Geometry Turned On!

Dynamic Software in Learning, Teaching, and Research

Geometry Turned On!

Dynamic Software in Learning, Teaching, and Research

James R. King
Doris Schattschneider
Editors

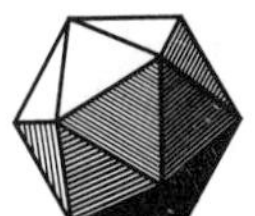

Published by
THE MATHEMATICAL ASSOCIATION OF AMERICA

ISBN 0-88385-099-0
Library of Congress Catalog Number 97-70509
Printed in the United States of America
Current Printing
10 9 8 7 6 5 4 3 2 1

MAA Notes Series

The MAA Notes Series, started in 1982, addresses a broad range of topics and themes of interest to all who are involved with undergraduate mathematics. The volumes in this series are readable, informative, and useful, and help the mathematical community keep up with developments of importance to mathematics.

MAA Notes

9. Computers and Mathematics: The Use of Computers in Undergraduate Instruction, *Committee on Computers in Mathematics Education, D. A. Smith, G. J. Porter, L. C. Leinbach, and R. H. Wenger,* Editors.
11. Keys to Improved Instruction by Teaching Assistants and Part-Time Instructors, *Committee on Teaching Assistants and Part-Time Instructors, Bettye Anne Case,* Editor.
13. Reshaping College Mathematics, *Committee on the Undergraduate Program in Mathematics, Lynn A. Steen,* Editor.
14. Mathematical Writing, by *Donald E. Knuth, Tracy Larrabee, and Paul M. Roberts.*
15. Discrete Mathematics in the First Two Years, *Anthony Ralston,* Editor.
16. Using Writing to Teach Mathematics, *Andrew Sterrett,* Editor.
17. Priming the Calculus Pump: Innovations and Resources, *Committee on Calculus Reform and the First Two Years,* a subcommittee of the Committee on the Undergraduate Program in Mathematics, *Thomas W. Tucker,* Editor.
18. Models for Undergraduate Research in Mathematics, *Lester Senechal,* Editor.
19. Visualization in Teaching and Learning Mathematics, *Committee on Computers in Mathematics Education, Steve Cunningham and Walter S. Zimmermann,* Editors.
20. The Laboratory Approach to Teaching Calculus, *L. Carl Leinbach et al.,* Editors.
21. Perspectives on Contemporary Statistics, *David C. Hoaglin and David S. Moore,* Editors.
22. Heeding the Call for Change: Suggestions for Curricular Action, *Lynn A. Steen,* Editor.
23. Statistical Abstract of Undergraduate Programs in the Mathematical Sciences and Computer Science in the United States: 1990–91 CBMS Survey, *Donald J. Albers, Don O. Loftsgaarden, Donald C. Rung, and Ann E. Watkins.*
24. Symbolic Computation in Undergraduate Mathematics Education, *Zaven A. Karian,* Editor.
25. The Concept of Function: Aspects of Epistemology and Pedagogy, *Guershon Harel and Ed Dubinsky,* Editors.
26. Statistics for the Twenty-First Century, *Florence and Sheldon Gordon,* Editors.
27. Resources for Calculus Collection, Volume 1: Learning by Discovery: A Lab Manual for Calculus, *Anita E. Solow,* Editor.
28. Resources for Calculus Collection, Volume 2: Calculus Problems for a New Century, *Robert Fraga,* Editor.
29. Resources for Calculus Collection, Volume 3: Applications of Calculus, *Philip Straffin,* Editor.
30. Resources for Calculus Collection, Volume 4: Problems for Student Investigation, *Michael B. Jackson and John R. Ramsay,* Editors.
31. Resources for Calculus Collection, Volume 5: Readings for Calculus, *Underwood Dudley,* Editor.
32. Essays in Humanistic Mathematics, *Alvin White,* Editor.
33. Research Issues in Undergraduate Mathematics Learning: Preliminary Analyses and Results, *James J. Kaput and Ed Dubinsky,* Editors.
34. In Eves' Circles, *Joby Milo Anthony,* Editor.
35. You're the Professor, What Next? Ideas and Resources for Preparing College Teachers, *The Committee on Preparation for College Teaching, Bettye Anne Case,* Editor.
36. Preparing for a New Calculus: Conference Proceedings, *Anita E. Solow,* Editor.

37. A Practical Guide to Cooperative Learning in Collegiate Mathematics, *Nancy L. Hagelgans, Barbara E. Reynolds, SDS, Keith Schwingendorf, Draga Vidakovic, Ed Dubinsky, Mazen Shahin, G. Joseph Wimbish, Jr.*
38. Models That Work: Case Studies in Effective Undergraduate Mathematics Programs, *Alan C. Tucker,* Editor.
39. Calculus: The Dynamics of Change, *CUPM Subcommittee on Calculus Reform and the First Two Years, A. Wayne Roberts,* Editor.
40. Vita Mathematica: Historical Research and Integration with Teaching, *Ronald Calinger,* Editor.
41. Geometry Turned On: Dynamic Software in Learning, Teaching, and Research, *James R. King and Doris Schattschneider,* Editors.
42. Resources for Teaching Linear Algebra, *David Carlson, Charles R. Johnson, David C. Lay, A. Duane Porter, Ann E. Watkins, William Watkins,* Editors.
43. Student Assessment in Calculus: A Report of the NSF Working Group on Assessment in Calculus, *Alan Schoenfeld,* Editor.

These volumes can be ordered from:
MAA Service Center
P.O. Box 91112
Washington, DC 20090-1112
800-331-1MAA FAX: 301-206-9789

Contents

III. Dynamic Visualization in History, Perception, Optics, and Aerodynamics

IV. The Worlds of Dynamic Geometry: Issues in Design and Use

Preface: Making Geometry Dynamic

Dynamic is the opposite of "static." *Dynamic* also connotes action, energy, even hype. Dynamic geometry* is active, exploratory geometry carried out with interactive computer software. The papers in this volume will convince you, we think, that dynamic geometry is full of action, energy, and, yes, even hype—the hype of excited individuals (students, teachers, researchers) who can't help but communicate their enthusiasm as they discuss the many implications of the software.

Mathematicians all know the power that a figure can provide—often a quick sketch or a diagram can make everything clear. We say "I see" and mean "I see and understand." In geometry, figures seem to be essential for most descriptions and proofs. Yet mathematicians also know the danger in relying on figures—inevitably, extra assumptions are made (suggested by a sketch), special cases are missed (omitted from a sketch), or absurd results are derived (from an inaccurate sketch). A classic case of the last occurence is the following oft-cited "theorem" and proof.[1]

Theorem. *All triangles are isosceles.*

Proof: Let ABC be a triangle with l the angle bisector of A, m the perpendicular bisector of BC cutting BC at midpoint E, and D the intersection of l and m. From D, draw perpendiculars to AB and AC, cutting them at F and G, respectively. Finally, draw DB and DC. Figure 1 shows a sketch of the situation.

$\triangle ADF \cong \triangle ADG$ (aas), so $AF \cong AG$ and $DF \cong DG$. $\triangle BDE \cong \triangle CDE$ (sas), so $BD \cong CD$. This implies $\triangle BDF \cong \triangle CDG$ (hyp.-leg), so $FB \cong GC$. Thus $AB \cong AF + FB = AG + GC \cong AC$, and so $\triangle ABC$ is isosceles. QED

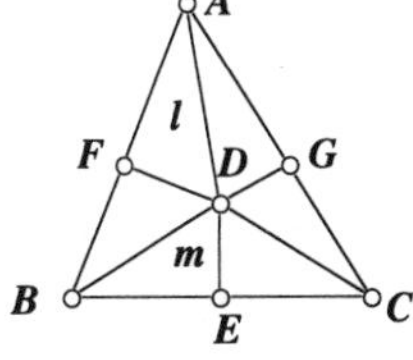

FIGURE 1

What is wrong with this proof? Absolutely nothing is wrong with the chain of reasoning—the conclusion must follow if the figure is accurate. But even though the sketch is plausible, the configuration cannot occur as drawn. D, the intersection of l and m, is never inside $\triangle ABC$.

This fallacious proof has been offered by some as a reason to ban the drawing of any figures! But they miss the point.

Quite to the contrary, this example offers compelling reasons to emphasize the need for acurately constructed figures, rather than quick sketches, especially when they are to be used in deductive aguments. As E. P. Northrop put it, "It shows how easily a logical argument can be swayed by what the eye sees in the figure and so emphasizes the importance of drawing a figure correctly, noting with care the relative positions of points esential to the proof. Had we at the start actually constructed—

*Here, and through this volume, the term "dynamic geometry" is used in its broadest sense. Although Key Curriculum Press has trademarked the phrase, our authors' use of the phrase does not refer to any particular computer software or published software support materials.

by means of rule and compass—the angle bisector and the various perpendiculars, we should have saved ourselves a good deal of trouble."[2] The "proof" also underscores the need for a variety of constructions, rather than just one. (Authors who discuss this particular "proof" display a variety of inaccurate sketches of various cases as well as accurate ones obtained by construction.)

Even careful constructions can be inaccurate, as anyone who has labored with a wobbly compass or a thick lead can testify. Complicated constructions are time-consuming and difficult to make precise. The time (and tedium) required to make a variety of figures by hand to illustrate various cases of a geometric configuration is beyond what almost anyone is willing to spend.

With dynamic geometry software, the proof above would be most unlikely. With the software, an investigator could make a single accurate sketch of the geometric configuration described in which the constraints on l (angle bisector of A), m (perpendicular bisector of BC), D (intersection of l and m), F and G (feet of perpendiculars from D to AB and AC, respectively) are specified. Then the shape of $\triangle ABC$ could be deformed at will with all the constraints being preserved in the newly created configuration. Almost instantaneously, a (seemingly infinite) family of accurately drawn configurations would be produced by manipulating a vertex of the elastic figure. The investigator would see that the only time that l and m do not intersect outside $\triangle ABC$ is when l and m coincide (and $B = F, C = G$), and that is exactly the case in which $\triangle ABC$ is isosceles. Moreover, in all but that special case, the equation in the last line of the "proof" is false. Figure 2 captures only a few of the different configurations that occur.

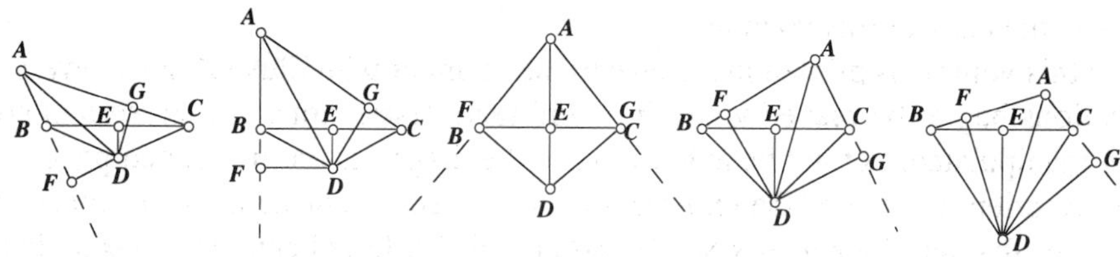

FIGURE 2

The ability to have students use a computer to construct accurate geometric configurations that can be easily altered to new configurations with the same constraints had its genesis less than 15 years ago with the *Geometric Supposers*.[3] This software, originally written for Apple II microcomputers, pioneered the idea of an environment in which constructions could be made on a specified figure (a triangle, for instance), and then replayed to produce a different case of the construction with the same constraints. The focus shifted from students laboriously making constructions by hand to verify a stated fact in a text (for which there seemed little reason to produce a proof) to a focus on students carrying out experiments, quickly producing many accurate sketches from which they conjectured properties that seemed to be "always" true. The only way to ensure that the observations were always true was to produce a proof. In response to teachers' requests, Dan Chazen and Richard Houde wrote a pamphlet based on their 5 year's experience using the new software. This described how they and other teachers "taught students to behave like working mathematicians who conjecture and prove within a community of learners," and answered "nittty-gritty questions about how the approach might be carried out."[4]

Today there are new generations of personal computers in the schools (and on teachers' and researchers' desktops and laps). The dynamic geometry software that has been developed takes advantage of the new hardware (with vastly increased memory, speed, and mouse interface for graphics) and builds on the early ideas. The most commonly used dynamic geometry software is *The Geometer's Sketchpad, Cabri Geometry II, Geometry Inventor*, and the successor to the Supposers, the *Geometric SuperSupposer*. These all provide means for accurate construction (and measurement) of geometric configurations (of points, segments, rays, lines, arcs, circles) and have the ability to replay a construction. What is most exciting about the new software, however, is that unconstrained parts of the configurations (arbitrary segments or points, for example, that are not dependent on any other objects) are moveable—they can literally be grabbed with a cursor (using the mouse) and be dragged or stretched—and as they move, all other objects in the configuration automatically self-adjust, preserving all dependent relationships and constraints. In doing this, a smooth sequence of instances of the configuration can be viewed, morphing contnuously from the first drawing to the final one that remains when the dragging is stopped. If the configuration is a general one (beginning with an "arbitrary" line, circle, or polygon, for instance) there can be no question that there are many (even an infinite number) of cases, and the visual evidence of invariant properties (such as concurrence of lines, collinearity of points, or ratios of certain measurements) is compelling.

In addition to construction capabilities, the programs allow certain affine transformations (translations, reflections, rotations, dilations) to act on figures or their parts. These transformations can be determined by "fixed" data or by "dynamic" data, such as a rotation by a fixed angle (e.g., 45°) or by a moveable angle that is in some sketched triangle, for instance. The software allows loci to be traced as parts of figures are dragged freely or guided along a specified circle or segment. The programs also have the special feature of being able to "record" the sequence of steps that leads to a particular configuration and then be able to use that construction as a "macro" tool in producing new configurations. For example, a script or macro that produced a tangent to a circle from a point outside the circle might be employed several times in producing a complicated figure. The software has components that link synthetic geometric constructions to analytic equations, coordinate representations, and graphs. Conics are also implemented in at least one program. Information on all of the software discussed by our authors is at the back of this volume.

Although this volume is printed in a conventional manner with illustrations in every article, most of the illustrations beg to be played with. You will read descriptions of how certain configurations behave when manipulated, but not be able to tweak the diagrams on the printed page. We want you to be able to experience some of the explorations described by our authors. To make that possible, dynamic sketches that use *Geometer's Sketchpad* or *Cabri II* have been made available by several of the authors and are posted on a Web page maintained by the Mathematics Forum. The URL is

http://forum.swarthmore.edu/dynamic/geometry_turned_on

This page can also be accessed by visiting the MAA web site: www.maa.org.

For those who do not have access to the software, demonstration copies of the software can be downloaded and used to open the sketches.

Overview of Contents

Dynamic geometry software has had a profound effect on classroom teaching wherever it has been introduced. Although not originally intended by its developers, it has also become an indispensible research tool for mathematicians and scientists. The papers in this volume give a good idea of the ways in which the software can be used, and some of the effects it can have. Authors also make clear that the software raises various questions for teaching and research, and its continuing evolution raises questions on the design of the software itself. Many of the papers here expand on presentations that were given at the annual meeting of the MAA in San Francisco in January 1995 in a special session devoted to dynamic geometry.

In an attempt to give an overview of the contents of the papers, we address the basic question "What is dynamic geometry software good for?"

Accuracy of construction. Dynamic geometry software provides an *accurate constructor* for any ruler-and-straightedge construction in Euclidean geometry, any configuration that can arise by applying affine transformations (isometries and dilations) to a Euclidean construction, or any locus of an object (or set of objects) that arises when part of a construction is moved along some path. Reliance on the accuracy of the software's geometric sketches and measurements is so basic that it is not discussed explicitly in any article, but it is implicit in every discussion. (Accuracy is, of course, limited to the tolerences of internal computation, screen display, allowed numerical display, and printer fidelity. And accuracy is occasionally diminished by the choice of heuristic for an algorithm or by pesky bugs that are bound to occur in such complex software.) Arnold and Bernadette Perham offer several suggestions for investigations of constructions that exhibit the collinearity of 3 points or exhibit invariance of the ratio of two segments. Tony Hampson describes how the software easily implements the techniques for perspective drawing and dramatically displays the principle of parallel lines converging at vanishing points, even as the viewpoint is changed. In every article, the sketches are produced by software; these attest to the author's reliance on it as a faithful constructor.

Visualization. As a demonstration tool in the classroom, dynamic geometry software can help students *see* what is meant by a general fact. Douglas Brumbaugh tells how an animated triangle got the attention of middle school students and demonstrated that the area of a triangle is determined by its base and height. Jim Morrow observes, "Dynamic geometry software can be used to aid the process of visualization in all mathematics classes, not just in the study of geometry. Students can construct, revise, and continuously vary geometric sketches. While visualization in itself is a powerful problem-solving tool, the capacity for students to make instantaneous and precise variations to their own visual representations adds a new dynamic dimension whose implications are only beginning to be understood." His several examples illustrate how the abstract concepts of variable, function, and invariance take on meaning through visualization. Al Cuoco and Paul Goldenberg observe "By allowing students to investigate continuous variation directly (without intermediary algebraic calculation), dynamic geometry environments can be used to help students build mental constructs that are useful (even prerequisite) skills for analytic thinking." Cathy Gorini shows how the software can help students in calculus see (and vary continuously) relationships between quantities in min-max applications and related-rates problems.

Exploration and Discovery. In a traditional geometry course, students are told definitions and theorems and assigned problems and proofs; they do not experience the discovery of geometric relationships, nor invent any mathematics. As Schwartz and Yerushalmy observed, "This constitutes a kind of satire on the nature of mathematical thinking and the way new mathematics is made."[5] Dynamic geometry software is perfectly suited for exploration and discovery—either guided, or completely open-ended (a.k.a. research). Tim Garry observes that it allows students to "test their own mathematical ideas and conjectures in a visual, efficient and dynamic manner and—in the process—be more fully engaged in their own learning." The range of investigation and the amount of guidance provided varies greatly with the level and experience of the students (or researchers).

Our authors provide a variety of of examples of investigations for students and in many cases, discuss the process and the outcomes of the tasks. For example, Tim Garry had his students investigate the ratio of perimeter to diameter for regular polygons and also explore questions concerning circles tangent to a given circle. Zhonghong Jiang and Edwin McClintock had preservice teachers find the shortest path (with restrictions) that connects towns A and B separated by one or more rivers and also find various constant ratios that arise in a construction of a triangle within a triangle. Students using the software often discover surprising things that are not in any text, and not known to the teacher. As Kathy Boehm points out, "Any teacher using... dynamic geometry software needs to be prepared for students to ask unexpected questions. I have had to answer, 'I don't know; let me get back to you,' many times." Michael Keyton describes the process of his students moving from guided discovery of well-known properties of families of quadrilaterals to open-ended investigation of new families of quadrilaterals (with newly-invented names) proposed by the students themselves. Fadia Harik gives her students (often teachers) tasks with deliberately ambigious directions and many possible solutions, designed to prompt questions and guide the process of formulating hypotheses.

Although it may seem a cliché, dynamic geometry software *empowers*. Students can get hooked pursuing open-ended problems. Michael Keyton reports that some of his students have come back after graduating to tell him some new result on an unfinished investigation begun in his class. Although ninth-grade student Ryan Morgan's discovery using dynamic geometry software received national attention[6], there are many more students who have experienced a similar thrill of discovery. And not only students. Douglas Hofstadter gives a riveting account here of his own pursuit of a geometric problem that tantalized him. Mike de Villiers describes how, with dynamic geometry software, he could follow the irresistible lure of "what if?" to find some new results related to known theorems.

Proof. While dynamic geometry software cannot actually produce proofs, the experimental evidence it provides produces strong conviction which can motivate the desire for proof. Dan Bennett was tantalized by a 40-year-old unsolved problem from the *American Mathematical Monthly* for which there was compelling computer evidence of its truth; he was determined to find a proof (and succeeded). Mike de Villiers discusses the synergy between using dynamic geometry software and proving the conjectures that spring from its use. Conviction is necessary for undertaking the (often difficult) search for a proof, he contends, and in addition, the software may even give insight into geometric behavior that can help with a proof. Subtle geometric relationships may be not at all obvious, but be revealed in experimenting with dynamic figures. Some teachers have been reluctant to use the software because they fear that visually convincing evidence will replace proofs of theorems. Our authors describe the opposite situation: when students make their own conjectures based on their explorations with the software, they know it is not enough to stop with the evidence; they need to devise a proof. Zhonghong Jiang and Edwin McClintock record for us the process by which their students arrived at proofs of their conjectures.

Transformations. Dynamic geometry software can *transform* figures in front of your eyes. Isometries and similarities are important examples of functions. In witnessing the action of these transformations moving and scaling figures, students see that functions are not synonymous with symbolic formulas. Doris Schattschneider points out that students can visually test fundamental properties of composition such as commutativity and inverses, as well as other abstract concepts encountered in the theory of groups. Jim Parks describes how students can identify a transformation by producing a sequence of points in its orbit.

Ross Finney's 1970 article "Dynamic Proofs of Euclidean Theorems"[7] (written long before the current dynamic geometry software was even an idea) begins, "Simple observations about [affine] transformations of the plane lead to elegant proofs of unusual Euclidean theorems." Proofs such as Finney's can be brought to life with the software—to show that one figure is the transformed image of another, one only needs to perform the transformation to see if the two figures coincide. Jim King's article is devoted to proofs that rely on similarities. Mike de Villiers also gives proofs that depend on the action of transformations.

Loci. It is virtually impossible for most people to imagine a point moving in a configuration (in which other several parts may also may be moving) and be able to describe the locus of the point's path as it travels. Dynamic geometry software, with its built-in feature to trace the locus of any specified object is ideally suited to show how a locus is generated and to reveal the shape of its traced path. Except for the most simple loci (circles, as the locus of points equidistant from a fixed center, and perhaps the conics), this rich subject has been avoided in most geometry texts. In fact, most of the classic curves arose as loci, and one must go to long out-of-print books to find these nonanalytic descriptions of them. Classic locus problems and intriguing generalizations of these, as well as surprising new loci, are discussed by our authors. Hampson's students investigate conics. Cuoco and Goldenberg describe various locus experiments—they create generalized ellipses, explore a locomotion problem, and solve some optimization problems by means of loci. Heinz Schumann and David Green describe some classic and non-classic curves that are constructed as loci and also provide some useful applications of loci. John Olive describes how the Joukowski transformation (which uses inversion in a circle) produces airfoil shapes as loci.

Simulation. Dynamic geometry software's special features of dragging, animation (of points on line segments or on circles), tracing loci, and random point generation provide many opportunities to simulate a surprising variety of situations. Our authors present imaginative examples: moving robot arms (Jim Morrow), a sine tracer (Tim Garry), a 2-dimensional random walk machine (Tim Garry), and an airfoil tracer (John Olive). David Dennis and Jere Confrey simulate a mechanical linkage devised by Rene Descartes (1637) that produces points in a geometric series and then pair those points with points in an arithmetic series to produce logarithmic curves. In this simulation, the curves can "flex and bend as the arithmetic and geometric sequences are manipulated," since the manipulation effectively changes the base of the logarithm. Ben Backus describes some of the special simulations devised to help in the teaching of optics and in vision research. Susan Addington and Stuart Levy use a 3D interactive viewer, *Geomview*, to produce a simulation of what is seen in an Ames room, a distorted room-size box that tricks our perception. Fadia Harik uses *The Physics Explorer: One Body* to simulate the action of a billiard ball bouncing off the sides of a square table.

Microworlds. Dynamic geometry software produces an environment in which Euclidean geometry can be explored. Three papers in our collection, all by designers of this type of software, discuss new environments that can be created with software. Jean-Marie Laborde (*Cabri-géomètre*) and Nick Jackiw (*The Geometer's Sketchpad*) each discuss ways in which their differing software can create a "Poincaré world" of hyperbolic geometry. Jackiw also discusses how other microworlds can be created through the use of scripts which produce new "tools" that replace the Euclidean tools and allow exploration fully within a new geometry. Allen and Trilling describe a program (*GéoSpécif*) under development that adds a new dimension to the dynamic geometry programs currently in use. Their program takes as input the specifications for a construction and then automatically produces the construction (if it is possible); using this, students can explore to discover various dependent relationships among the parts of the construction. All these software developers acknowledge the challenges that dynamic geometry software presents—there is a delicate balance between what might be desired and what is feasible, and there is an ongoing dialogue as to what should be the design of software that can best enhance learning for all.

Notes

1. This fallacious proof has been in the literature for over 100 years; it is probably even older. It appears, with many "cases" discussed, in the following sources: *Mathematical Recreations and Essays*, by W.W. Rouse Ball and H.S.M. Coxeter, 13th ed., Dover 1987 (1st edition 1892); *Riddles in Mathematics*, by E.P. Northrop, Van Nostrand, 1944; *Fallacies in Mathematics*, by E.A. Maxwell, Cambridge [Eng.] University Press, 1959; and *Mistakes in Geometric Proofs*, by IA.S. Dubnov, Heath, 1963.
2. Northrop (see 1. above), page 101.
3. Schwartz, J. and Yerushalmy, M. designers, *The Geometric Supposers*. Pleasantville, NY: Sunburst Communications, 1983–1991.
4. Chazen, D. and Houde, R. (1989). *How to Use Conjecturing and Microcomputers to Teach Geometry*, National Council of Teachers of Mathematics.
5. Schwartz, J.L. and Yerushalmy, M. (1986). "Using Microcomputers to Restore Invention to the Learning of Mathematics," *Contributors to Thinking*, D. Perkins and R. Nickerson, eds. Lawrence Erlbaum Associates, Hillsdale, NJ, pp. 293–298.
6. Watanabe, T., Hanson, R., and Nowosielski, F.D., "Morgan's Theorem," *Mathematics Teacher*, 89 (May 1996) 420–423.
7. Finney, R. (1970). "Dynamic Proofs of Euclidean Theorems," *Mathematics Magazine* 43, 177–185.

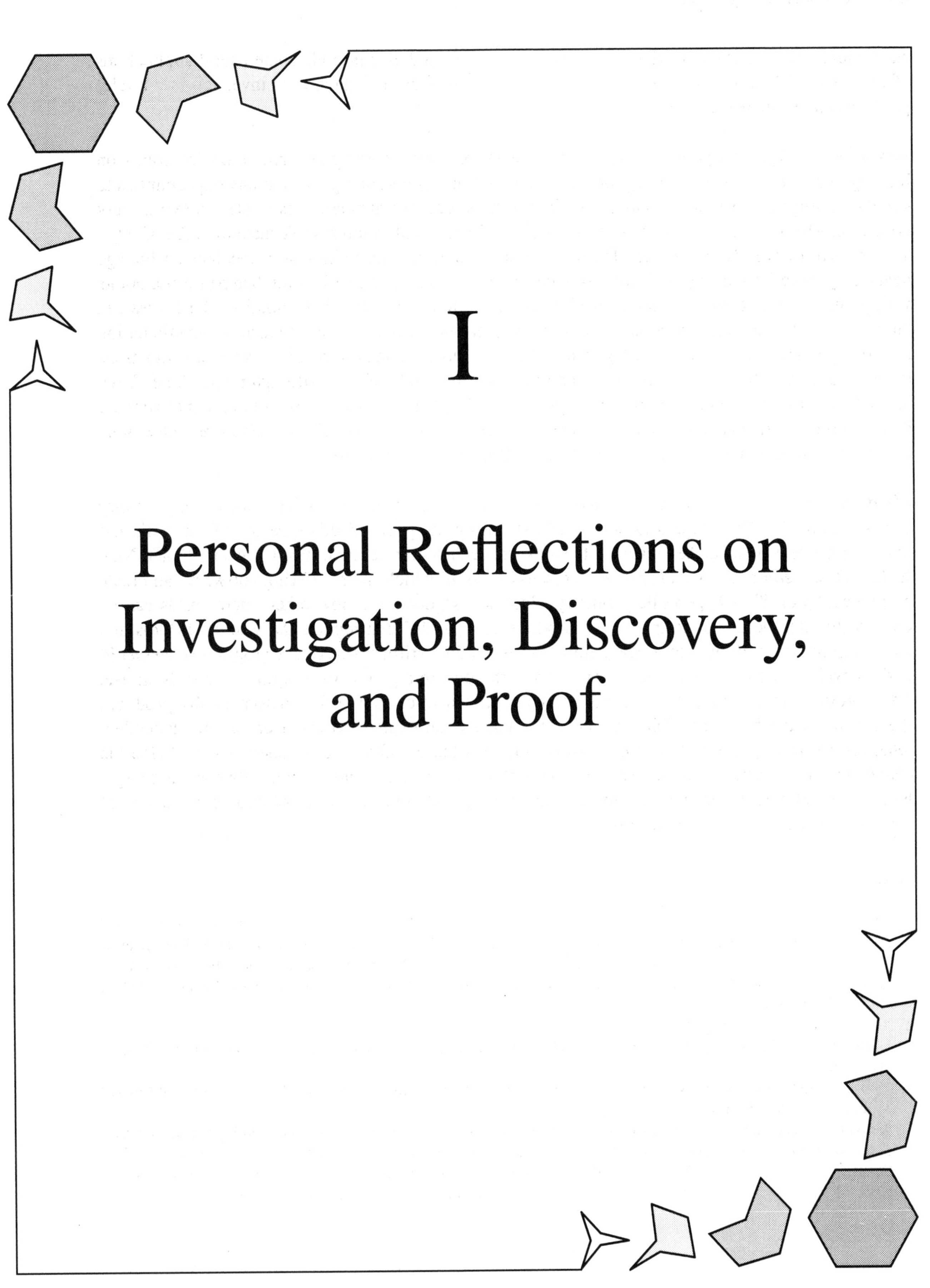

I

Personal Reflections on Investigation, Discovery, and Proof

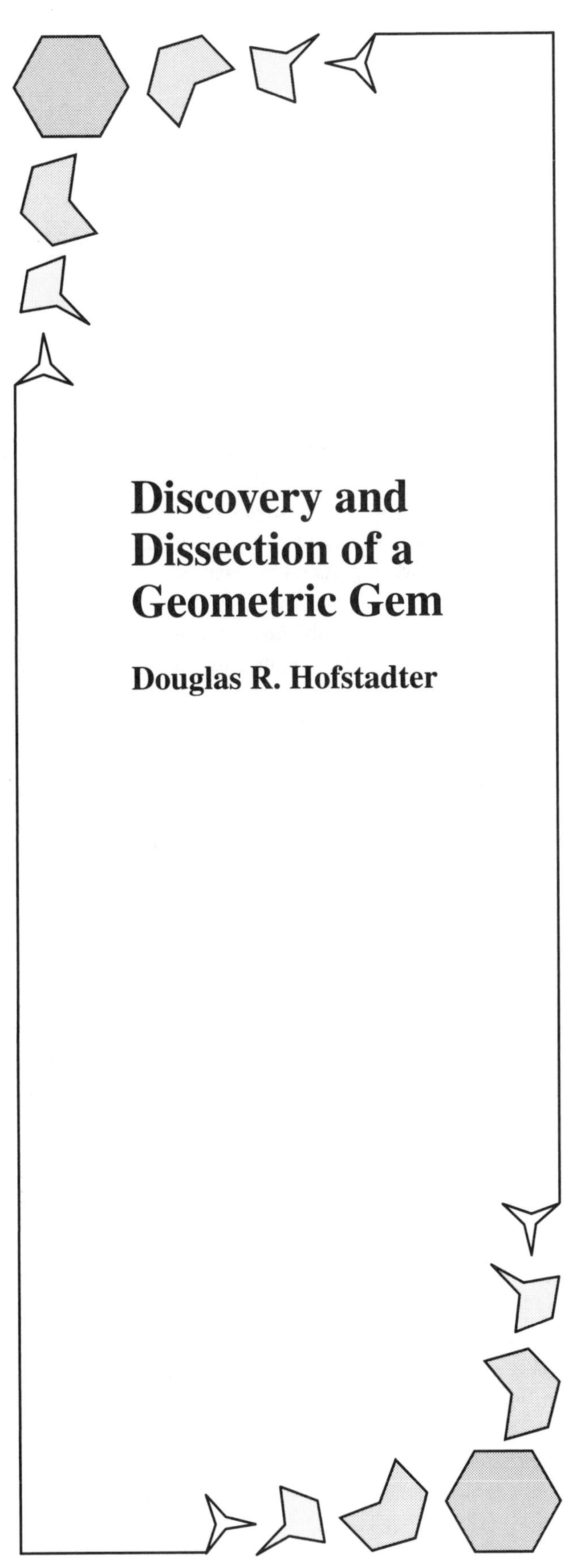

Discovery and Dissection of a Geometric Gem

Douglas R. Hofstadter

Bewitched... by Circles, Triangles, and a Most Obscure Analogy

Although many, perhaps even most, mathematics students and other lovers of mathematics sooner or later come across the famous Euler line, somehow I never did so during my student days, either as a math major at Stanford in the early sixties or as a math graduate student at Berkeley for two years in the mid-sixties (after which I dropped out and switched to physics). Geometry was not by any means a high priority in the math curriculum at either institution. It passed me by almost entirely.

Many, many years later, however, and quite on my own, I finally did become infatuated—nay, *bewitched*—by geometry. Just plane old Euclidean geometry, I mean. It all came from an attempt to prove a simple fact about circles that I vaguely remembered from a course on complex variables that I took some 30 years ago. From there on, the fascination just snowballed. I was caught completely off guard. Never would I have predicted that Doug Hofstadter, lover of number theory and logic, would one day go off on a wild Euclidean-geometry jag! But things are unpredictable, and that's what makes life interesting.

I especially came to love triangles, circles, and their unexpectedly profound interrelations. I had never appreciated how intimately connected these two concepts are. Associated with any triangle are a plentitude of natural circles, and, conversely, so many beautiful properties of circles cannot be defined except in terms of triangles.

Not surprisingly, some of the most important points of a triangle are the centers of various natural circles associated with it. Three such circles are the circumcircle, which is the smallest circle that the triangle will fit inside[1], and which thus passes through all the triangle's vertices; the incircle, which is the largest circle that will fit inside the triangle, and which is thus tangent to all three of its sides; and the nine-point circle, a circle that somehow manages to pass through the midpoints of all three sides, the feet of all three altitudes, and three further notable points. The centers of these circles are the circumcenter O, the incenter I, and the nine-point center P.

Two other famous special points are the centroid G (also known as the "barycenter"), which is the center of gravity of the triangle (meaning that the triangle would balance perfectly if it were supported by a pin located precisely at its centroid), and the orthocenter H, which is where all three altitudes cross. The Euler line or, as it more properly ought to be called, the Euler segment (which term I will use henceforth) connects four out of these five most special of special points. To be specific, the circumcenter, the centroid, the

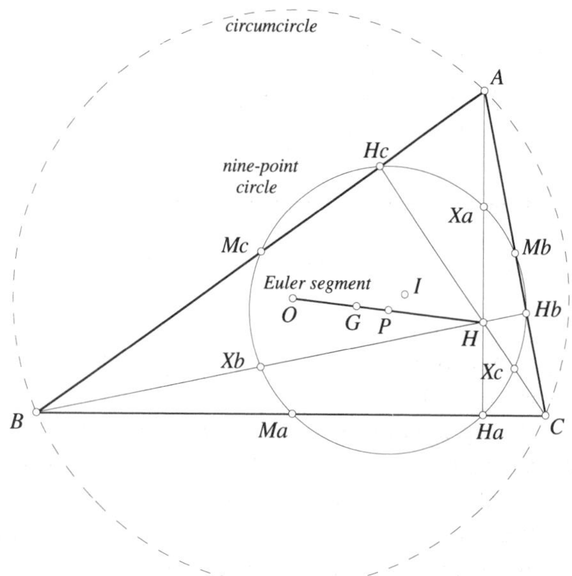

FIGURE 1. The nine-point circle passes through *Ma, Mb, Mc* (*ABC*'s midpoints); *Ha, Hb, Hc* (the feet of *ABC*'s altitudes); and *Xa, Xb, Xc* (the midpoints of the segments connecting *H* to each of *ABC*'s vertices).

nine-point center, and the orthocenter all lie on a single line. (Poor little neglected incenter!)

Figure 1 shows a typical triangle *ABC* with its circumcircle, its nine-point circle, its three midpoints, its three altitudes, and its Euler segment *OGPH*. The Euler segment runs from *ABC*'s circumcenter O to its orthocenter H. Precisely one-third of the way from O to H, the segment passes through *ABC*'s centroid G; at exactly the halfway point it passes through P, the center of the ninepoint circle. Figure 1 also shows last, but not least, the poor forgotten incenter I, somehow left out of the party.

Although I loved the Euler segment, I was deeply puzzled as to why the incenter I had been excluded from it, and felt that the incenter surely had to have its own special way of relating to these four points, or else, perhaps, its own coterie of special friends (although which ones they might be, I had no hunch about). Unresolved questions like this can lure one on, ever more deeply, into the study of special points and their unexpected hidden patterns. In any case, I was certainly hooked by these questions.

As I grew more involved with triangles, I started to see a metaphorical connection between my love for their special points and a mathematical love I had had from childhood: the love for special points on the number line, of which the quintessential examples are, of course, π and e. Perhaps the most exciting aspect of math for me was learning of equations that showed secret links among such numbers,

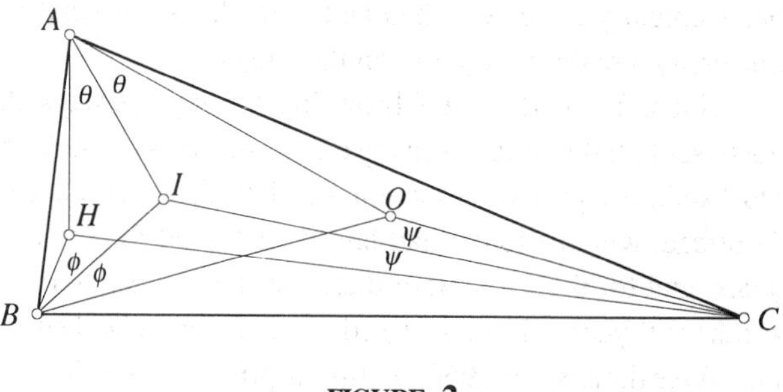

FIGURE 2

such as Euler's equation:

$$e^{i\pi} = -1$$

When I first saw this, perhaps at age 12 or so, it seemed truly magical, almost other-worldly. One can even draw an analogy between this equation, which relates four important numbers in a most astonishing way, and the Euler segment, which relates four important triangular points in a most astonishing way.

One day I made a little discovery on my own, which can be stated in the following picturesque way: If you are standing at any vertex and you swing your gaze from the circumcenter to the orthocenter, then, when your head has rotated exactly halfway between them, you will be staring straight at the incenter. More formally (see Figure 2), the bisector of the angle formed by the two lines joining a given vertex with the circumcenter and with the orthocenter passes through the incenter. (A more technical way of characterizing this property is to say that O and H are "isogonic congugates".) It wasn't too hard to prove this, luckily.

This discovery, which I knew must be as old as the hills, was a relief to me, since it somehow put the incenter back in the same league as the points I felt it deserved to be playing with. Even so, it didn't seem to play nearly as "central" a role as I felt it merited, and I was still a bit disturbed by this imbalance, almost an injustice.

Seeking to quench my avid thirst for geometrical insight, I went to a couple of superb technical bookstores with row after row of math books—but even there, all I found on geometry was a handful of rather thin volumes. In any case, I bought all the relevant books I could find, of which my favorite was *Geometry Revisited,* by H.S.M. Coxeter and Samuel Greitzer. None of these books on its own was anywhere close to definitive, but I nonetheless drank from them all with great gusto. A few of these books referred to a long-out-of-print volume by Julian Lowell Coolidge, called *Treatise on the Circle and the Sphere*, published in 1916. I didn't know whether this book or any of the other old-timers that were occasionally cited would have anything much to say beyond what my modern books collectively had told me, but eventually I decided I had better go check out what my university's math library had on the subject.

Soon, I found myself browsing through perhaps the dustiest of all the library's many dusty shelves—those in the old-fashioned-geometry section—and there I came across Coolidge, which I found had lain undisturbed for some 12 years, and for 9 years before that, and then before that, yet another 10 years. In other words, it had been checked out only three times since 1960. To my surprise, it was quite a big tome—some 600 pages jam-packed with beautiful diagrams and theorems. A moment's skimming was enough to tell me that this was a treasure trove of geometric gems. Without further ado, I checked it out and joyfully took it home.

Browsing through Coolidge's Chapter One (a small book in itself, rich enough to make me feel very humble), I came across something that almost took my breath away. There was apparently a second segment that not only was *reminiscent* of the Euler segment, but in fact was deeply *analogous* to it. This segment, which seemed, strangely, to have no name, ran from the incenter I to another special point that I already knew and loved, the *Nagel point* N. Figure 3 illustrates the construction of the Nagel point of a triangle ABC.

Like the Euler segment, the anonymous segment passed through the centroid G as well as through a point S that was the center of another very interesting circle, the Spieker circle (which is the incircle of ABC's median triangle). Moreover, the relative positions of these points along the segment were the same as in the Euler line: G was one-third of the way from I to N, and S fell exactly halfway between I and N (see Figure 4).

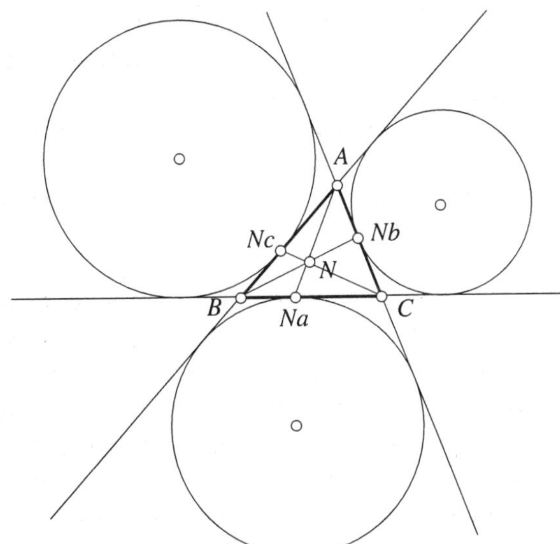

FIGURE 3. An excircle of triangle ABC is a circle outside ABC that grazes all three of its sides (considered as infinite lines). Three such circles exist, and, when their grazing-points are joined with the opposite vertices, the joining-lines turn out to all meet in a single point: the Nagel point N.

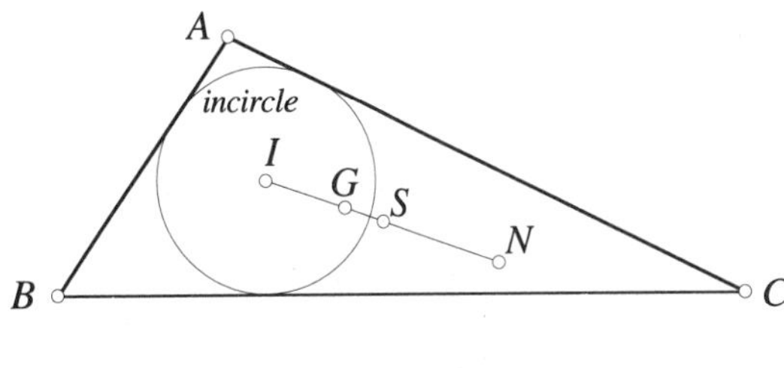

FIGURE 4

To cap it all off, Coolidge listed a long and systematic set of parallels between the Spieker circle and the nine-point circle, some of the high points of which are given below (translated into crisper modern terms from Coolidge's slightly verbose "turn-of-the-centurese"):

The nine-point circle...	*The Spieker circle...*
is the circumcircle of ABC's median triangle;	*is the incircle of ABC's median triangle;*
has radius one-half that of ABC's circumcircle;	*has radius one-half that of ABC's incircle;*
is the circumcircle of the triangle whose vertices are the midpoints of the segments that join ABC's vertices with its orthocenter;	*is the incircle of the triangle whose vertices are the midpoints of the segments that join ABC's vertices with its Nagel point;*
passes through the points where ABC's sides are cut by the lines that join ABC's vertices with its orthocenter (i.e., the feet of ABC's altitudes)	*is tangent to the sides of ABC's median triangle where that triangle's sides are cut by the lines that join ABC's vertices with its Nagel point.*

Figure 5 shows the Spieker circle and nine-point circle of

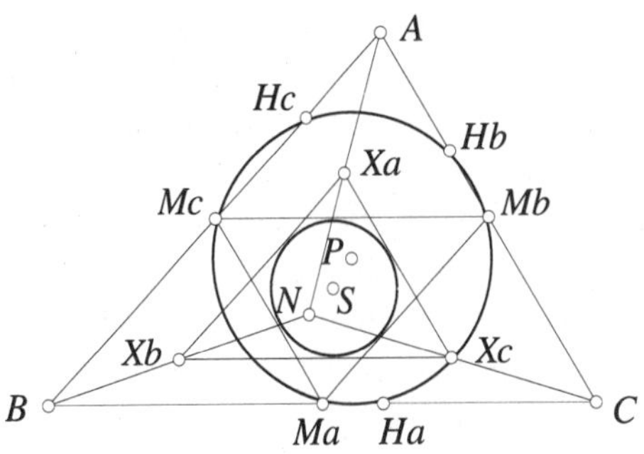

FIGURE 5

triangle ABC and illustrates the properties listed in this comparison.

Altogether, this systematic set of correspondences between properties of the Euler and Nagel segments (including the correspondences between properties of the nine-point and Spieker circles) constituted one of the most remarkable and complex mathematical analogies I had ever run across. And to my great pleasure, it restored the honor of the incenter, while also elevating the Nagel point to a level of respect much higher than I had previously accorded it.

I wondered to myself, "Why does this fantastic second segment have no standard name? Why is it not routinely mentioned in the same breath as the Euler segment? Why are the two of them not treated by geometers as precisely equal companions?" Even Coolidge didn't go much beyond the mere act of describing this segment. At the end of Chapter One, he did go so far as to suggest that there probably are other circles analogous to the Spieker circle that remain to be discovered, but that was about it.

Surely, I thought, there is more to it than this. In mathematics, such a striking and intricate analogy can't just happen by accident! There's got to be a reason for it. But there was no discussion in Coolidge of why the parallels were so perfect and so systematic. In the whole library I found only one other book that mentioned this segment, again leaving it nameless and giving less information on it than Coolidge.

I was baffled. Why was this companion segment—which I began calling the Nagel segment, after the discoverer of its outlier endpoint—so neglected? Was it truly less important than the Euler segment? Or was it just that it had been discovered at a time when people were beginning to lose interest in this kind of geometry? I could not help but mull this over, and the image of these two segments, each one lopsidedly cutting the other into two pieces, reverberated through my head intensely.

In order to gain a deeper intuitive feel for these things, I ambled into my study, plunked myself down in front of my trusty Macintosh, fired it up, and double-clicked on the icon labeled "Geometer's Sketchpad". Within a couple of minutes at most, I had constructed a picture of a triangle ABC with its two associated segments, as in Figure 6. It looked excellent, but this picture was not the destination—it was just the starting-point.

I clicked on point C and started "dragging" it around the screen. As I did so, everything else that depended on point C, whether point or line, started moving in synchrony, perfectly maintaining the geometric relationships established by my construction. I could now watch the dynamic way in which the Euler and Nagel batons swiveled around simultaneously as the triangle defining them changed. This "dynamogram"

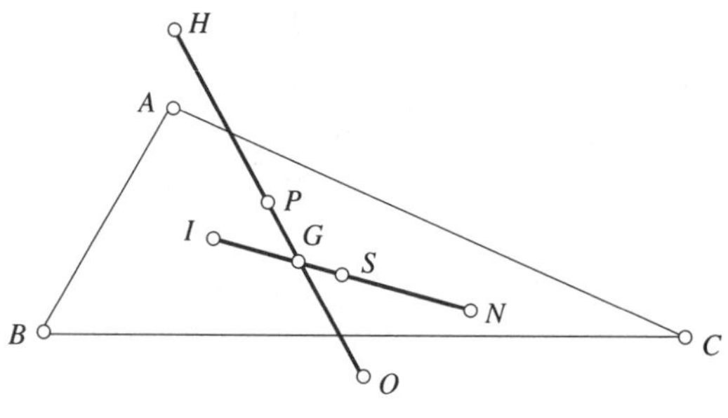

FIGURE 6

was revelatory in a way that no static image could possibly be. I certainly feel fortunate and grateful to have this wonderful program at my fingertips, and I truly wish some of the old-time geometers, such as Euclid, Apollonius, Pappus, Menelaus, Desargues, Euler, Ceva, Poncelet, Steiner, von Staudt, Brocard, Nagel, Gergonne, Spieker—and let's not forget Coolidge, of course!—could have seen it.

I looked for patterns, questing after comprehensibility. But, although my dynamogram was pretty and fascinating, my eyes didn't pick much up at first. Figure 7(A) shows the two crisscrossing segments by themselves, without triangle ABC.

Recall that in the tight analogy between the segments, O maps onto I and N maps onto H. It seemed therefore very natural to construct the lines OI and NH, each of which links counterpart points together. After all, these lines, if built, would constitute a concrete physical realization of the abstract analogy—a lovely idea, irresistible to me.

It took but a moment to construct them, and the instant they flicked onto my screen, I saw something most promising—they appeared to be parallel! To test this hopeful hypothesis, I tweaked triangle ABC and found that always the OI-NH parallelism stayed true. Moreover, I found that the two midpoints, P and S, when joined, added a third parallel line to the first two Figure 7(B)). I felt as if I had stumbled on a stupendous connection between the segments!

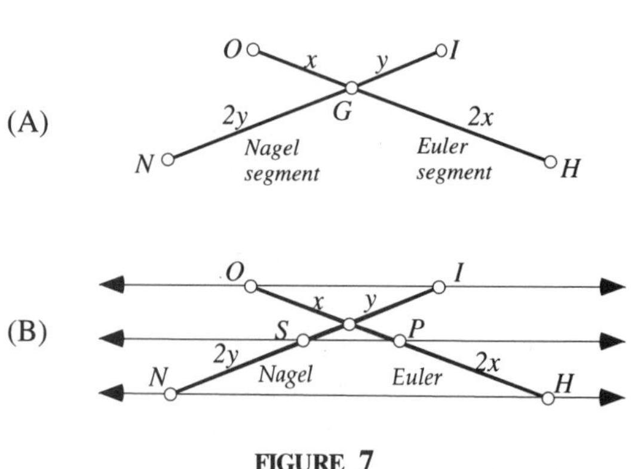

FIGURE 7

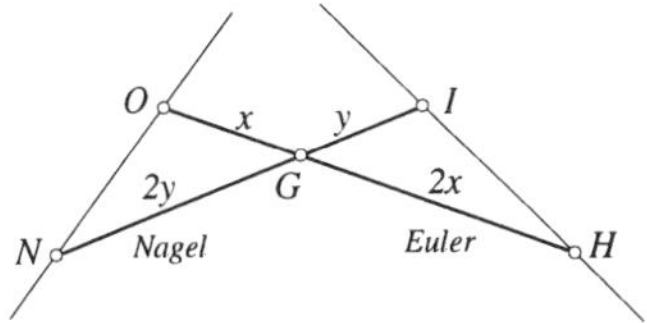

FIGURE 8

A couple of minutes' thought deflated me rather devastatingly, however. I soon realized that this supposedly "profound insight" was in fact a triviality: any two line segments that cut each other in identical length-ratios will always have parallel lines defined by their tips. My parallelism had nothing whatsoever to do with the fact that I was dealing with two Deeply Significant Segments—it was just a simple consequence of similar triangles. I felt ashamed of myself for my premature elation.

The next thing I did was to make the only other two lines that remained, *HI* and *NO* (Figure 8). But, unlike *OI* and *NH*, these two lines didn't have any strong *reason* pushing for building them.

Disappointingly, they turned out to be neither parallel nor perpendicular, and after watching them for a little while, I concluded they were of no interest. But where to look, then, for insight into what lay behind this tantalizing analogy? I was at a loss.

Bedazzled... by a Sparkling Crystal

Disappointed at having made essentially no progress on the mystery of the interrelationship of the two analogous segments, I gradually started letting the matter slide into "background mode". And so a few days later, I found myself idly thinking about a completely different and rather simple geometric question:

> *Suppose somebody gives me a point G and asks for the general recipe for making triangles having G as centroid. Can I simply "splay out" from G three segments of arbitrary lengths in arbitrary directions, letting their tips define the vertices?*

That, of course, I knew instantly was nonsense. There had to be some constraints on the segments that defined the three vertices—after all, there is a relationship between a triangle's centroid and its vertices! I remembered that the segment from the centroid to any vertex forms a part of one of the triangle's three medians (a segment that links a vertex with the opposite side's midpoint). And then I remembered that medians always cut each other in a characteristic way, namely into subsegments whose length-ratio is $1:2$. So I wrote numbers indicating the relative lengths of the subsegments, as in Figure 9.

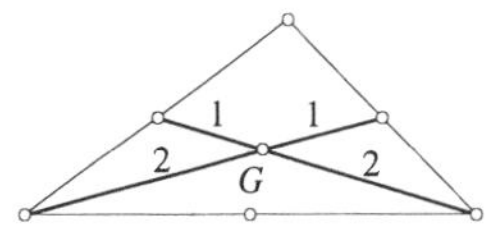

FIGURE 9

The instant I put those numbers down on the paper, something clicked in my mind. This was a turning point in the whole process, for in the crisscrossing heavy lines of this picture I suddenly recognized something familiar—my two fundamental segments crisscrossing each other, slicing each other up in that lopsided $1:2$ way. Unexpectedly, the little puzzle had brought me back to my earlier quest via the back door! Looking at one picture and seeing another—that was what everything had hinged on.

This new vision meant that I could interpret my two segments as medians of a hidden triangle, with *N* and *H* as two of its vertices. And of course, since any triangle has *three* medians, this meant there was one more segment, which would complete the trio of which the Euler and Nagel segments were now seen to be simply the first two. Excitedly, I sketched the third median by joining the midpoint of the segment *NH* to *G* and extending this segment by twice its length. The next step was to draw the triangle with these three medians (Figure 10). I was starting to tingle now, because it seemed to me that since *two* of the medians were known segments having very important endpoints and midpoints (not to mention sublime properties), it almost had to be the case that this *third* median's endpoints and midpoint would also have absolutely fundamental properties.

Was I on the verge of a significant discovery? It certainly felt that way. My new triangle gave me a strong sense of symmetry and closure—the sense of bringing something beautiful but unfinished to its inevitable completion. Since two of its sides' midpoints were named by vowels, I called the new midpoint "*U*", and the new vertex became "*T*". I also gave the name "*V*" to the midpoint of the new median. This point was the counterpart to the nine-point center *P* and the Spieker center *S*. So my new segment—the third median and the analogue of segments *IGSN* and *OGPH*—was

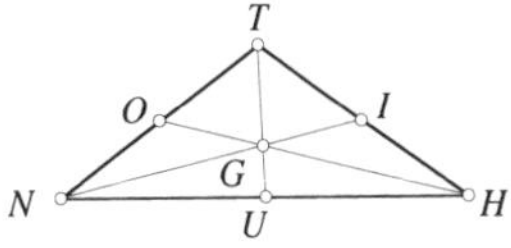

FIGURE 10

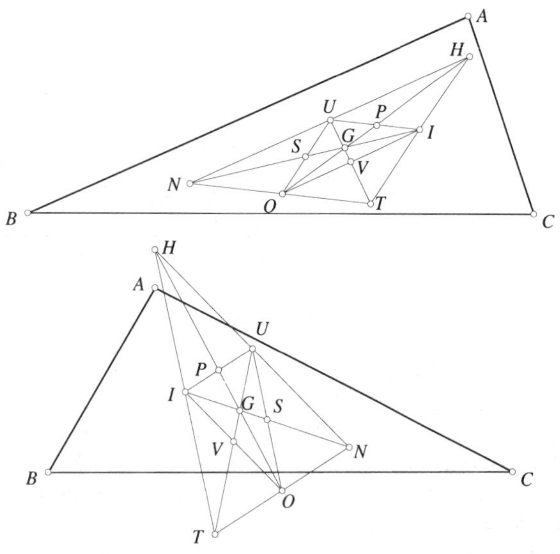

FIGURE 11

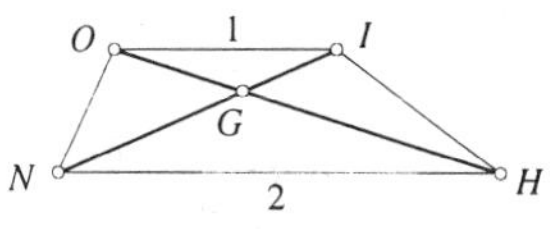

FIGURE 12

UGVT. Betraying my rather juvenile excitement, I gave the triangle as a whole the corny name "Magic Triangle". Figure 11 shows the Magic Triangles for two different triangles *ABC*.

The irony did not escape me that two of my Magic Triangle's sides (*HI* and *NO*) were lines that I had earlier rejected as meaningless. At the earlier time, it had not occurred to me that the lines' intersection-point might be of interest.

At about this point, I started feeling a little self-conscious about the hokey name "Magic Triangle", and I found a more poetic name that seemed to say a little more and to fit very nicely: *hemiolic crystal*. With its crisscrossing medians, its six external points, and its four internal points, it looked at least a little bit like a crystal. The term "hemiolic" came from the musical notion of *hemiola*, which refers to the ambiguity inherent in a six-beat measure: should it be heard as three groups of two notes each, or as two groups of three notes each? Here, should we think of the six points forming the exterior of this crystal as two sets of three points (*HNT* and *OIU*)? Or should we rather think of them as three sets of two points (*OH, IN,* and *UT*)?

All this excitement was predicated on something entirely unsure: that my new third segment was in some sense meaningful. I was convinced it would have to be, but I still didn't know the first thing about its new points T, U, and V! I could imagine three possibilities:

(1) All three of my supposedly new points T, U, and V turn out to be known entities. If this is the case, fine—because I would have revealed an unsuspected unity among them (the fact that they belong to a third segment much like the Euler and Nagel segments).

(2) At least one of my new points *T, U,* and *V* is a new discovery and turns out to have interesting and novel properties. This is even better than (1), because then not only have I discovered a sublime high-level trio and a great new segment, but also one or more nifty new special points!

(3) None of my new points *T, U,* and *V* has any interesting property at all. On one level, this would seem pretty dismal, but just think—on another level, the absence of meaning would be so strange that it would in itself be fascinating.

All in all, then, I felt secure no matter how things might turn out—provided, that is, that the whole idea hadn't already been discovered, in which case I would just be a jackie-come-lately.

Certainly Coolidge hadn't known about the new segment or the crystal in 1916, for his book doesn't mention either one at all. In fact, though, he came pretty close to finding them—he explicitly talks about the "trapezoid" defined by the points *OIHN* (Figure 12), but he doesn't see that it naturally beckons one to complete it by putting a little triangle on its short top-side.

And Roger Johnson, a disciple of Coolidge's who came out with his own scholarly treatise on circles and triangles in 1929, briefly describes the unnamed second segment, but doesn't mention any connections between it and the Euler segment. So at least up 'til 1929, my crystal was certainly completely unsuspected. "And," I thought to myself, "how much deep exploring of triangles has there been since 1929?"

Later, I found a more recent book by Nathan Altshiller-Court—*College Geometry: An Introduction to the Modern Geometry of the Triangle and the Circle,* published in 1952. This extremely comprehensive treatise mentioned (and did not name) the Nagel segment but didn't connect it with the Euler segment, much less hint at the existence of a third, related segment.

All this gave me quite a bit of hope that I was the first to see this gem. Was I truly the first person in decades to find a major new property of the triangle?

The day after I had come up with the notion of the crystal itself, I made an all-out effort to find its meaning—that is, to "decipher" its new points. My first attempt focused on V, because it seemed the easiest. My idea was that, just like

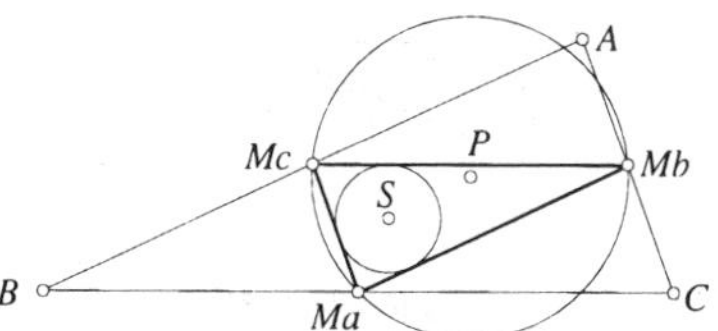

FIGURE 13

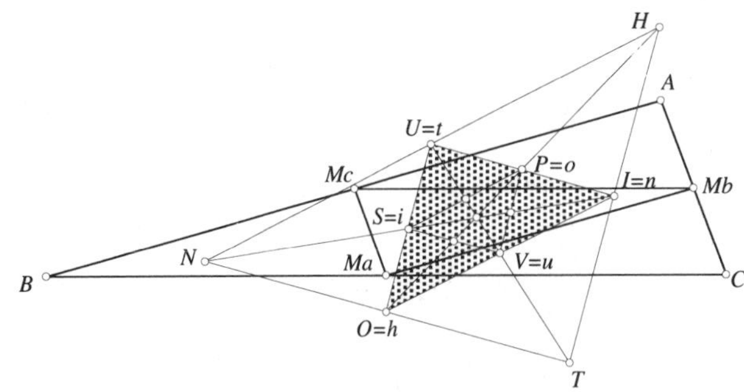

FIGURE 14. The hemiolic crystal of triangle *ABC* is *HUNOTI*; that of its medial triangle *MaMbMc* is *hunoti*. Since the medial triangle is half the size of *ABC* and centered on the same point (G) but rotated 180° with respect to it, its hemiolic crystal is likewise half the size of *HNT*, centered on the same point as it, and rotated 180° with respect to it. This establishes a one-to-one correspondence between the points *OVIPUS* and the points *hunoti*, in that order.

P and S, its counterparts on the other two medians, V ought to be the center of some spectacular circle that managed to jump through several hoops at once. So on the screen I drew a circle of variable size centered on V and let it shrink and grow while carefully looking for coincidences of any sort—simultaneous tangencies here and there, simultaneous passages through various points, simultaneous whatever! But I saw nothing. It was somewhat shocking, but I took it in stride. After all, it would be nice if V had beautiful properties but ones so subtle that nobody before me had ever noticed them. I still had high hopes that V would in the end turn out to have lots of meanings.

As a result of this mild setback, I went back to thinking a bit more about P and S, since it was on them that my analogy was based. On reflection, it seemed to me that both P and S really belonged more to *ABC*'s medial triangle (the triangle formed by the midpoints of *ABC*'s sides) than to *ABC* itself, because P is the center of that triangle's circumcircle and S is the center of its incircle. Semi-symbolically stated, P is the O of the medial triangle, and S is the I of the medial triangle (see Figure 13).

When I thought about this in connection with the medians of my crystal, the analogy seemed clearly to be telling me this: "V is the U of the medial triangle." But what did this mean? It seemed to me to suggest that the meaning of V would be elucidated only after I elucidated the meaning of U, and so this pushed me to switch the focus of my quest to the two new points directly on the hemiolic crystal (T and U), rather than the single new point inside it.

I made a small and rather simple discovery at this point, which was that by drawing the segments connecting O, I, and U inside the hemiolic crystal of *ABC*, I was thereby constructing the hemiolic crystal of *ABC*'s medial triangle. In rather opaque language, I had found the following result:

> OVIPUS–*hunoti* Theorem. The hemiolic crystal of *ABC*'s medial triangle is the medial triangle of *ABC*'s hemiolic crystal.

It sounds kind of grandiose, but really it is quite trivial. Figure 14 illustrates this.

At this point, I went off on a long and completely fruitless wild-goose chase, motivated by a very jumbled-up analogy that I made in a confused moment, involving my new points T and U and several of the old points of the hemiolic crystal. At the end of this wasted time I got extremely discouraged about the probability of my crystal's meaningfulness—so discouraged, in fact, that I was even led to questioning whether I had correctly understood the passage in Coolidge where I had learned about the so-called "Nagel segment".

So I went back to Coolidge and rechecked the passage in which he describes the anonymous Euler-like segment. To my relief, I found I had gotten it exactly right. Very carefully, I hunted through Coolidge for any possible hints of things to look for in old or new dynamograms. It was then that I first noticed the rather important role played by two "auxiliary triangles"—one involving *ABC*'s orthocenter H, the other involving *ABC*'s Nagel point N. Each auxiliary triangle was constructed in the same way. You took a point X (H in one case and N in the other), connected it in turn with A, B, and C, and then bisected those segments. The three midpoints defined X's auxiliary triangle.

Coolidge was whispering a secret, but to my ears his analogical message was loud and clear: *Construct the auxiliary triangle belonging to T*! Clearly, it was time to wake up *Geometer's Sketchpad* again. I first constructed a triangle *ABC* and its hemiolic crystal *HUNOTI*, then the auxiliary triangle *TaTbTc* belonging to T. In its original form, this picture was very cluttered. I couldn't make head or tail of what was going on (Figure 15).

It seemed I would simply have to do some judicious pruning—namely, I would have to "hide" a bunch of points and lines. So, I started stripping away line after line, point

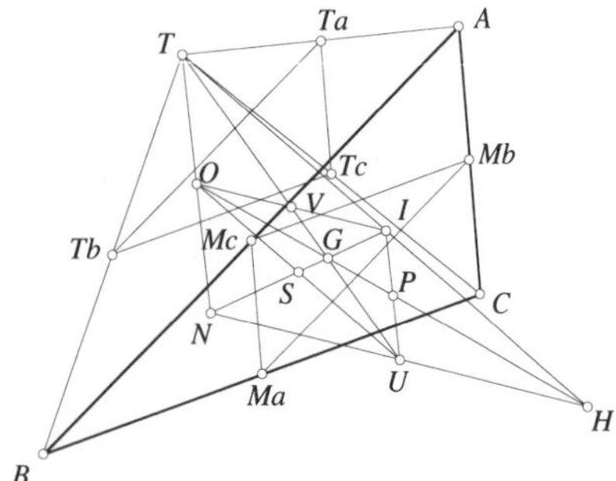

FIGURE 15

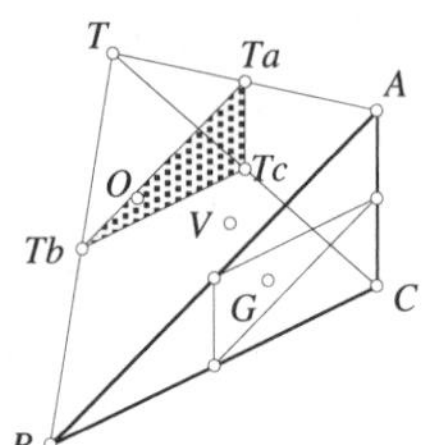

FIGURE 16

after point, in an attempt to boil the picture down to its essence. I pared the image down very gradually, winnowing out a point here, a line there, until it started seeming that I had reached roughly the right level of simplicity. I also highlighted T's auxiliary triangle by shading it, since that was supposed to be the real focus of my perception (Figure 16).

I kept on looking for something interesting. And, as I randomly dragged a vertex of ABC around, a small coincidence caught my eye—the fact that point O kept staying inside the shaded region. It seemed inconsequential, but I wondered, "Is this really always so? Why would that be the case?" So I kept on watching it as ABC changed shape, and indeed it never left the shaded region. I noticed that, whenever *TaTbTc* got long and pointy, O seemed to cling very close to one of the longer sides, and to slide down toward the pointy end. This rang a bell. I recognized this trait—it's a characteristic of the Nagel point of a triangle! (How did I know this? Easy—I had spent some time watching Nagel points sliding around inside their home triangles, during an earlier stage of my geometry binge.) At last, I was picking up a meaningful message: O seemed to be acting like the Nagel point of T's auxiliary triangle.

Of course, this needed some confirmation. To make sure that my intuition was right, I quickly added some new construction lines to my dynamogram, which defined the actual Nagel point of T's auxiliary triangle. This new point, N', landed smack on top of O, and, as I moved things around, their co-incidence never changed at all. I was in like Flynn!

For a moment, I was a bit worried that this, too, might prove to be yet another false epiphany—another piece of geometrical trivia—but, as I considered it carefully, I came to the firm conclusion that this was a meaningful and unexpected finding. So now I knew at least something definite and new about T:

> *Triangle ABC's circumcenter O is the Nagel point of the auxiliary triangle of T.*

It wasn't exactly the most perspicuous and exciting property of all time, but I felt that I now had genuine confirmation that the new points of the hemiolic crystal were not meaningless. This was enough for one day, and I went to bed with a feeling of great satisfaction.

The next morning, however, I felt distinctly uneasy with the current stage of my discovery. For a segment that was supposed to be on a par in importance with the Euler segment, my new segment was surely not acting very important. The only piece of information I had about it was a rather obscure fact about one of its endpoints. Surely, there must be more to the new points than just that one teeny theorem!

By the way, note that I just referred to my screen-based observation as a "fact" and a "theorem". Now any red-blooded mathematician would start screaming bloody murder at me for referring to a "fact" or "theorem" that I had not proved. But that is not my attitude at all, and never has been. To me, this result was so clearly true that I didn't have the slightest doubt about it. I didn't need a proof. If this sounds arrogant, let me explain. The beauty of *Geometer's Sketchpad* is that it allows you to discover instantly whether a conjecture is right or wrong—if it's wrong, it will be immediately obvious when you play around with a construction dynamically on the screen. If it's right, things will "stay in synch" right on the button no matter how you play with the figure. The degree of certainty and confidence that this gives is downright amazing. It's not a proof, of course, but in some sense, I would argue, this kind of direct contact with the phenomenon is even more convincing than a proof, because you really see it all happening right there before your eyes. None of this means that I did not want a proof. In the end, proofs are critical ingredients of mathematical knowledge, and I like them as much as anyone else does. I just am not one who believes that certainty can come *only* from proofs.

My unproven "theorem" could be tersely phrased as follows, using an obvious notation to symbolize the notion of "auxiliary triangle":

$$O \text{ is the } N\text{-point of the } T\text{-}\Delta.$$

Since my feeling about the crystal was that it is a highly symmetric structure, it seemed natural to wonder whether analogous statements might not hold for the points on the other two sides—*HUN* and *TIH*. Thus I was led, purely by a sense of elegance, analogy, and symmetry, to make the following two rather bold speculations:

U is the H-point of the N-Δ.
I is the T-point of the H-Δ.

I didn't have a great deal of confidence in either of them, because, although formally they made a symmetric trio with the first one, when you looked at their meanings, they said amazingly different things. Here are all three, spelled out more completely:

(1) *O, the circumcenter of ABC, is the Nagel point of T's auxiliary triangle.*
(2) *U is the orthocenter of the auxiliary triangle belonging to ABC's Nagel point.*
(3) *I, ABC's incenter, plays the T role for the auxiliary triangle of ABC's orthocenter H.*

These statements seem to have nothing to do with one another! Still, I felt there was at least a sporting chance that this guess might pan out, so I pitter-patted down the hall in my bedroom slippers, turned on the old Mac, and clicked on my faithful "verification engine", *Geometer's Sketchpad.*

In the twinkling of an eye, I had made a new dynamogram showing *ABC*, its Nagel point N, and its U point. I then constructed N's auxiliary triangle. The question was, did U look like that triangle's orthocenter? It looked at least plausible on the screen, but I needed proof—"eyeball proof", that is. And so I constructed the three altitudes of N's auxiliary triangle. The first one ran straight through U, skewering it perfectly. So did the second, and so did the third. Bingo! Moreover, when I moved the vertices of *ABC*, everything stayed completely right on target (Figure 17). There was no doubt that the first of my two analogy-based speculations was true! It was quite a stunning moment for me.

The question remained, what about the other speculation? I wasn't going to deny myself the pleasure of actually seeing its truth on the screen. So I made the dynamogram, and played around with it. To my astonishment, the two points that were supposed to be coincident with each other were moving around completely independently of each other. No relationship at all! Could it be that my third and final statement was wrong while the other two were right?

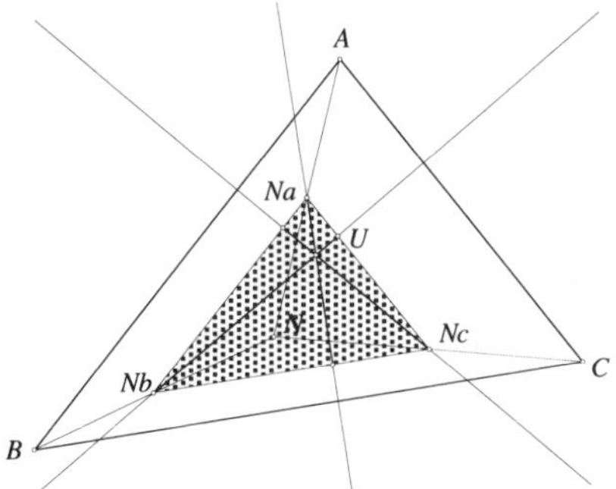

FIGURE 17. Eureka! U is the H of N's auxiliary triangle!

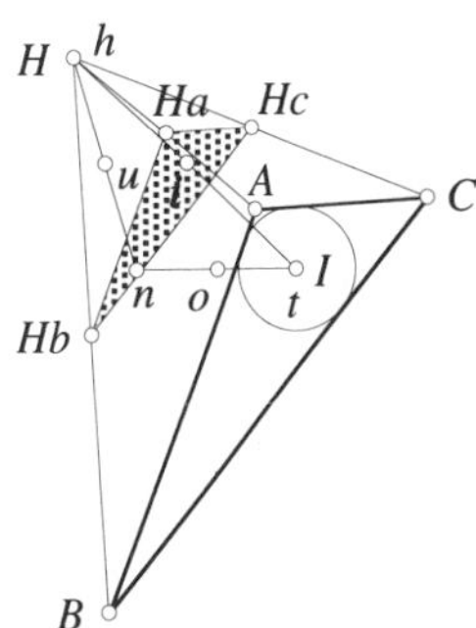

FIGURE 18

I had a moment or two of self-doubt, but then my ever-strong belief in beauty and symmetry—a kind of "inner compass"—resurged and started to gain the upper hand. I thought to myself, "Don't be silly. The theorem is of course true. Let me try again." And in trying it again, I discovered the slight error I'd committed. Indeed, the correctly-built new dynamogram fully verified the third member of the trio, and victory was mine (Figure 18).

It strikes me that a key ingredient here was the fact that I had enough self-confidence to trust my "inner compass" more than what I saw before me on the screen. If I hadn't, I might well have been stopped in my tracks, and never made the third discovery that closed the circle—or the triangle, to be more accurate.

I decided to call my cyclic trio of results the "Garland Theorem", as it represents a "garland" of linked results. I was now extremely satisfied: I had found enough to convince me that the hemiolic crystal was a nontrivial new idea with at least a few elegant and nonintuitive (or non-obvious) properties. So, at the conclusion of my second day, this was how things stood.

The next morning, I felt far less satisfied. What I had found was just a set of three little curiosities. Big deal! I needed something much more impressive. I also felt I needed proofs. Never having been the completely analytic do-it-in-your-head type, I went back to *Geometer's Sketchpad*, to be in contact with the phenomena themselves in a very concrete way. The obvious thing to do, it seemed, was to take an auxiliary triangle such as H's, and to take *ABC*, and to exhibit their full hemiolic crystals together on the screen. To be sure, this might make a hugely messy screenful, but

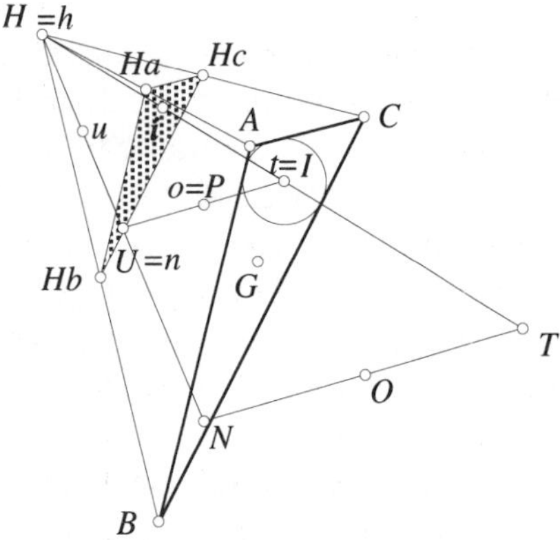

FIGURE 19

then again, it might not. No harm trying. Figure 19 shows what I found.

The diagram was surprisingly unmessy—in fact, as simple as it could possibly have been. First of all, *ABC*'s orthocenter H coincides with the orthocenter h of H's auxiliary triangle (this is provable in a snap). This, together with the obvious fact that the auxiliary triangle is half the size of *ABC* and oriented parallel to it, implied that the two hemiolic crystals, one big and one small, must fit snugly together, their sides perfectly superimposing. Thus two of the midpoints of the big one had to coincide with two vertices of the small one, and this implied the three further point coincidences $n = U$, $o = P$, $t = I$. Rather than being baffling, it all made sense in a very straightforward manner. So not only had I now discovered some new facts, I also had a clear understanding of why these facts were true. In other words, as new results started pouring in, so did proofs of them! I couldn't have asked for more.

Two more diagrams—one involving the auxiliary triangle of N and one involving the auxiliary triangle of H—gave me precisely analogous results, and for precisely analogous reasons: the snug nesting of a half-size triangle inside the full-size one. Since each diagram gave me four results where the day before I had had just one, I had almost effortlessly multiplied the content of the Garland Theorem by a factor of four, so that it now ran this way:

> *The "Garland" Theorem.* For any point X, define X's auxiliary triangle, denoted "X-Δ", by joining X with each of A, B, and C, and connecting the midpoints of those segments. The following relations then hold among the points of *ABC*'s hemiolic crystal:
>
> $(1h)$ H is the H-point of the H-Δ.
>
> $(1n)$ N is the N-point of the N-Δ.
>
> $(1t)$ T is the T-point of the T-Δ.
>
> $(2h)$ U is both the H-point of the N-Δ and the N-point of the H-Δ.
>
> $(2n)$ O is both the N-point of the T-Δ and the T-point of the N-Δ.
>
> $(2t)$ I is both the T-point of the H-Δ and the H-point of the T-Δ.
>
> $(3h)$ P is the O-point of the H-Δ.
>
> $(3n)$ S is the I-point of the N-Δ.
>
> $(3t)$ V is the U-point of the T-Δ.

And its proof was now clear, to boot. I was getting to the point where it felt like the hemiolic crystal really was a significant new idea in the geometry of the triangle. Even if my discovery turned out not to be new, these two days had been without a doubt the most exhilarating mathematical experience I had had in 30 years. I was truly thrilled—nay, bedazzled—by the hemiolic crystal I had unearthed. And I had a hunch that what I had seen so far was still just the tip of the iceberg.

Bewildered . . . by the Meaning of it all

Math-making being as elusive as it is, I feel it could be instructive to make a few comments about this particular act of geometry-making, an act that I was fortunate enough to be able to watch from a ringside seat as an ongoing process in my own mind. First of all, as I hope to have made clear, I am not much of a geometer. I am just a geometry amateur, and something of a bumbling amateur, at that. More often than I like to admit, I have "false epiphanies" in which I mistake trivia for profundities, and I have a rather primitive idea of how to construct proofs. Compared to a titan like H.S.M. Coxeter, I have but a minuscule storehouse of knowledge. So why in the world was I the lucky one to have found this gem? Although some luck was certainly involved, I don't think it was a complete accident.

First of all, I am apparently just a bit more taken than are most geometers with the mystery of the special points of a triangle. In fact, I doubt that most would use the word "mystery" in this connection. Yet in these points and their beautiful patterns I sense something almost mystical about the "essence of triangularity". To most geometers, it probably doesn't feel quite that compelling. That in itself is symptomatic of a significant attitude-disparity. And my unswerving fascination has led me on, something like an infatuation or even an obsession, to muse for months on end about special points and their interrelationships. I didn't call the first section of this essay "Bewitched" for nothing! So that's the first point.

Secondly, I am someone who for years has been in love with analogies, and with the concept of analogy itself. Not only do I use analogies in my thinking and my writing all the time, I am also constantly jotting down other people's analogies, wondering how they arise, pondering what makes some of them good and others bad, and musing about what they say about the subcognitive mechanisms of the mind. For almost 15 years, my professional research has been largely about making a computer model of how people make analogies and use them in the creative process. So you can imagine that, when I came across the analogy between the Euler and Nagel segments, I was electrified. It was not just a beautiful analogy, but a beautiful analogy smack-dab in the midst of a field that I was intoxicated with already. It was a mind-boggler of an analogy, and to add to that, I was baffled as to why nobody but nobody had mentioned it in any books on geometry since Roger Johnson's 1929 treatise. This struck me as so irrational and narrow-minded that it also served as a kind of goad to my curiosity. Was there something I was missing? Was the Nagel segment, despite its seeming fundamentality, just a trivial sidekick to the Euler segment? Was it worthy of nothing but a footnote (if even that) when the Euler segment was treated like a glamorous movie star? I couldn't figure this out, and the image of these two "twin" segments, each one cutting the other in that strange, lopsided way, just etched itself into my brain.

These two factors together added up to something unique, I guess—this special-points-of-a-triangle analogy par excellence started swirling around in my head over and over again. I couldn't let it loose, or, rather, it wouldn't let me loose. It was the analogical unity of the two segments that caught my imagination, much more than either segment on its own. Coolidge's systematic, point-by-point listing of parallels between them certainly was a critical element, and, I have to say, so was the bizarre fact that he himself voiced no curiosity as to why there was this deep and beautiful analogy at the heart of triangularity. The fact that this obvious, salient question went completely unasked was almost as much a source of mystery to me as was the analogy itself.

One further key factor that mustn't be overlooked is the fortuitous existence and tremendous power of *Geometer's Sketchpad*. Somehow, this program precisely filled an inner need, a craving, that I had, to be able to see my beloved special points doing their intricate, complex dances inside and outside the triangle as it changed. And my own personality welcomed a computer program to explore mathematics, and felt that it afforded visions of geometric truth, an attitude that perhaps would be a little bit less accepted by a traditional mathematician. In short, living in the 1990s and having a Macintosh and enjoying computers was also part of it.

I have to admit, there is one last crucial factor that allowed me, of all people, the privilege of making this discovery (if discovery it is): the incredible downplaying and neglect, by several generations of mathematicians the world over, of the "anonymous" segment so much like Euler's. So, to mathematicians everywhere, for not looking in this direction, I hereby express my great debt of gratitude. Thank you!

Putting on my cognitive-scientist's hat now, I would like to point out something that struck me as I reviewed this chronicle of my discovery process—namely, the large number of analogies, good and bad, that figured critically in it. Here, then, is a list of the main analogies that I think served as guiding or misguiding forces, with a brief comment on each one:

(1) My vague, intuitive feeling that special points in a triangle are very much like special constants on the real line, such as e and π. Since I have deeply loved such constants from childhood, this analogy was in part responsible for getting me so hooked on geometry.

(2) Coolidge's systematic mapping between properties of the Euler segment and properties of the Nagel segment. This was a bolt out of the blue.

(3) Seeing the three parallel lines HN, OI, and PS as a physical instantiation of the Euler/Nagel analogy—in other words, a meta-analogy that maps a visible geometric diagram onto an abstract analogy.

(4) Looking at the crisscrossing-medians diagram that arose in a simple puzzle, and recognizing in it the much more profound Euler/Nagel diagram.

(5) Mapping the new point V onto the known points P and S because of their analogous positional roles in their respective segments, and concluding that V ought to be the center of some important circle associated with ABC.

(6) The scramble-brained analogy that led me so far astray for several hours one day that I eventually felt compelled to return to Coolidge's book for confirmation of my sanity—where I then chanced upon another key analogy—namely. . .

(7) The idea of constructing T's auxiliary triangle, arrived at by analogy with two other constructions described in Coolidge.

(8) Looking at the behavior of a certain point marked "O" in a certain dynamogram, and seeing it as Nagel-point-like behavior with respect to a triangle it was inside.

(9) Seeing the three letters "O, N, T" not just as standing for concepts involved in an abstract relationship, but as symbolizing one side of the crystal.

(10) Jumping from a discovery attached to the *NOT* side of the crystal to the idea that maybe two further analogous results would hold, attached to the *HUN* and *TIH* sides of the crystal.

Any number of people have looked at a picture of crisscrossing medians cutting each other in their characteristic way, yet not seen the Euler and Nagel lines in it. The difference was, of course, that I was rather obsessed with the Euler/Nagel connection, so I came to the medians picture with a highly biased eye. These cases exemplify *seeing something as something else.* They are the kinds of things that cognitive scientists interested in the deep underpinnings of creativity ought, in my opinion, to study very carefully.

The story of the hemiolic crystal is by no means a closed book. Indeed, I see many fascinating avenues to explore. One has to do with the fact that you can take an unlabeled picture of a hemiolic crystal and legally label its points in four different ways. (It might seem that there should be six ways to label it, but, as it turns out, there is a constraint: the *UT* median cannot be as long as the Euler median *OH*. This limits the possible labelings to four.) Thus one has *HUNOTI, HITONU, NUHITO,* and *NOTIHU*. All the "cousin" crystals belong to cousin triangles *ABC*, $A'B'C'$, $A''B''C''$, and $A'''B'''C'''$—and so the obvious question is: How are all four cousin triangles related to one another?

But the biggest remaining mystery for me concerns the meanings of the points U and T. Although they play beautifully symmetric roles in the Garland Theorem, I have so far been unable to find any concise and catchy characterizations for them on their own. Just what is the T-point? What is the U-point? Mysteries beckon, mysteries call, mysteries ever lure me on. . .

A few weeks ago, flush with excitement about the earliest of these discoveries, I penned a letter to the great geometer H.S.M. Coxeter, hoping to see whether my ideas were new and of merit. After briefly describing my newfound passion for geometry and, recounting my discovery of the hemiolic crystal and its properties, I concluded with the following lines: "I will never be quite the same, after having drunk so deeply from the infinite well of geometry. My life is in some central way forever changed, thanks to the mysteries and beauties of triangles and circles." And so it is.

Endnote

1. Actually, this characterization holds only for acute triangles.

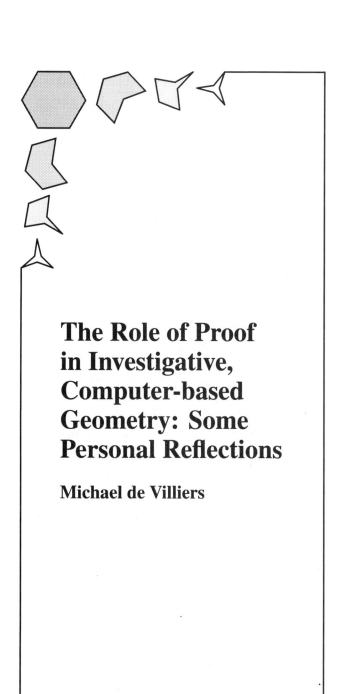

The Role of Proof in Investigative, Computer-based Geometry: Some Personal Reflections

Michael de Villiers

There is a tendency amongst many mathematicians to report only their final results in a neatly organized fashion, not discussing and reflecting much upon the processes of discovery/invention and proof. This tends to give a distorted perspective of mathematical creativity as being always purely deductive.

In what follows, I intend to present some personal discoveries I made recently with the aid of dynamic geometry software. At the time of the investigation, I believed these results to be original and was not aware that two had been published elsewhere. In all these examples, actual conviction preceded the eventual proofs, and the purpose is to reflect on the need and function of proof in such cases. Readers are also encouraged to first investigate the results with their own dynamic constructions before reading the proofs. A personal model of how new discoveries are sometimes made in mathematics, as well as that model's relevance to mathematics education, will also be briefly discussed.

Example 1

The false impression is sometimes created that mathematicians are only problem solvers who spend most of their time trying to solve *already given* problems. However, mathematicians continually create their own new problems by asking questions, making hypotheses, and testing them.

A problem-posing heuristic I have often found useful is to ask "*what—if?*" questions when coming across any mathematical result. Consider for example Figure 1, which shows a convex quadrilateral $ABCD$ with equilateral triangles ABP, BCQ, CDR, and DAS constructed on the sides so that the first and third are exterior to the quadrilateral, while the second and the fourth are on the same side of sides BC and DA as is the quadrilateral itself. Then quadrilateral $PQRS$ is a parallelogram.

When I first came across this result in Yaglom [1962, p. 39], I immediately wondered *what would happen if,* instead

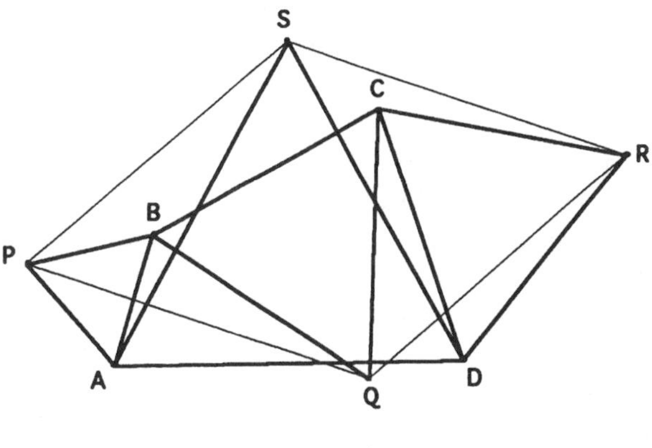

FIGURE 1

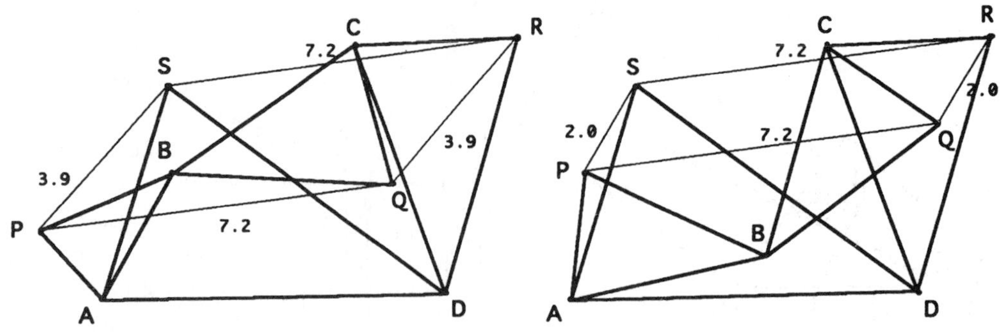

FIGURE 2

of equilateral triangles, *similar* triangles PBA, QBC, RDC, and SDA were constructed on the sides. After quickly constructing a dynamic version of the figure with *Cabri* by using a macro-construction for the similar triangles on the sides as shown in Figure 2, I found that $PQRS$ remains a parallelogram. By measuring opposite sides and dragging the vertices of the figure around the screen, it can be seen that this result appears to be true in general. In fact, I used the property checker of *Cabri* to check whether both pairs of opposite sides were parallel, and each time it gave the message: "*this property is* ***true in a general position***.[1]"

Although I was initially very skeptical about the property checker on *Cabri,* I have yet to "*catch*" it out and have consequently learnt to trust it to a very large degree when investigating new or unknown conjectures like that above. Armed with conviction that the generalization was indeed true,[2] I then proceeded with the task of constructing a deductive proof. Why did I still feel a need to prove the above result if I was already convinced of its truth?

Firstly, it is important to point out that it is precisely because I was convinced of its truth that I felt challenged to find a deductive proof, not because I doubted the result. Why? Well, here was a result that was obviously true, and I was intrigued to try to find out *why* it was true. If the above construction had not shown $PQRS$ to be a parallelogram, I would certainly not have wasted my time trying to find a proof. I therefore experienced the search for and eventual construction of a deductive proof in this case as an intellectual challenge, definitely not as an epistemological exercise in trying to establish its "*truth*."

The proof I devised is a "dynamic" one, that is, it uses transformations that act on the figure. I later learned that a dynamic proof of the result had been given earlier by Ross Finney [1970]. (His paper cites a reference from 1881 for the result.) Unlike Finney, who in 1970 had no access to dynamic geometry software, I could witness the transformations of my figure on a computer screen and visually verify the claims made about their actions.

Example 2

Some years ago I came across Van Aubel's theorem in Martin Gardner's book *Mathematical Circus* [1981, pp. 176–179], namely, that the centres of squares on the sides of any quadrilateral $ABCD$ form a quadrilateral $EFGH$ with equal and perpendicular diagonals (see Figure 3).[3] Again I wondered what would happen if, instead of squares on the sides, one constructed similar rectangles or rhombi on the sides. It was however not until recently that I had an opportunity to investigate these questions.

After some initial experimentation on the arrangement of the similar rectangles and rhombi on the sides, I discovered the following two generalizations of Van Aubel:

(1) If similar *rectangles* are constructed on the sides of any quadrilateral as shown in Figure 4, then the centres of these rectangles form a quadrilateral with *perpendicular* diagonals.

(2) If similar *rhombi* are constructed on the sides of any quadrilateral as shown in Figure 5, then the centres of these rhombi form a quadrilateral with *equal* diagonals.

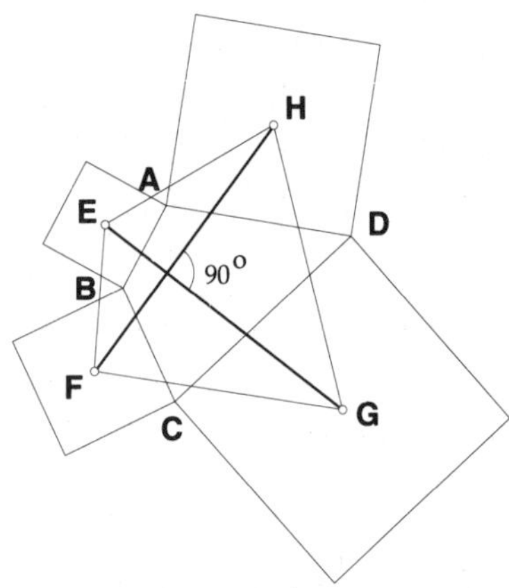

FIGURE 3

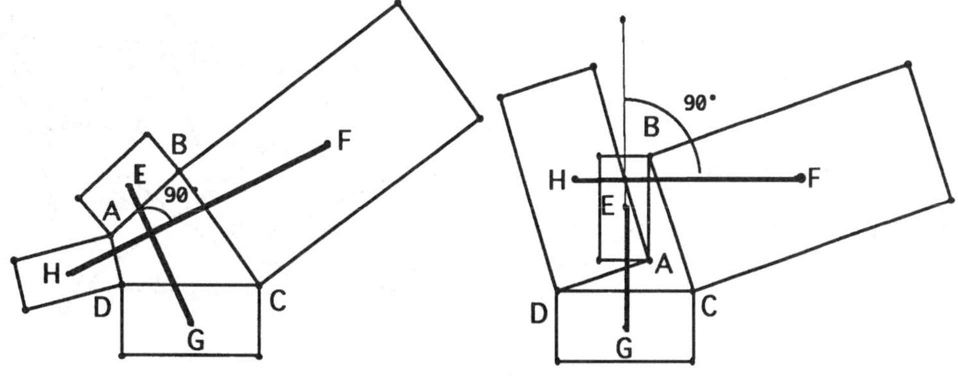

FIGURE 4

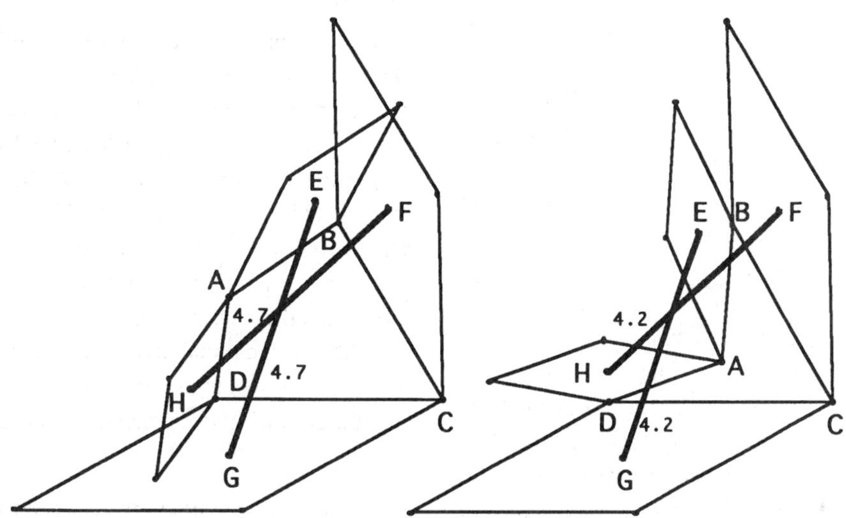

FIGURE 5

Again in both cases, it was very easy to click and drag any of the vertices of $ABCD$ around the screen to see whether EG remains perpendicular to HF in the first case, and whether EG remains equal to HF in the second case. Again I used the property checker of *Cabri* to verify that both results were indeed "*true in general.*"[4]

As before, this experimental confirmation motivated me to start looking for proofs. As was the case previously, I did not really experience a need for further certainty, but rather of *explanation* (why were they true?) and of *intellectual challenge* (can I prove them?). Interestingly, when the base quadrilateral $ABCD$ is a parallelogram, the quadrilateral $EFGH$ is a rhombus in the first case and is a rectangle in the second case. Also note that Van Aubel's theorem is the intersection of the above two results, just as squares fit the intersection of the definitions of rectangles and rhombi.

Proofs. The proofs of (1) and (2) below utilize two Lemmas concerning transformations. The first Lemma is proved in Yaglom [1962, pp. 34–35], and the second in Yaglom [1968, pp. 42–43]; both Lemmas can be found in De Villiers [1996]. A spiral transformation denoted (k, z) is the sum (composition) of a dilation with scale factor k and a rotation through angle z about a fixed centre.

Lemma 1. *Let α and β be angles. If $\alpha + \beta \neq 360°$, then the sum of two rotations with centres O_1 and O_2 through angles α and β, respectively, is a rotation through angle $\alpha+\beta$ around a centre O such that $\angle O_1 O O_2 = 180° - \frac{1}{2}(\alpha + \beta)$. If $\alpha + \beta = 360°$, the sum of two rotations with centres O_1 and O_2 through angles α and β, respectively, is a translation if there are no fixed points and the identity otherwise.*

Lemma 2. *The sum of two spiral similarities $(k, 90°)$ and $(\frac{1}{k}, 90°)$ around centres O_1 and O_2 is a halfturn around a centre O which is the vertex of an isosceles triangle $O_1 O O_2$ with $OO_2 = OO_1$ and angles $\angle OO_1O_2 = \angle OO_2O_1 =$* $\arctan\left(\frac{1}{k}\right)$.

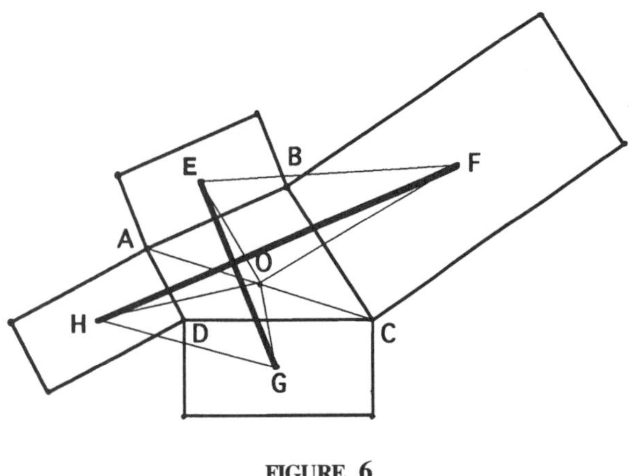

FIGURE 6

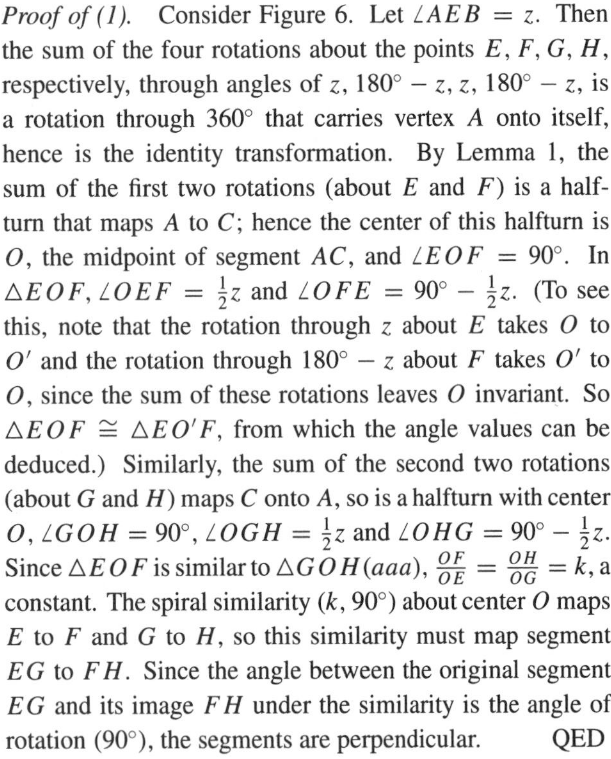

Proof of (1). Consider Figure 6. Let $\angle AEB = z$. Then the sum of the four rotations about the points E, F, G, H, respectively, through angles of z, $180^\circ - z$, z, $180^\circ - z$, is a rotation through 360° that carries vertex A onto itself, hence is the identity transformation. By Lemma 1, the sum of the first two rotations (about E and F) is a halfturn that maps A to C; hence the center of this halfturn is O, the midpoint of segment AC, and $\angle EOF = 90^\circ$. In $\triangle EOF, \angle OEF = \frac{1}{2}z$ and $\angle OFE = 90^\circ - \frac{1}{2}z$. (To see this, note that the rotation through z about E takes O to O' and the rotation through $180^\circ - z$ about F takes O' to O, since the sum of these rotations leaves O invariant. So $\triangle EOF \cong \triangle EO'F$, from which the angle values can be deduced.) Similarly, the sum of the second two rotations (about G and H) maps C onto A, so is a halfturn with center O, $\angle GOH = 90^\circ$, $\angle OGH = \frac{1}{2}z$ and $\angle OHG = 90^\circ - \frac{1}{2}z$. Since $\triangle EOF$ is similar to $\triangle GOH(aaa)$, $\frac{OF}{OE} = \frac{OH}{OG} = k$, a constant. The spiral similarity $(k, 90^\circ)$ about center O maps E to F and G to H, so this similarity must map segment EG to FH. Since the angle between the original segment EG and its image FH under the similarity is the angle of rotation (90°), the segments are perpendicular. QED

Proof of (2). Consider Figure 7. From the similarity of the rhombi we have $\frac{EB}{EA} = \frac{FB}{FC} = \frac{GD}{GC} = \frac{HD}{HA} = k$, a constant. The sum of the four spiral similarities $(k, 90^\circ)$, $(\frac{1}{k}, 90^\circ)$, $(k, 90^\circ)$, and $(\frac{1}{k}, 90^\circ)$ about the points E, F, G, H carries vertex A onto itself; hence this sum is the identity transformation. By Lemma 2, the sum of the spiral similarities about E and F is a halfturn; since it maps A to C, the center of the halfturn is O, the midpoint of segment AC. Also, O is the vertex of isosceles triangle EOF with $\angle OEF = \angle OFE = \arctan(\frac{1}{k})$ and $\angle EOF = 180^\circ - 2\arctan(\frac{1}{k})$. Similarly, the sum of the two spiral similarities about G and H is a halfturn

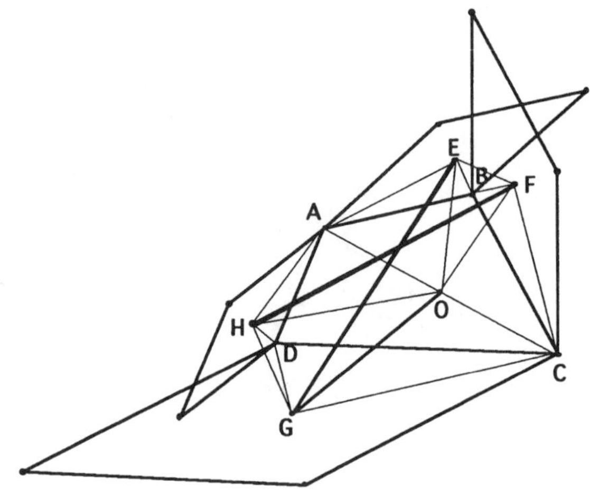

FIGURE 7

that maps C onto A, so has the same center O and $\angle GOH = 180^\circ - 2\arctan(\frac{1}{k})$. Since isosceles triangles EOF and GOH have equal vertex angles, a rotation about O through the vertex angle $\angle EOF = \angle GOH$ maps E to F and G to H. This same rotation must map $\triangle OEG$ onto $\triangle OFH$. Thus corresponding segments EG and FG are equal. QED

It should be pointed out that the above two results can be proved by using vectors, for example, generalizing the approach by Kelly [1966] (see De Villiers [1996]). However, these vector proofs are less explanatory (to me personally) than those given above. It is also possible to generalize the approach of DeTemple and Harold [1996] using the Finsler–Hadwiger theorem to include further dual properties regarding the two general configurations (see De Villiers [in press]). I only recently learnt that the similar rectangles generalization has already appeared in Neuberg (1894).

Example 3

In Ross Honsberger's excellent book *Mathematical Gems III* the reader is introduced to the so-called "*equilic quadrilateral*", namely a quadrilateral $ABCD$ with one pair of opposite sides equal, say $AD = BC$, which are inclined at 60° to each other. (The latter condition might also be stated in the form $\angle A + \angle B = 120^\circ$). Then one of the engaging results he proves is the following:

"If $ABCD$ is an equilic quadrilateral and equilateral triangles are drawn on AC, DC, and DB, away from AB, then the three new vertices, P, Q, and R are collinear" (see Figure 8).

As before, I again wondered what would happen if $ABCD$ was any quadrilateral with opposite sides equal and

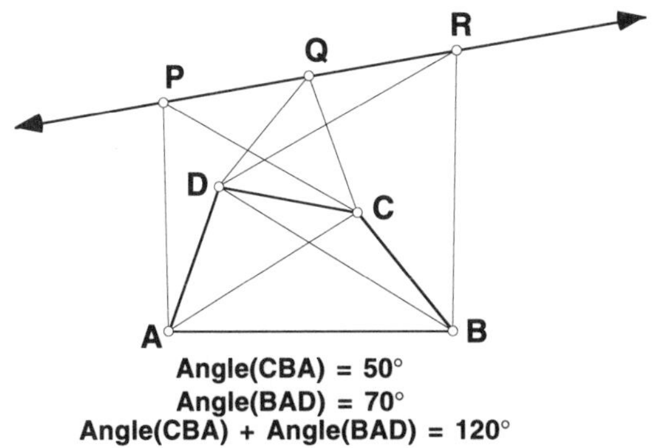

FIGURE 8

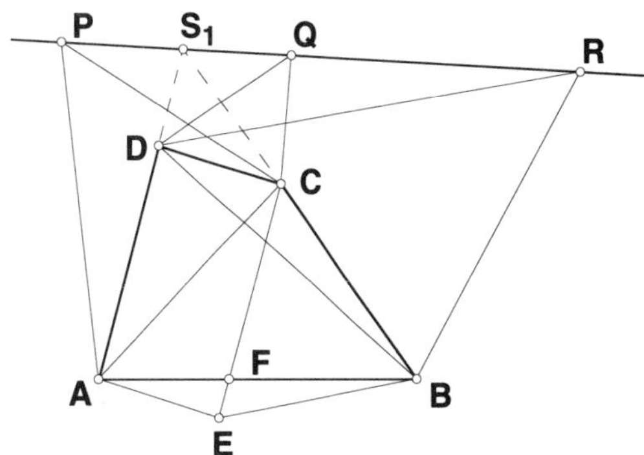

FIGURE 10

triangles PAC, QDC, and RDB similar to each other. Would P, Q, and R then still be collinear?

By investigating these questions with *The Geometer's Sketchpad*, I discovered the following interesting generalization:

"Given any quadrilateral $ABCD$ with $AD = BC$, if similar triangles PAC, QDC, and RDB are constructed on AC, DC, and DB, away from AB so that $\angle APC = \angle ASB$, where S is the intersection of AD and BC extended, then P, Q, and R are collinear" (see Figure 9).

Furthermore, I found that the point S is also collinear with the other three points. Using a dynamic construction with *Sketchpad* as shown in Figure 9, and varying either angle A or B, or the shape of the similar triangles, it was easy to see that the result was true in general. As before, this conviction was based on the continual experimental confirmation of the result, as well as the absence of counterexamples, and provided the motivation to start looking for a deductive proof, an activity I would hardly have embarked upon had I doubted the result. (In such an event, I would have been looking for counterexamples rather than a proof).

Interestingly, after one or two unsuccessful attempts at proving this result, I then noticed, while manipulating the configuration, that $\angle ACB = \angle APQ$, thereby enabling me to construct a proof. This shows how investigation by dynamic software can also sometimes assist in the eventual construction of a proof. Basically, the following proof involves first showing $\angle ACB = \angle APQ$ and $\angle ADB = \angle QRB$, and then that PQ and QR have the same direction. It should be noted that the condition that $\angle APC = \angle ASB$ may also be alternatively stated as $\angle PAC + \angle PCA = \angle A + \angle B$ or $\angle APC = 180° - \angle A - \angle B$.

Proof. Consider Figure 10. Connect P with Q and Q with R. Construct CE parallel and equal to DA as shown. Call the point F where CE cuts AB. Since $ADCE$ is a parallelogram, $\angle CAE = \angle DCA$ (alternate) and $\angle CFB = \angle A$ (corresponding). In triangle CFB we therefore have $\angle ECB = 180° - \angle A - \angle B = \angle APC$.

From the similarity of triangles PAC and QDC, we have $\angle PCA = \angle QCD$. Therefore $\angle PCA + \angle PCD = \angle QCD + \angle PCD$ which implies that $\angle DCA = \angle QCP$, and therefore

$$\angle CAE = \angle QCP \tag{1}$$

From the similarity of triangles PAC and QDC we also have $\frac{PC}{AC} = \frac{QC}{DC}$. But $DC = AE$ since $ADCE$ is a parallelogram, and therefore

$$\frac{PC}{AC} = \frac{QC}{AE} \tag{2}$$

By (1) and (2), triangles AEC and CQP are similar, which

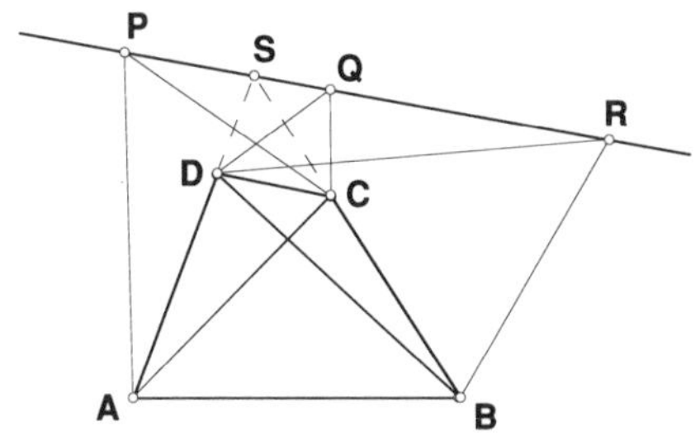

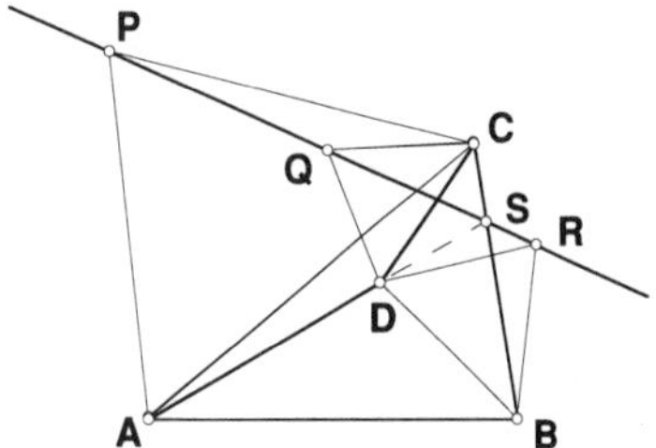

FIGURE 9

implies $\angle ACE = \angle CPQ$. Since $\angle APC = \angle ECB$ (shown earlier), we have $\angle ACE + \angle ECB = \angle APC + \angle CPQ$, and so $\angle ACB = \angle APQ$. By constructing DG parallel and equal to CB, we can prove in a similar manner that $\angle ADB = \angle QRB$.

A counterclockwise rotation of size $\angle PAC$ around A carries C to a point C' on the line through AP, and B to B'. But, since $\angle ACB = \angle APQ$, we have $\angle AC'B' = \angle APQ$ and, therefore, $C'B'$ is parallel to PQ. Thus PQ is inclined to BC at an angle of size $\angle PAC$. Similarly, from a clockwise rotation of size $\angle DBR$ around B that takes A to A' and D to D', we have that $A'D'$ is parallel to QR, that is, QR is inclined to AD at an angle of $-\angle DBR$.

Since AD and BC are inclined towards each other at an angle of $180° - \angle A - \angle B = 180° - \angle PAC - \angle DBR$, rotating BC through $\angle PAC$ and AD through $-\angle DBR$ (in appropriate directions), aligns $B'C'$ and $A'D'$ in the same direction, and we therefore have that PQ and QR also line up in the same direction, i.e. P, Q, and R are collinear.

Now construct $\angle QS_1D = \angle QCD$ with S_1 on QR. Connect S_1 with C. We now prove $\angle DS_1C = 180° - \angle A - \angle B$ and that S_1CB and S_1DA are straight lines, hence S_1 and S are the same point.

From the construction we have $QDCS_1$ is a cyclic quadrilateral (equal angles on same chord). Thus $\angle DS_1C = \angle DQC = 180°$. Also $\angle S_1CD = \angle PQD$.

In triangle PQC, we have $\angle PQD = 180° - \angle CPQ - \angle QCP - \angle DQC = \angle S_1CD$. But $\angle BCD = \angle DCA + \angle ACE + \angle ECB$ and $\angle DCA = \angle QCP$ (proved earlier), $\angle ACE = \angle CPQ$ (proved earlier) and $\angle ECB = \angle DQC$ (construction). Therefore $\angle S_1CD + \angle BCD = 180°$ and S_1CB is a straight line. Similarly we can prove that S_1DA is a straight line. Therefore S_1 and S are the same point, namely the intersection of AD and BC. (Note: if S falls on QP, we simply construct $\angle QS_1C = \angle QDC$ and show in the same manner that S_1 and S are the same point). QED

Looking Back. Carefully looking back at the above proof, I realized that I had never used the property that $AD = BC$; in other words, the result was immediately generalizable to *any* quadrilateral! This illustrates the indispensable value of an explanatory proof which enables one to generalize a result by the identification of the fundamental properties upon which it depends. This point is discussed further in De Villiers [1990].

The Psychology of Mathematical Discovery and Proof

What follows here is a personal model of how new discoveries may sometimes be made in mathematics and is based mainly on the kind of explorations I have done in elementary plane geometry over the past ten years or so. There is no intention to present it as a model which encompasses the complex totality and rich diversity of mathematical discovery and proof.

Logically, mathematics is assumed to be based upon the following fundamental axiom: "*Something is true (T), if and only if, it can be (deductively) proved (P).*" However, from a psychological perspective, it is more useful to represent it in the following equivalent, but different logical forms:

(a) the forward implication ($T \Rightarrow P$): if something is true, then it can be proved.
(b) the converse ($P \Rightarrow T$): if something has been proved, then it is true.
(c) the inverse ($T' \Rightarrow P'$): if something is false, then it cannot be proved.
(d) the contrapositive ($P' \Rightarrow T'$): if something cannot be proved, then it is false.

It is unfortunate that in textbooks and teaching only the converse ($P \Rightarrow T$) is usually conveyed; in other words, that we must first prove results, before we can accept them as true. However, in actual mathematical research, the forward implication ($T \Rightarrow P$), its inverse ($T' \Rightarrow P'$), and contrapositive ($P' \Rightarrow T'$) often play a far greater role in motivating and guiding our actions.

For example, suppose we were to make a conjecture and then test some cases. If the conjecture is not supported by these cases, we reject it as false and according to the inverse do not even bother trying to prove it. On the other hand, if it is supported by these cases, we might start to believe it is true, which according to the forward implication then gives us the encouragement to start looking for a proof. If after a while we are not successful in producing a proof, we might start doubting the validity of the conjecture (according to the contrapositive), and then consider some more cases, after which the whole process is repeated.

This process of conjecturing, testing, refuting, proving, and reformulating can sometimes go through several cycles and is represented in Figure 11. Two famous historical examples which spanned many decades and went through several cycles are the Euler–Descartes theorem and Cauchy's theorem about the continuity of the limit of any convergent series of continuous functions.

In the above model, conviction is not seen as the exclusive prerogative of proof. To the contrary (as shown by our earlier examples), conviction often precedes proof and is probably far more frequently a prerequisite for finding a proof. Why would we spend months or years trying to prove certain conjectures if we weren't already reasonably convinced of their truth?

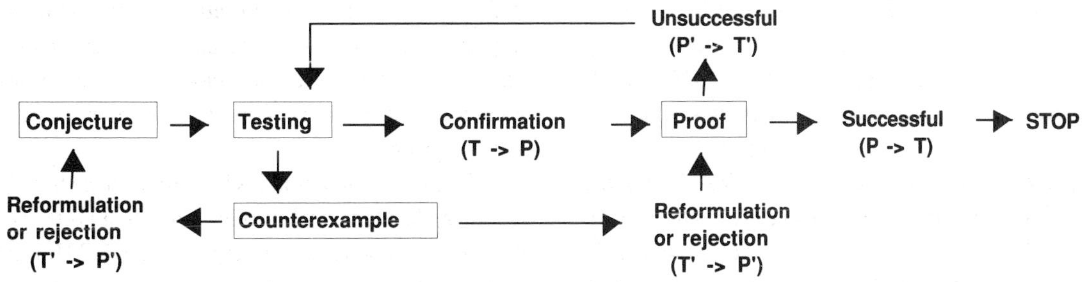

FIGURE 11

The following quotes by some well-known mathematicians also support these ideas:

> "... having verified the theorem in several particular cases, we gathered strong inductive evidence for it. The inductive phase overcame our initial **suspicion** and gave us a strong **confidence** in the theorem. Without such **confidence** we would have scarcely found the courage to undertake the proof which did not look at all a routine job. When you have satisfied yourself that the theorem is **true**, you start **proving** it." (bold added)
>
> —George Pólya [1954, pp. 83–84]

> "The mathematician at work makes vague guesses, vizualizes broad generalizations, and jumps to unwarranted conclusions. He arranges and rearranges his ideas and becomes **convinced** of their **truth** long before he can write down a logical proof. The conviction is not likely to come early—it usually comes after many attempts, many failures, many discouragements, many false starts ... **experimental** work is needed ... thought-experiments. When a mathematician wants to prove a theorem about an infinite-dimensional Hilbert space, he examines its finite dimensional analogue, he looks in detail at the 2- and 3-dimensional cases, he often tries out a particular numerical case, and he hopes that he will gain thereby an insight that pure definition juggling has not yielded." (bold added)
>
> —Paul Halmos [1984, p. 23]

> "Actually the mathematician does not rely upon rigorous proof to the extent that is normally supposed. His creations have a meaning for him that precedes any formalization, and this meaning gives the creations an existence or reality ipso facto. ... Great mathematicians know before a logical proof is ever constructed that a theorem must be **true**..." (bold added)
>
> —Morris Kline [1982, pp. 313–314]

Although public conviction through the medium of mathematical journals usually relies heavily on rigorous proof, personal conviction usually depends on a **combination** of intuition, quasi-empirical verification, and the existence of some form of logical (but not necessarily rigorous) proof. In fact, a very high level of conviction may sometimes be reached even in the absence of a proof. For instance, in their discussion of the "*heuristic evidence*" in support of the still unproved twin prime pair theorem and the famous Riemann Hypothesis, Davis and Hersh [1983: 369] conclude that this evidence is "*so strong that it carries conviction even without rigorous proof.*"

Recent years have seen an explosion in the use of computers in many areas of mathematics. For example, Pollak [1984:12] describes the increased use of computers as experimental aids as follows:

> "... we find ourselves examining on the machine a collection of special cases which is too large for humans to handle by conventional means. The computer is encouraging us to practice unashamedly and in broad daylight, certain customs in which we indulge only in the privacy of our offices, and which we never admitted to students: experimentation. To a degree which never appears in the courses we teach, mathematics is an experimental science... The computer has become the main vehicle for the experimental side of mathematics".

An interesting development has been the design of a computer program called *Graffiti* which generates its own hypotheses in graph theory and tests them (see Kolata [1989]). According to the designer of the program, Fatjlowicz from Houston University, *Graffiti* has already led to five published articles and at least 20 mathematicians are presently working on proving several other *Graffiti* results.

Another striking example of the dramatic influence of computer experiments on research in pure mathematics is the sudden burst of developments in areas such as fractal geometry, dynamical systems, and chaos. Due to the complexity of the problems in these domains, it was only really with the advent of the computer that mathematicians could start exploring a sufficient number of examples to develop their intuition and to see patterns that could lead to conjectures and ultimately to theorems.

Computers are also increasingly being used as a means of verification. A famous example is the computer proof of the four-colour problem in 1976 by Appel and Haken. Recently Branko Grünbaum [1993] used *Mathematica* to verify some geometric conjectures. He presents an interesting argument that the probability of his findings being false are, for all practical purposes, zero. Very briefly, the argument goes as follows. If we try to prove such geometric assertions by using analytic geometry, we wind up with some algebraic relationships which have to be shown to be identities. However, as pointed out by Davis [1977], an algebraic identity can be conclusively established with a single numerical check by using algebraically independent transcendental numbers. Although computers cannot actually operate with transcendental numbers, a series of experiments with points selected at random achieves much the same result. In other words, if experiment after experiment with randomly selected points reaffirms the same result, the probability of the result being false effectively becomes zero. (Since the property checker of *Cabri* is a trade secret, I do not know exactly how it works, but it probably works along similar lines.) Particularly relevant are the following remarks by Grünbaum [1993, p. 8]:

> "Do we start trusting numerical evidence (or other evidence produced by computers) as proofs of mathematics theorems? ... if we have no doubt—do we call it a theorem? ... I do think that my assertions about quadrangles and pentagons are theorems ... the mathematical community needs to come to grips with the possibilities of new modes of investigation that have been opened up by computers."

Several more examples of the increasing use of the computer, not only as an exploratory tool in different areas, but also as a means of verification, are given in a recent controversial article by Horgan [1993] in *Scientific American*. In addition to the fact that he ignores numerous areas in mathematical research where computers are not (yet?) particularly useful, he proceeds from the narrow viewpoint that the **only** function of proof is that of the establishment of mathematical truth and concludes that *"proof is dead."* However, as will be briefly discussed in the next section, a deductive proof may also be useful for many other reasons.

Proof as a Means of Explanation, Discovery and Self-Realization

Although I have often achieved confidence[5] in the general validity of a conjecture by seeing its truth displayed while objects undergo continuous transformation across the screen (or using the property checker), this provides no personally satisfactory *explanation of **why** it may be true.* It merely confirms that it is true, and, even though the consideration of more and more examples may increase my confidence even more, it gives no psychological satisfactory sense of illumination. There is no **insight** or **understanding** into how it is the consequence of other familiar results. It has been my experience that the more convinced I become, the more motivated I also become to find out why it is true. Davis and Hersh [1983, p. 368] remark that, despite the convincing heuristic evidence in support of the Riemann Hypothesis, one may still have a burning need for explanation:

> "It is interesting to ask, in a context such as this, why we still feel the need for a proof.... It seems clear that we want a proof because... if something is true and we can't deduce it in this way, this is a sign of a lack of understanding on our part. We believe, in other words, that a proof would be a way of understanding **why** the Riemann conjecture is true, which is something more than just knowing from convincing heuristic reasoning that it **is** true."

Recently Gale [1990: 4] also emphasized that the function of Lanford and other mathematicians' proofs of Feigenbaum's experimental discoveries in fractal geometry was that of explanation and not that of verification at all. Thus, in most cases when the results concerned are intuitively self-evident and/or they are supported by convincing quasi-empirical or computer evidence, the main purpose of proof for mathematicians is certainly not that of verification, but rather that of explanation.

Admittedly not all proofs are explanatory; sometimes we have to be satisfied with simply a verification proof, but the ideal is generally to try to arrive at some form of satisfactory explanation. I can recall many times when I have spent hours, days, and weeks trying to redraft my own non-explanatory proofs into explanatory ones. Of course, I am not always successful, as some results seem to lend themselves more easily to certain classes of proofs that may be considered relatively *non-explanatory*—indirect, mathematical inductive, analytic geometric, and vector proofs (compare Hanna [1989]). It is certainly unthinkable to do without such powerful proof techniques, but I would normally turn to them only as a last resort. Robert Long [1986, p. 616] writes:

> "Proofs that yield insight into the relevant concepts are more interesting and valuable to us as researchers and teachers than proofs that merely demonstrate the correctness of a result. We like a proof that brings out what seems to be essential. If the only available proof of a result is one that seems artificial or contrived it acts as an irritant. We keep looking and thinking."

Also to Manin [1981, p. 107] and Bell [1976, p. 24], explanation is a criterion for a "*good*" proof. It is "*one which makes us wiser*", and it is expected "*to convey an insight into* **why** *the proposition is true.*"

Another very important function of proof is that of **discovery**. The production of a proof that identifies its underlying explanatory properties, can sometimes lead to a further unanticipated generalization, as we mentioned in Example 3. I have often experienced the value of proof as a means of discovery, finding results which would not have been likely through more or less "*blind*" experimentation (see De Villiers [1991 and 1996].

Mathematicians know that proving something is an intellectual challenge that can be compared to the physical challenge of completing an arduous marathon or triathlon. In this sense, proof serves the function of **self-realization** and **fulfillment**. Proof is a testing ground for the intellectual stamina and ingenuity of the mathematician. To paraphrase Mallory's famous comment on his reason for climbing Mount Everest: "*we prove our results because they're there.*"

We also point out that a proof may also fulfill several other functions such as *systematization, communication, memorization,* and *algorithmization.* (For more details, see De Villiers [1990], Renz [1981], and Van Asch [1993]).

Students and Proofs

Although most students seem to have no further need for conviction once they explore geometric conjectures in dynamic geometry environments like *Cabri* or *Sketchpad*, it is not difficult to solicit further curiosity by asking them **why** they think a particular result is true. Challenge them to try to *explain* it. Students quickly admit that inductive/experimental verification merely confirms; it gives no satisfactory insight or understanding. They seem to be willing to then view a deductive argument as an attempt at explanation, rather than verification.

To present the fundamental function of proof as explanation and discovery requires that students be introduced early to the art of problem posing and allowed sufficient opportunity for exploration, conjecturing, refuting, reformulating and explaining as outlined in Figure 11 (compare Chazan [1990]). Dynamic geometry software strongly encourages this kind of thinking. It is powerful as a means of verifying true conjectures and also extremely valuable in constructing counterexamples to false conjectures.

Movshovitz-Hadar [1988a and b] argues similarly (but not identically) for stimulating presentations of results that solicit the surprise and curiousity of students so that they are susceptible to responsive proofs which leave them with "*an appreciation of the invention, along with a feeling of becoming wiser.*"

We should also be quite honest in telling our students that we as mathematicians often prove results simply because of the intellectual challenge involved. We should not try to present a fairy tale of always wanting to obtain "*certainty*". We should also try to give more attention to the communicative aspects of proof by actually negotiating with our students the criteria for acceptable evidence, explanations and/or arguments. As anyone with a bit of experience in actual research will testify, the systematization function of proof (the arrangement of a series of results in a strictly axiomatic-deductive form) comes to the fore only at a very advanced stage, and should be withheld in any introduction to proof.

Endnotes

1. The English translation from the French is poor and should actually read "*this property is true in general*". Presumably, the property checker is based on the mathematical theory described in Davis [1977]. The property checker of *Cabri* is also able to construct and display a counterexample if the property is not true in general.
2. I used a macro-construction for the construction of the similar triangles on the sides which does not allow for changes to the constructed shape; in other words, their basic *shape* remains *fixed* during the transformation of the base quadrilateral. So strictly speaking, the property checker of *Cabri* in this case shows only that the result is true in general (i.e., for **any** quadrilateral) for these *particular* similar triangles. One could however easily construct other similar triangles on the sides with another macro-construction and repeat the experiment. Although I was sufficiently convinced at this stage, I could have created a dynamic configuration in *Sketchpad* that allows for the dynamic transformation of not only the base quadrilateral, but also the similar triangles on the sides.
3. I have just recently been informed by a Dutch mathematician that the "Von Aubel" in Kelly (1966) is a spelling error. This error was unfortunately taken over by Gardner (1971; 1981; 1992). The correct version is "Van Aubel."
4. As in the previous case, the respective shapes of the similar rectangles and rhombi on the sides were fixed. (The rectangles were constructed so that their lengths were twice their widths.) I could have repeated the experiments with other sim-

ilar rectangles or rhombi, or used *Sketchpad* to create dynamic configurations which allow for dynamic changes to either the base quadrilateral or the rectangles or rhombi on the sides.

5. I am aware of a few cases with *Cabri* and *Sketchpad* where certain constructions do not work out correctly. For example, in *Cabri* the traditional construction of a common tangent to two circles produces two points on the one circle (instead of only one). Using the checking facility shows that the line segments from the center of that circle to the two points are both perpendicular to the tangent; therefore, the two points are actually coincident. (In fact, one cannot distinguish the two points visually even when zooming in on them.) Similar problems also occasionally arise in *Sketchpad.* I am aware of a relatively complicated construction where two circles which should end up being tangent to each other are not. However, from a research point of view such a *false negative* is probably not too problematic as one may still suspect the truth of such a result provided the error seems very small, and continues to remain so under transformation. It would be more serious to find *false positives* (results continually confirmed by dynamic geometry software which are actually false). So far, however, I have not yet come across any such results and have learnt to trust dynamic software to a very large degree.

Acknowledgement. This research forms part of the Spatial Orientation and Spatial Insight (SOSI) Project funded by the Foundation for Research Development (FRD), Pretoria, South Africa.

Bibliography

Bell, A.W. (1976). A study of pupils' proof-explanations in mathematical situations. *Educational Studies in Mathematics,* 7, 23–40.

Chazan, D. (1990). Quasi-empirical views of mathematics and mathematics teaching. In G. Hanna, and I. Winchester, (Eds.) *Creativity, thought and mathematical proof.* Toronto: OISE.

Coxeter, H.S.M. and Greitzer, S.L. (1967). *Geometry revisited.* New York: Random House. Washington, DC: Mathematical Association of America.

Davis, P.J. (1977). Proof, completeness, transcendentals and sampling. *Journal Assoc. Comp. Machin.*, 24, 298–310.

Davis, P.J. and Hersh, R. (1981). *The Mathematical Experience.* Boston: Birkhauser (1983) Great Britain: Pelican Books.

DeTemple, D. and Harold, S. (1996). A Round-up of Square problems. *Mathematics Magazine*, 16, 15–27.

De Villiers, M.D. (1990). The role and function of proof in mathematics. *Pythagoras,* 24, 17–24.

—— (1991). Vertical line and point symmetries of differentiable functions. *Int. Journ. Math. Ed. Sci. and Technol.*, 22(4), 621–644.

—— (1996). *Some adventures in elementary geometry.* Durban: Univ. Durban-Westville.

—— (in press) Dual Generalizations of Van Aubel's Theorem.

Finney, R.L. (1970). Dynamic Proofs of Euclidean Theorems. *Mathematics Magazine*, 43, 177–185.

Gale, D. (1990). Proof as explanation. *The Mathematical Intelligencer*, 12(1), 4.

Gardner, M. (1979). *Mathematical Circus.* New York: Knopf. (1981) Gt. Britain: Chaucer Press. (1992) Washington, DC: Mathematical Association of America.

Grünbaum, B. (1993). Quadrangles, pentagons, and computers. *Geombinatorics*, 3, 4–9.

Halmos, P. (1984). Mathematics as a creative art. In Campbell, D. and Higgens, J. (1984). *Mathematics: people, problems, results.* Vol. II, 19–29. Belmont: Wadsworth.

Hanna, G. (1989). Proofs that Prove and Proofs that Explain. *Proceedings of the 13th International Conference on the Psychology of Mathematics Education*, Paris, 45–51.

Hersh, R. (1993). Proving is convincing and explaining. *Educational Studies in Mathematics*, 24, 389–399.

Horgan, J. (1993). The death of proof. *Scientific American*, Oct., 74–82.

Kelly, P.J. (1966). Von Aubel's Quadrilateral Theorem. *Mathematics Magazine*, (39) 35–37.

Kline, M. (1982). *Mathematics. The Loss of Certainty.* Oxford: Oxford University Press.

Kolata, G.B. (1989). A program that makes conjectures—mathematicians meet computerized ideas. *New York Times,* 18 June 1989.

Long, R.L. (1986). Remarks on the history and philosophy of mathematics. *The American Mathematical Monthly,* 93, 609–619.

Manin, Y.I. (1981). A Digression on Proof. *The Two-year College Mathematics Journal*, 12(2), 104–107.

Movshovitz-Hadar, N. (1988a). Stimulating presentations of theorems followed by responsive proofs. *For the Learning of Mathematics*, 8(2), 12–19; 30.

—— (1988b). School mathematics theorems—An endless source of surprise. *For the Learning of Mathematics*, 8(3), 34–40.

Neuberg, J. (1894). Sur Quelques Quadrilat'eres Sp'eciaux. *Mathesis*, Tome IV, 268–271.

Pollak, H.O. (1984). The effects of technology on the mathematics curriculum. In A.I. Olivier, (Ed.) *The Australian Experience: Impressions of ICME 5.* Centrahil: AMESA.

Pólya, G. (1954). *Mathematics and Plausible Reasoning. Induction and Analogy in Mathematics.* Vol. 1. Princeton: Princeton University Press.

Renz, P. (1981). Mathematical proof: What it is and what it ought to be. *The Two-Year College Mathematics Journal.* 12(2), 83–103.

Van Asch, A.G. (1993). To prove, why and how? *Int. J. Math. Educ. Sci. Technol.*, 24(2), 301–313.

Yaglom, I.M. (1962). Trans. Alan Shields. *Geometric Transformations I.* New York: Random House. Washington, DC: Mathematical Association of America.

—— (1968). Trans. Alan Shields. *Geometric Transformations II.* New York: Random House. Washington, DC: Mathematical Association of America.

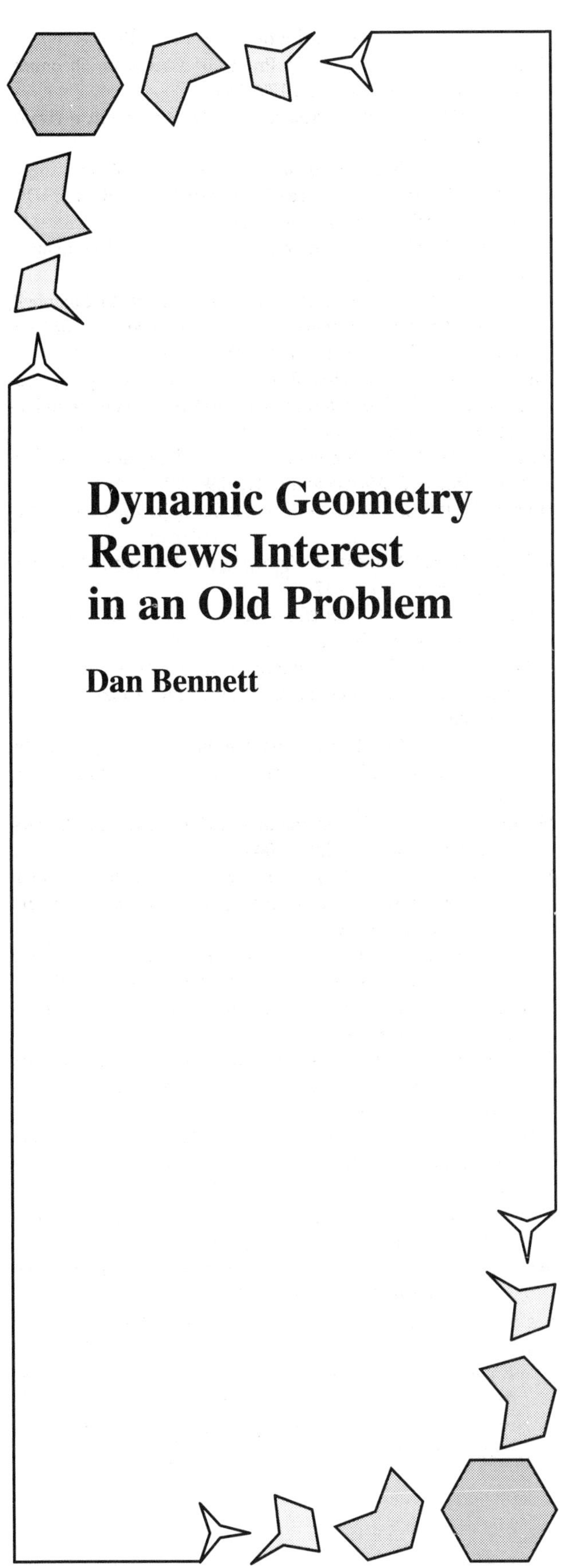

Dynamic Geometry Renews Interest in an Old Problem

Dan Bennett

Much has been said and written about computers as harbingers of the death of proof. (See Horgan, for example.) There are at least two ways in which computers, it is argued, have at best changed the role of proof in mathematics and at worst have rendered proof irrelevant: 1) Computers can actually do proofs, and 2) in many cases, a computer can supply as much empirical evidence as you could possibly require to persuade you of the truth of a conjecture. It is this second impact that dynamic geometry has most been a party to: you don't need a proof to be convinced of the truth of something that you can see with your own eyes. This alleged discord between computers and proof is of particular concern to those of us who work in the field of secondary school geometry, where the role of proof in the curriculum was being questioned before dynamic geometry came along. In this paper, I relate an experience that I hope demonstrates how dynamic geometry and proof can, in fact, exist in harmony.

Before I move to my example, let me summarize a counter-argument to those who might argue that dynamic geometry renders proof irrelevant. Michael de Villiers names five roles of proof: verification, explanation, communication, discovery, and systematization. He also offers, in passing, a possible sixth role of aesthetics or self-realization. Without remaking all his points here, suffice it to say that he makes a strong case that there are more (and better) reasons to prove things than simply to verify their truth. Dynamic geometry has an impact on the role of proof as verification; a proposition whose truth may not be obvious from a few static figures might be verified by looking at the continuous set of cases offered by dynamic geometry. So now we need other motivation, besides verification, to prove things. I would say that high school students have always needed better reasons to prove things, given that many things we ask them to prove are obvious even without dynamic geometry. And, if they're not obvious, the simple fact that they're "in the book" is reason enough not to doubt them.

I think for me, and I imagine for many students, the strongest motivation for proof is de Villier's sixth role. Stated differently, I'd say I most often do proofs for the challenge and the satisfaction that comes from discovering them. Stated more crassly, one role of proof is ego gratification. It was with that motivation, armed with a tool for doing dynamic geometry (*The Geometer's Sketchpad*), that I set out to prove a proposition that was first posed in 1953. I wanted to prove it because in 1993 professional mathematicians were calling it an unsolved problem and talking about it in the relatively new context of computers and geometry.

The following problem was posed by Josef Langr as problem E 1050 in MAA's *American Mathematical Monthly*,

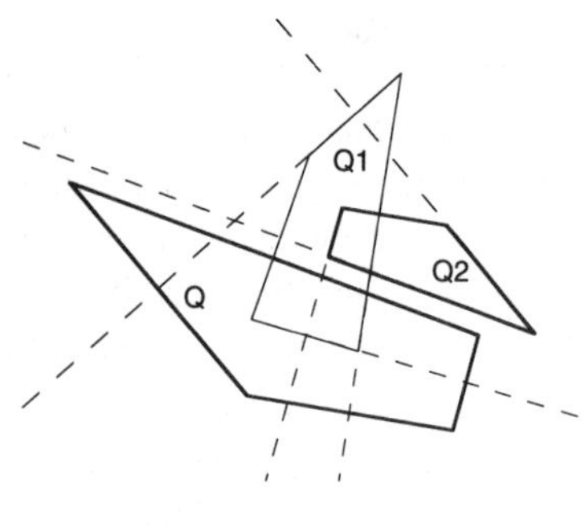

FIGURE 1

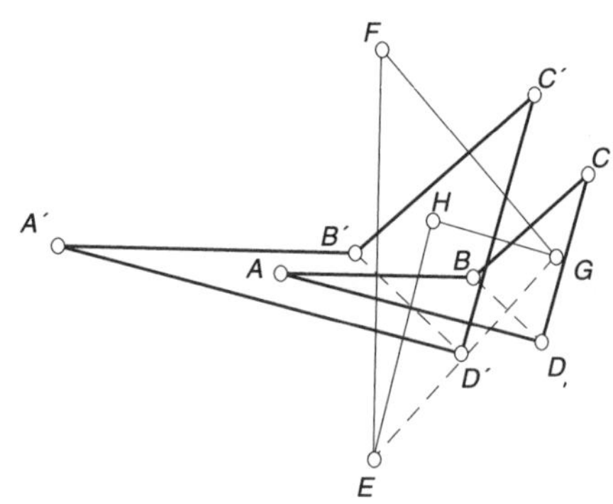

FIGURE 2

vol. 60, 1953: The perpendicular bisectors of the sides of a quadrilateral Q form a quadrilateral Q_1, and the perpendicular bisectors of the sides of Q_1 form a quadrilateral Q_2. Show that Q_2 is similar to Q and find the ratio of similitude. (See Figure 1.)

Branko Grünbaum wrote about the problem in 1993, demonstrating that the truth of the proposition could hardly be doubted, despite its not having been proven. He programmed *Mathematica* to generate enough examples to show that not only were Q and Q_2 similar, but they were homothetic (that is, each is the image of the other under a dilation). He went on to pose several more questions about the quadrilaterals and analogous problems for other polygons that might be explored empirically on computers or even proven by them. He concluded by describing the problem of computers and proof as "a serious mathematical, philosophical and practical problem that needs to be addressed."

I first heard of the problem from a colleague who'd seen Doris Schattschneider demonstrate it at a meeting of *Sketchpad* User's Group in January, 1994. She presented it as an example of an interesting problem made more interesting by dynamic geometry. She showed how by using *Sketchpad* one need perform only a single construction that could then be manipulated to show the whole class of figures that satisfied the constraints of the problem. Even more than with Grünbaum's static examples, this dynamic example made it obvious that the proposition was true. What really piqued Schattschneider's audience's interest, though, was her allusion to the fact that in 40 years the proof had not been published.

That's what piqued *my* interest in the problem—I just couldn't believe the proof could be that hard. As it turns out, it wasn't. Once I had the proof, it seemed likely that the 40-year absence of a published proof had more to do with a lack of interest in the problem than with the difficulty of a proof. So, instead of being a story of a hard nut finally cracked by a computer, this is a story of a computer used as a tool for doing an elementary proof of an interesting problem—a problem given new life by virtue of the fact that it's fun to look at dynamically.

What follows is my proof that quadrilateral Q is homothetic to quadrilateral Q_2. Quadrilateral Q is assumed to be not concyclic; that is, its vertices do not lie on a circle.

> In quadrilateral Q (*ABCD*), *EF* is the perpendicular bisector of *AB*, *FG* is the perpendicular bisector of *BC*, *GH* is the perpendicular bisector of *CD*, and *HE* is the perpendicular bisector of *DA*. Taken in order, the intersections of the perpendicular bisectors form quadrilateral Q_1 (*EFGH*). The perpendicular bisectors of the sides of Q_1 form quadrilateral Q_2 ($A'B'C'D'$). (See Figure 2.)
>
> Draw diagonal *BD*. The plan is to show that $\triangle ABD$ is similar to $\triangle A'B'D'$. The same argument can be used for other pairs of triangles formed by diagonals, so transitivity will give us all angles equal and all sides in the same proportion.
>
> AB is perpendicular to EF is perpendicular to $A'B'$ (given). So AB is parallel to $A'B'$.
>
> AD is perpendicular to EH is perpendicular to $A'D'$ (given). So AD is parallel to $A'D'$.
>
> Next, it's necessary to show that BD is parallel to $B'D'$. Because E is the intersection of the perpendicular bisectors of AB and AD, it is the circumcenter of $\triangle ABD$; thus it is equidistant from B and D. Point G is the circumcenter of $\triangle DBC$, so it is also equidistant from B and D. Since points E and G are both equidistant from B and D, EG is the perpendicular bisector of BD. (See Figure 3.)
>
> Similarly, B' is the circumcenter of $\triangle EFG$, and D' is the circumcenter of $\triangle EHG$. So $B'D'$ is the perpendicular bisector of EG.
>
> Since BD is perpendicular to EG is perpendicular to $B'D'$, BD is parallel to $B'D'$.

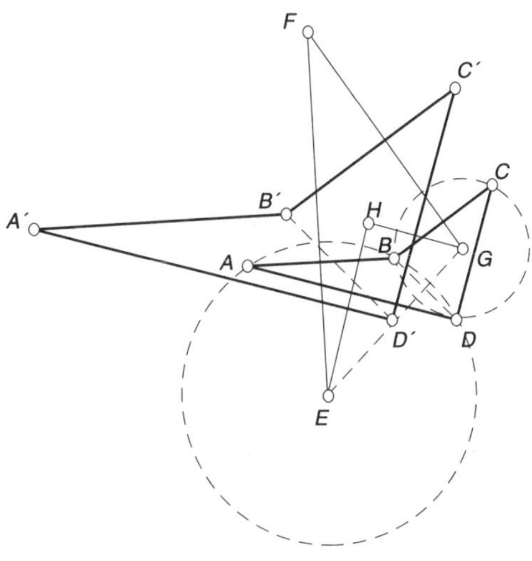

FIGURE 3

Since AB is parallel to $A'B'$, AD is parallel to $A'D'$, and BD is parallel to $B'D'$; $\triangle ABD$ is similar (and homothetic) to $\triangle A'B'D'$.

The same argument can be applied to other triangles formed by diagonals, so all angles in quadrilateral $ABCD$ are equal to corresponding angles in quadrilateral $A'B'C'D'$, and corresponding sides are proportional.

Therefore, quadrilateral $ABCD$ is similar to quadrilateral $A'B'C'D'$.

So it turns out that there is an elementary proof of the theorem about perpendicular bisector quadrilaterals—a proof a bright high school student could follow, or perhaps even come up with. What good did dynamic geometry do in constructing the proof? I wish I could say that I gleaned some insight while moving parts of the figure around. I've certainly had that experience doing dynamic geometry, and I even think I've come to be able to do rudimentary dynamic geometry in my head. But, in fact, the opposite was true in this particular problem. My main use of dynamic geometry was to create the figure in the first place and then manipulate it until I could see everything. You might have wondered why Figure 2 shows such a distorted, nonconvex quadrilateral. If you play with the problem using a dynamic geometry tool, you'll find that the quadrilaterals behave somewhat wildly. There are few cases in which all the quadrilaterals are of reasonable size, perpendicular segments intersect one another, and no parts of the figure overlap one another in deceptive ways. In convex and self-intersecting quadrilaterals, the dilation ratio is negative, so the correspondences in the similar figures are harder to keep track of. The proof doesn't rely on any of this specialness, but it's certainly easier to see the relationships in this configuration. I doubt I would ever have had the patience to construct enough figures with compass and straightedge to get one that was readable. Ironically, once I had a figure I liked on *Sketchpad*, I printed it out and studied the static figure while I thought about the proof! Whatever value moving figures have for visualization, we mustn't underestimate the power of dynamic geometry for simply creating good static ones!

Still, that's a rather modest claim to make for dynamic geometry. We'd like more to think of dynamic geometry as a tool for insight, rather than just a nifty construction tool. It's hard to analyze where insight comes from, and in this little proof it's even hard to say what the insights are. (They're hardly earth shaking.) But I think I can credit dynamic geometry for giving me ways of thinking and seeing, before I attempted this problem, that lent me some insights into the problem. The strategy of drawing diagonals and looking for similar triangles, as obvious as that seems, was in the front of my mind because of *Sketchpad* work I'd been doing previously with special quadrilaterals classified by properties of their diagonals. Seeing the vertices of Q_1 as circumcenters, which is the actual key to the problem, came to me, I think, because of past experience playing with circumcenters dynamically. It would be only a slight exaggeration to say that thanks to *Sketchpad* I've seen the circumcenter of every triangle in the universe. I have no doubt that experience helped me look for and spot the key circumcenters in this problem.

Dynamic geometry also gives an important instant insight into the second part of the problem: It's apparent as soon as you drag that there is not a simple, single ratio of similitude. (See Figure 4.) Moreover, it "feels" as you drag as if the ratio of similitude is most influenced by changes in the angles, leading one to pursue a trigonometric solution to the problem.

I didn't get anywhere with the ratio of similitude, but Doris Schattschneider has found many nice relationships for special-case quadrilaterals. For example, when Q is a parallelogram, Q, Q_1, and Q_2 are all similar. Quadrilateral

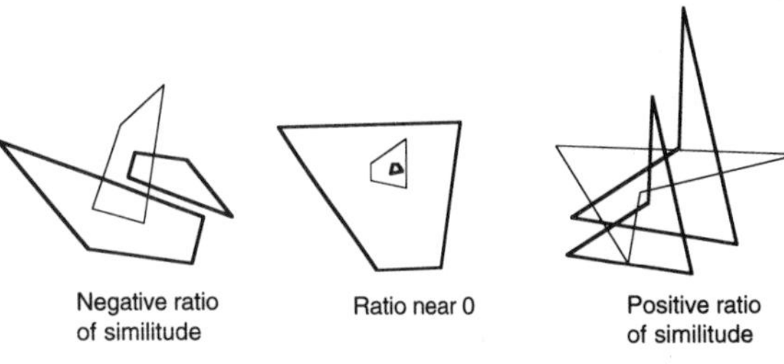

FIGURE 4

Q_1 is gotten from Q by a spiral similarity of 90 degrees about the center of Q with ratio of similitude equal to the tangent of the acute angle of the parallelogram. Q_2 is gotten from Q_1 by applying the same spiral similarity to Q_1. This implies that the scale ratio of the dilation (which is a product of the two spiral similarities) is the negative of the square of the tangent of the acute angle of Q. (Note that this ratio is negative.) In particular, when Q has a 45 degree angle, then Q, Q_1, and Q_2 are congruent. Schattschneider also found a more general (and more complex) result for quadrilaterals with one pair of parallel sides.

Grünbaum's frequent collaborator G.C. Shephard used trigonometry to prove the theorem, and he found the ratio of similitude for the general case. The "feeling" one gets from playing with the figure—that the ratio has to do only with angles and not with side lengths—is confirmed by Shephard. The ratio is a complicated formula that involves trigonometric functions of the angles.

In Langr's statement, and in my proof of the problem, it was assumed that the perpendicular side bisectors of the first quadrilateral Q formed a quadrilateral Q_1, and that the perpendicular side bisectors of Q_1 formed a quadrilateral Q_2. Shephard's proof and a proof by James King (whose proof is equivalent to the one presented here) also deal with a degenerate case: When Q is concyclic, Q_1 collapses to a single point (and so Q_2 isn't defined). Again, dynamic geometry makes the truth of this proposition obvious. In a dynamic sketch you can collapse the two quadrilaterals to a point simply by dragging a vertex of Q until the circles in Figure 3 coincide, cirumscribing Q. There are two other possible degenerate cases to be considered: if Q_1 has three collinear points (Q_1 is called degenerate), then two of its side bisectors will be parallel, and so Q_2 will not exist. Also, if Q_1 is concyclic, then Q_2 will not exist. King also showed that if you begin with *any* nonconcyclic quadrilateral, then its perpendicular bisectors will form a nondegenerate quadrilateral that is not concyclic, so that Langr's theorem is true for any quadrilateral that is not concyclic. King's paper, "Quadrilaterals Formed by Perpendicular Bisectors" follows this article.

Finally, I think the greatest strength of dynamic geometry as a tool for tackling this problem is that it made a 40-year-old problem worth exploring again. Try it. It's really captivating to watch Q_1 dart around while Q and Q_2 stay always homothetic. Far from being a disincentive for a proof, dynamically exploring this interesting relationship motivated me and others to pursue a proof. For 40 years the problem went unsolved; then suddenly several people came up with proofs and other interesting results at virtually the same time. Jim King told me he came up with his proof on a plane ride. I ask, at the risk of maligning one of the esteemed editors of this volume, is it possible that, if he hadn't been traveling with his laptop computer, he'd have watched a bad movie or read a trashy novel instead? Dynamic geometry won't kill proof. It just makes it all the more fun.

Bibliography

de Villiers, M. "The Role and Function of Proof in Mathematics," *Pythagoras* 24 (1990), pp. 18–23. See also de Villiers, M. "The Role and Function of Proof in Dynamic Geometry: Some Personal Reflections," *Geometry Turned On*, pp. 15–24.

Grünbaum, B. "Quadrangles, Pentagons, and Computers," *Geombinatorics* 3 (1993), pp. 4–9.

Horgan, J. "Trends in Mathematics: The Death of Proof," *Scientific American*, October 1993, pp. 92–103.

King, J., "Quadrilaterals Formed by Perpendicular Bisectors," *Geometry Turned On*, pp. 29–32.

Langr, J., Problem E 1050, *American Mathematical Monthly* 60 (1953), p. 551.

Shephard, G.C. "The Perpendicular Bisector Construction," *Geometriae Dedicata* 56 (1995), pp. 75–84.

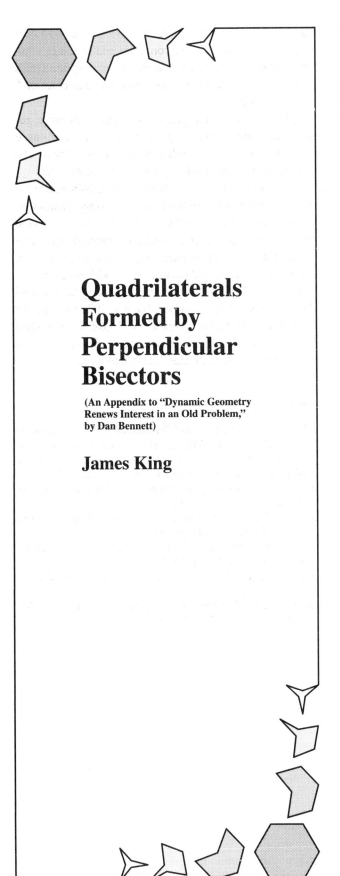

Quadrilaterals Formed by Perpendicular Bisectors

(An Appendix to "Dynamic Geometry Renews Interest in an Old Problem," by Dan Bennett)

James King

Dan Bennett's paper [1] gives a beautifully simple geometric proof of Langr's conjecture. His proof assumes that at each stage the four perpendicular side bisectors of a quadrilateral also form a quadrilateral. This is true (except for one special case), and the proof of this fact is the goal of this paper. The reasoning in our proof is not just a technical exercise but illustrates some ideas about line reflections and perpendicular bisectors which are interesting in their own right.

As Dan mentioned, I found the proof presented in his paper using *Sketchpad* on an airplane after hearing Doris Schattschneider describe the problem in a talk. The idea of using line reflections for treating the special cases described in this paper was in my mind because of *Sketchpad* explorations that I use in teaching transformational geometry; these will also be described below in section 4.

1. Quadrilaterals and the Langr Problem

Recall that the Langr Problem begins with a quadrilateral Q. One constructs the quadrilateral Q_1 as the intersection of the perpendicular bisectors of the sides of Q and then constructs quadrilateral Q_2 as the intersection of the perpendicular bisectors of the sides of Q_1. What must be considered is that in some cases the perpendicular side bisectors of a quadrilateral may not define a quadrilateral. Since we are considering a delicate point, it is a good idea to define exactly what we (and Langr, and Bennett) mean by a quadrilateral.

Definition 1. A ***quadrilateral*** $ABCD$ is a polygon whose vertices are distinct points A, B, C, D, no three of which are collinear.

In general, given four points E, F, G, H, four lines through the pairs EF, FG, GH, and HE can define degenerate "quadrilaterals" $EFGH$ like these:

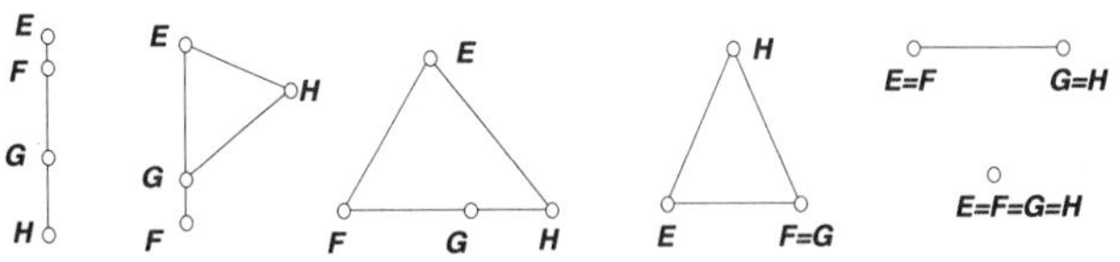

If Q_1 is a degenerate quadrilateral (has three collinear points), then Q_2 is not defined.

2. Concyclic Quadrilaterals

The most obvious case where Q_1 is not a quadrilateral is when Q can be inscribed in a circle; such a quadrilateral Q is called ***concyclic.*** In this case the sides of the quadrilateral are chords of a circle, so the perpendicular bisector of each side

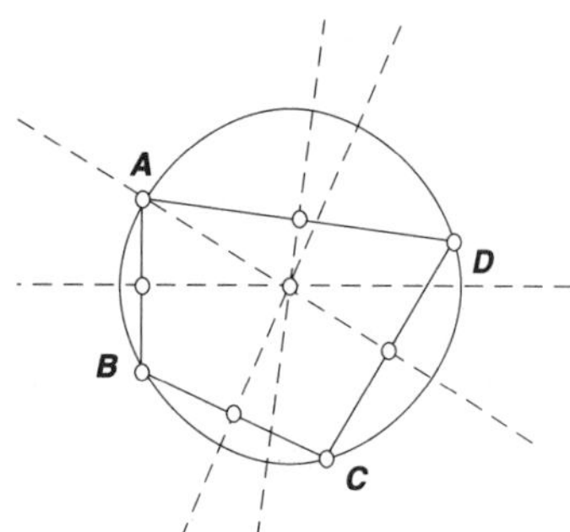

FIGURE 1

passes through the center of the circle (see Figure 1). Thus the perpendicular bisectors of concyclic quadrilaterals are concurrent. If Q is concyclic, then Q_1 is not a quadrilateral, but a point. In the sections that follow, we will show this is the only case in which Q_1 is degenerate.

3. Degenerate Quadrilaterals from Perpendicular Bisectors

Let the quadrilateral $Q_1 = EFGH$ be formed from the intersections of the perpendicular side bisectors of quadrilateral $Q = ABCD$ as in Bennett's Figure 3. Then the vertices of Q_1 are the circumcenters of the four triangles formed by A, B, C, and D. For example, E is the circumcenter of triangle DAB, and G is the circumcenter of triangle BCD.

We have seen that if Q is concyclic, then the four points E, F, G, Hcoincide. In what other ways can Q_1 fail to be a quadrilateral?

First, we consider the possibility that two (or more) of the vertices of Q_1 coincide.

Lemma 1. *If two or more of the points E, F, G, H coincide, then all four points coincide and the quadrilateral Q is concyclic.*

Proof. We may assume that $E = G$. Then by our remark above, triangle DAB is inscribed in a circle centered at E and triangle BCD is inscribed in a circle with the same center E. Since B and D are on both circles, the two circles coincide. Thus A, B, C, D all lie on the same circle, and Q is concylic. QED

The other possibility to consider is when three of the vertices of Q_1 are collinear.

Lemma 2. *If the points E, F, G, H are distinct, then no three of the points can be collinear, and so $Q_1 = EFGH$ is a quadrilateral.*

Proof. Suppose that E, F, G are collinear. Since the sides of Q_1 are perpendicular bisectors of Q, this means that two consecutive bisectors are the same line; in this case $EF = FG$. But this is impossible since it would mean that the lines DA and AB must be the same, which contradicts the fact that Q is a quadrilateral. QED

Lemmas 1 and 2 together give us the following result.

Proposition. *Given a quadrilateral Q which is not concyclic, the perpendicular bisectors of the sides form a quadrilateral Q_1.*

4. Perpendicular Bisector Quadrilaterals and Line Reflections

We have established that if Q is not concyclic, then the construction for the Langr problem produces Q_1, which is a quadrilateral. If we can show that Q_1 is also not concyclic, then Q_2 must also be a quadrilateral and Bennett's argument holds whenever Q is not concyclic. In this section we will prove the following result.

Theorem. *If Q_1 is a quadrilateral formed by the perpendicular bisectors of the sides of a quadrilateral Q, then Q_1 cannot be concyclic.*

The main idea in the proof of the theorem uses the properties of isometries defined as compositions of line reflections, applied to sets of lines used in the Langr construction. First we introduce some notation and properties of isometries.

If k is a line, we denote the reflection in line k by M_k (for mirror reflection in line k). We will use the following well-known facts about line reflections (see [2], for example).

- If a and b are lines intersecting at P, then the product (composition) $M_b M_a$ is a rotation with center P and angle of rotation equal to twice the directed angle between the lines a and b.
- If a, b, and c are lines concurrent at P, then the composition $M_c M_b M_a$ is a line reflection M_d, where d is a line through P.
- The product of two rotations with angles a and b, respectively, is a rotation through angle $a + b$ if $a + b$ is not a multiple of 360° and a translation otherwise.

To see what is special about sets of lines which are the perpendicular bisectors of the sides of a polygon, we first consider the simplest case of a triangle. In this case the perpendicular bisectors are always concurrent.

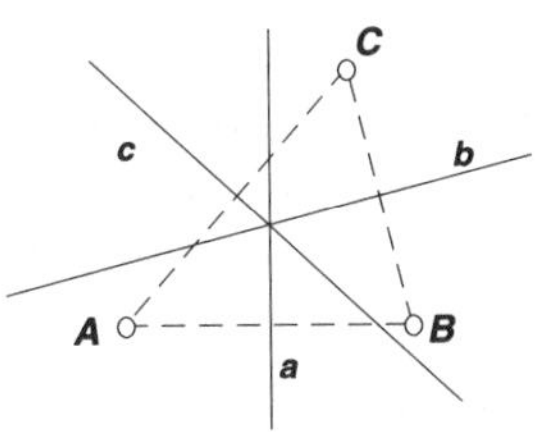

FIGURE 2

Question. Given three concurrent lines a, b, c, when are they the perpendicular bisectors of the sides of a triangle ABC?

Answer. There is always such a triangle ABC. Observe in Figure 2, where a, b, and c are the perpendicular bisectors of AB, BC, and CA, respectively, that A is carried to itself by the triple reflection: $A = M_c(M_b(M_a(A)))$. Since this triple reflection is equal to a single line reflection M_d, $A = M_d(A)$ if and only if A is on the line M_d. So take A any point on the line d and let $B = M_a(A)$ and $C = M_b(M_a(A))$ to form the desired triangle.

This exercise makes a nice dynamic geometry exploration. Start with the three concurrent lines a, b, c, and any point A. Form $B = M_a(A)$, $C = M_b(B)$ and $A' = M_c(C)$. Connect the four points, in order, with segments. Drag A to see when it coincides with A'. Observe that the trace of the midpoint of AA' is a line.

Now we consider the case of quadrilaterals. Given a quadrilateral $ABCD$, the isometry S_{ABCD} is defined to be the product of the reflections in the four sides of $ABCD$:

$$S_{ABCD} = M_{DA}M_{CD}M_{BC}M_{AB}.$$

Lemma 3. *Given a quadrilateral $ABCD$, S_{ABCD} cannot be the identity transformation.*

Proof. S_{ABCD} is the product $(M_{DA}M_{CD})(M_{BC}M_{AB})$ of two rotations by non-zero angles; the centers of rotation are B and D. S_{ABCD} cannot be the identity, since the point B is fixed by the first rotation and moved by the second, so B does not equal $S_{ABCD}(B)$. QED

Given a quadrilateral $ABCD$ let a, b, c, and d denote the perpendicular bisectors of AB, BC, CD, and DA, respectively. The isometry T_{ABCD} is defined to be the product of the reflections in these four perpendicular side bisectors:

$$T_{ABCD} = M_d M_c M_b M_a.$$

As in the case of a triangle, the isometry T_{ABCD} leaves fixed the vertex A.

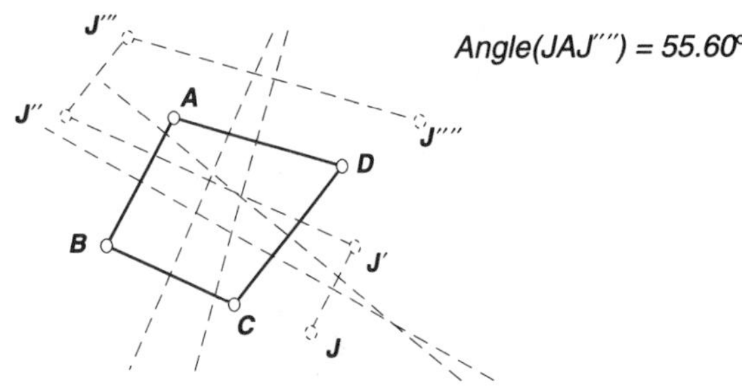

FIGURE 3

Lemma 4. *Given a quadrilateral $ABCD$, T_{ABCD} is either a rotation with center A or the identity.*

Proof. T_{ABCD} is the product $(M_dM_c)(M_bM_a)$ of two rotations and so must be a rotation or a translation. If T_{ABCD} is a non-zero rotation, then A must be its center. If T_{ABCD} is a translation, it must be the identity, since the only translation with a fixed point is the identity. QED

It is interesting to explore the action of T_{ABCD} with dynamic geometry software. Begin with a point J and follow its images under the line reflections. In Figure 3, $T_{ABCD}(J) = J''''$. If point J is dragged around, then the measure of angle JAJ'''' does not change. If one of the vertices of $ABCD$ is dragged so that the quadrilateral appears to be concyclic, then J'''' moves to coincide with J, as it should.

Lemma 5. *If $Q = ABCD$ is concyclic, then T_{ABCD} is the identity.*

Proof. If Q is concyclic, the perpendicular bisectors are concurrent at a point P. The isometry T_{ABCD} is the product of two rotations $(M_dM_c)(M_bM_a)$, each of which has its center at P, so T_{ABCD} fixes P. Since T_{ABCD} fixes both P and A, it is the identity mapping. QED

Before proving the theorem, we need one last lemma.

Lemma 6. *If $Q = ABCD$ is a quadrilateral, then either S_{ABCD} and T_{ABCD} are both nontrivial rotations with the same angle or both are translations (allowing the identity as a translation).*

Proof. Since a is perpendicular to AB and b is perpendicular to BC, the angle between a and b is the same as the angle between AB and BC. This means that the angle of rotation of the product M_bM_a is the same as the angle of rotation of $M_{BC}M_{AB}$. Similarly, the angle between c

and d is is the same as the angle between CD and DA, so the angle of rotation of the product $M_d M_c$ is the same as the angle of rotation of $M_{DA} M_{CD}$. Thus the angle of rotation of $T_{ABCD} = (M_d M_c)(M_b M_a)$ is the same as that of $S_{ABCD} = (M_{DA} M_{CD})(M_{BC} M_{AB})$. T_{ABCD} is a translation if and only if S_{ABCD} is also, since this occurs if and only if the sum of the angles in the products of the two rotations is a multiple of 360°. QED

Proof of the Theorem We begin with $Q = ABCD$ and assume that Q is not concylic. We construct $Q_1 = EFGH$ and wish to show that Q_1 is not concyclic. Suppose Q_1 is concyclic. Then by Lemma 5, T_{EFGH} is the identity and so by Lemma 6, S_{EFGH} must be a translation. But S_{EFGH} is the same transformation as T_{ABCD}, since the line reflections defining them are the same. Lemma 4 then implies that T_{ABCD} must be the identity. But this is a contradiction, since by Lemma 3 S_{EFGH} cannot be the identity if $EFGH$ is a quadrilateral. This implies that if Q is not concyclic then $Q_1 = EFGH$ cannot be concylic. We are done.

References

1. Bennett, D., "Dynamic Geometry Renews Interest in an Old Problem," *Geometry Turned On*, pp. 25–28.
2. Yaglom, I.M., trans. Alan Shields. *Geometric Transformations I.* New York: Random House. Washington, DC: Mathematical Association of America, 1962.

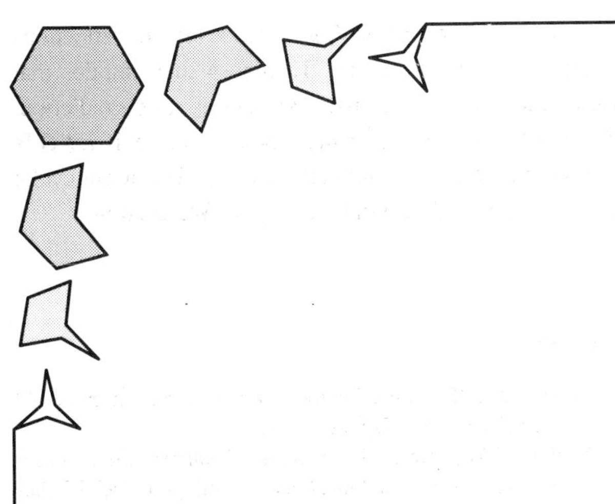

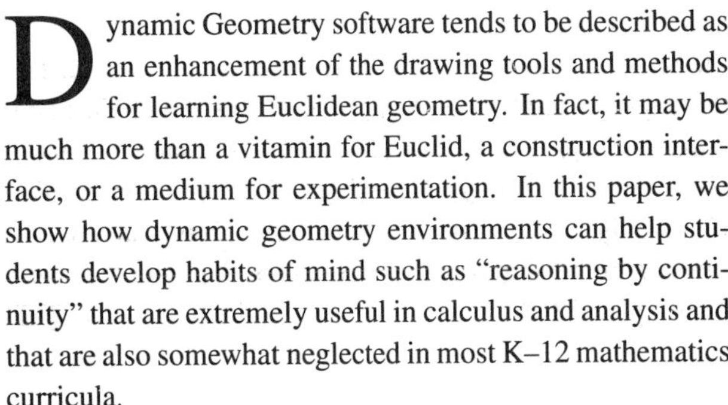

Dynamic Geometry as a Bridge from Euclidean Geometry to Analysis

Albert A. Cuoco
and
E. Paul Goldenberg

Dynamic Geometry software tends to be described as an enhancement of the drawing tools and methods for learning Euclidean geometry. In fact, it may be much more than a vitamin for Euclid, a construction interface, or a medium for experimentation. In this paper, we show how dynamic geometry environments can help students develop habits of mind such as "reasoning by continuity" that are extremely useful in calculus and analysis and that are also somewhat neglected in most K–12 mathematics curricula.

Dynamic Geometry in Analytic Thinking

The usual school approach to the solution of problems involving continuous variation goes something like this:

- Find an algebraic expression for the function that relates the paired quantities.
- Draw the Cartesian graph of this function.
- Use a combination of information gained from the graph and the algebraic model of the function to solve the problem.

The experience of many teachers is that questions about continuous change and the way in which two quantities vary in relation to one another get lost in the shuffle. Students focus on the algebraic manipulations that set up the problem, on the technology that draws the graph, and on interpreting the algebraic and visual information (both of which are viewed as static phenomena) that they have produced. The "Cartesian connection" between the algebraic expression and the graph also tends to get lost; students see the equation of a curve as a code for generating a picture rather than as a characterization of the coordinates of the points on the curve (see, e.g., [5]).

Think, for example, of the problem that asks students to find a rectangle of fixed perimeter that maximizes area. We'd *like* students to imagine a continuum of rectangles, all with the same perimeter. We'd like them to do a thought experiment starting with a long thin rectangle that gradually gains height until it is tall and narrow, and we'd like them to imagine that the area is changing in a continuous way throughout the process. We'd even like them to imagine a graph of area against the length of one side developing dynamically. But nothing in the usual school methods requires or even encourages such thought experiments.

Dynamic geometry software is a medium in which continuous variation is the primary investigatory tool. Students can model a situation and then see how it changes as some feature is dragged smoothly across the computer screen.

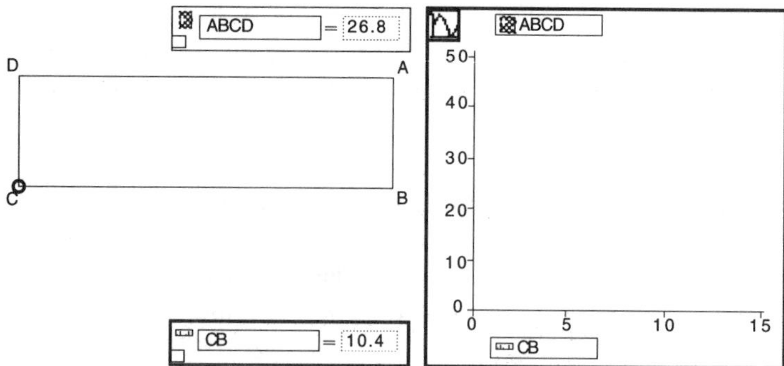

FIGURE 1. The setup.

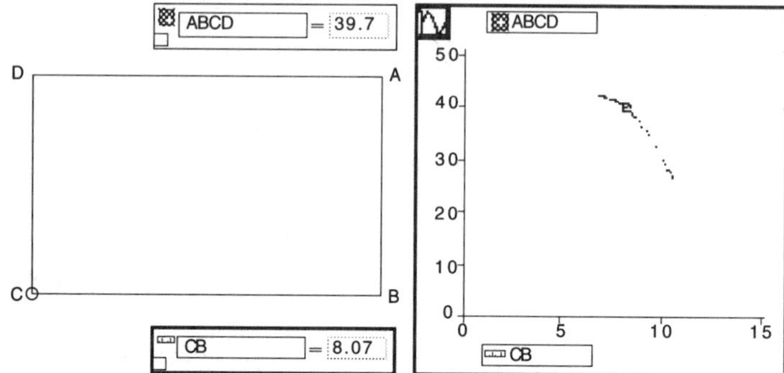

FIGURE 2. Dragging C.

Claim 1. By allowing students to investigate continuous variation directly (without intermediary algebraic calculation), dynamic geometry environments can be used to help students build mental constructs that are useful (even prerequisite) skills for analytic thinking.

Take, for example, the problem of finding a rectangle with fixed perimeter of maximum area. In dynamic geometry environments, students can build a rectangle that can be deformed but will maintain a constant perimeter. Such environments calculate area and length, so it is easy to ask the system for the area of any instance of the rectangle and to also ask it for the length of a side. Simply dragging the rectangle will produce the desired conjecture. In fact, "wiring" the area and length calculations to appropriate axes in the *Geometry Inventor* will even produce the desired Cartesian graph, without any recourse to algebraic symbols at all.

Figures 1–3 show three snapshots of what a student would observe. At the left is the rectangle that can be dynamically changed in shape by pointing and clicking. On the right is a window that shows a scatterplot of length-area pairs sampled from the rectangle as it changes, allowing the student to view area as a function of the variable edge length. In essence, students are building a function defined on a geometric construction.

As corner C of the rectangle is dragged, the rectangle changes shape but, because of the construction constraints,

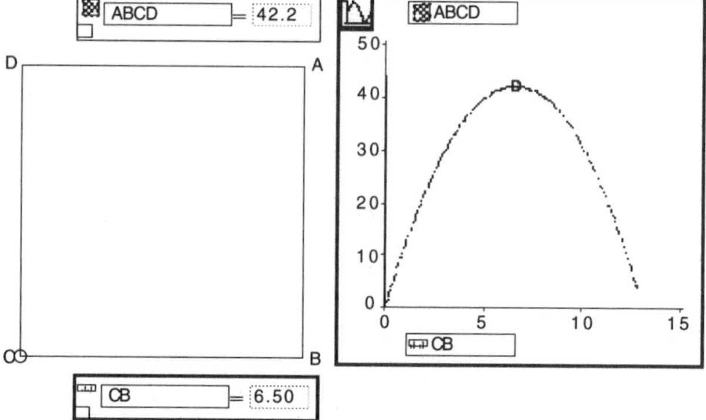

FIGURE 3. An experimental solution.

does not change perimeter. The graph develops dynamically. This gives students a kinesthetic feeling for the continuous variation of the system, much more than they'd feel by plotting points. Quickly, they can generate the entire graph, approximate the maximum value of the area, and solve the problem experimentally.

In addition to this potential for bringing continuous variation to the foreground of students' experience, we see other potential benefits in the uses of dynamic geometry environments.

Claim 2. Dynamic geometry software can be used to help students develop a much broader collection of techniques for visualizing continuous variation and the behavior of functions. Such environments have the potential to change the dominant role played by $\mathbb{R}$ to $\mathbb{R}$ functions and their associated Cartesian graphs in the current analysis curriculum.

Claim 3. Dynamic geometry environments can encourage students to develop mental images of functions that are especially suited to analysis.[1]

In [2], we describe three dominant ways functions are used in mathematics: as algorithms, mappings, and continuous covariation. In that paper, we describe several computational environments well suited to the "function as algorithm" and "function as table" images; dynamic geometry environments are equally well suited to developing the image of a function as a continuous dependence between two objects. This point of view is a prerequisite for a genuine grasp of the basic ideas in analysis.

There are several caveats about our claims:

- We are *not* calling for an elimination of algebraic calculations in analysis. What dynamic geometry allows us to do is to "separate the variables" for awhile so that students can concentrate on the essentially topological properties in a situation without getting bogged down in "the algebra."
- Dynamic geometry software allows students to study the behavior of functions defined on geometric constructions. We are *not* advocating that students' entire analysis experience be built around such functions. There are certainly other classes of functions that are important in the development of analytic thinking. Examples of functions that can be essentially non-geometric include functions whose independent variable is time, and functions that arise from the analysis of data.
- Any limitations with respect to the objects that can be built or manipulated (typically, constructions involving points, lines, circles, conics, and various geometric loci) should not be taken as *prescribing* the content of geometry, precalculus, or analysis experiences for students. The special power of dynamic geometry software lies not in the particulars of its content, but in its ability to allow direct manipulation (or "direct engagement" as discussed in [6]) of certain features of a construction.
- Computer experiments should be seen as precursors to, not replacements for, thought experiments. In effect, the dynamic geometry experiments are like "jump starts" for thought experiments.
- Our claims are currently conjectures. Investigating these conjectures is part of the work of a research project underway at EDC, "An Epistemology of Dynamic Geometry," that will investigate the extent to which dynamic geometry environments can help students develop the ways of thinking that are essential to analysis and that are discussed in this article and in [6].

Continuous Change in Euclid

Geometry has always been a staple of the high school mathematics curriculum, but, for the past few decades, it has been looked upon as somewhat disconnected from the rest of the secondary mathematics program. The current course contains a treatment of some of the classical Euclidean results ("the base angles of an isosceles triangle are congruent"), formulae for area and volume, an increasingly smaller discussion of stylized deductive proofs, and perhaps a smattering of material on coordinate geometry, triangle trigonometry, and rigid motions of the plane.

But there is another, much older, and perhaps more important, theme in elementary geometry: in addition to studying properties of two- and three-dimensional space, geometry built on the Euclidean tradition of using proportional reasoning to think about real numbers in a way that developed intuitions about continuously changing phenomena.

Thought experiments involving conic sections, locus problems, the behavior of geometric phenomena under continuous change, and mechanical devices (linkages, pin and string constructions, and so on) can give students experience with "moving points" and their paths. Ideas about continuity (with respect to the topology of the real line) were embedded in the works of many classical geometers; celebrated legends like Zeno's paradox show that movement and motion have long been on the minds of mathematicians. But, because continuity and motion were not explicit parts of the *Elements*, such topics were not explicitly included in most geometry courses in Western Europe and the United States. Such ideas did, however, become part of the pedagogy and folklore among many high school teachers. Dynamic geometry software is an effective medium for bringing this pedagogy and folklore into the mainstream geometry curriculum.

For example, there are several theorems that relate central angles to angles formed by two chords, angles formed by a tangent and a chord, and angles formed by two secants or tangents to the arcs they intercept in a circle. All of these

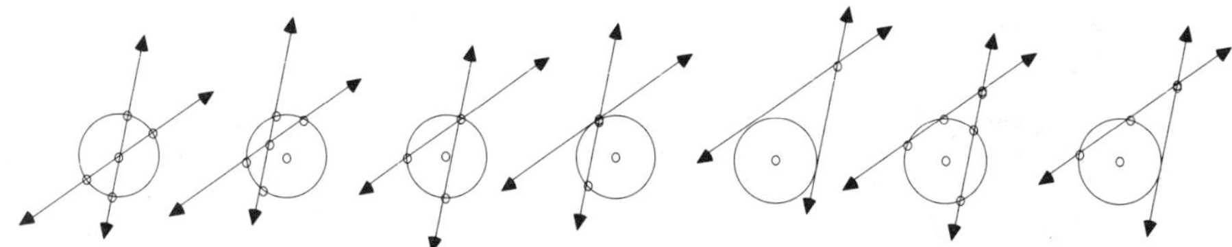

FIGURE 4. Seven cases of two lines intersecting a circle. Is this seven static theorems, or one dynamic one?

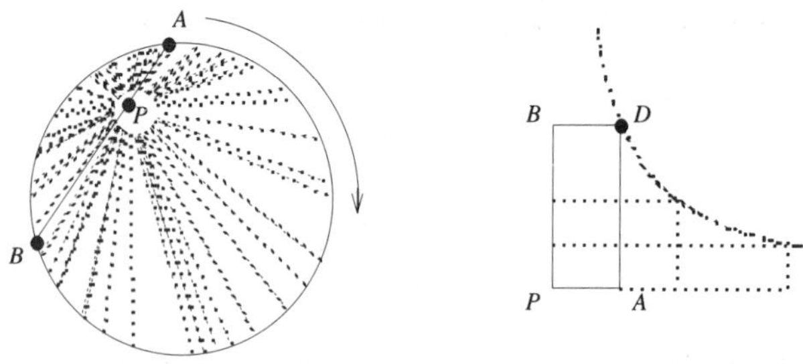

FIGURE 5. At the left, a circle with movable point A on it, fixed P, and chord $\overline{AB}$ through P. At the right, lengths PA and PB are copied on horizontal and vertical lines intersecting at P.

theorems can be related by a thought experiment that asks students to envision the vertex of the angle starting at the center of a circle and gradually moving to the exterior of the circle. This thought experiment can become a real experiment in most dynamic geometry software environments (see Figure 4).

A similar figure (and the accompanying construction in, say, *Geometer's Sketchpad*) can be used to motivate the "power of a point" theorems. Recall the definition: given a fixed circle and a point P not on the circle, if a line through P intersects the circle in A and B, then the "power of P" is $PA \cdot PB$, and this is independent of the line chosen. Theorems related to the power of a point that were formerly thought to be esoteric curiosities can be given practical application in dynamic geometry environments. For example, one can use the power of a point inside a circle to construct a family of rectangles with constant area (see [10] for more details). Figure 5 illustrates such a construction. As A is moved around the circle, the product $PA \cdot PB$, and therefore the area of the family of rectangles at the right, remains constant.

Back to the Future: Implications for Curriculum

In addition to investigating classical problems in geometry through the lens of continuous variation, dynamic geometry environments allow teachers and students to develop techniques for investigating situations that have until now required advanced methods. We discuss two broad categories.

1. Locus of a point. Dynamic geometry environments allow students to trace the path of a point that is subject to certain constraints. This "tracing point" moves as a result of the direct manipulation of one or more features of a sketch; in other words, it is a *function* of these features. When students construct such sketches, they are constructing computational models for functions, and the behavior of such functions can be *experienced* in a very kinesthetic[2] way when students experiment with their creations. In addition, some dynamic geometry software allows students to gather up the image points of such functions into a *set* that can be manipulated by other functions.

Take, for example, a *Sketchpad* construction of an ellipse.

In Figure 6, circles of radius CE and ED are constructed with centers A and B, respectively. As E slides back and forth along $\overline{CD}$, points F and G (the intersections of the two circles) trace out the upper and lower halves of an ellipse with foci at A and B and whose major axis has length CD. The *construction* of this sketch is driven by the fact that the positions of F and G are determined by that of E. And the actual *drawing* of the ellipse is accomplished by moving E continuously along $\overline{CD}$. In other words, students who construct this sketch are modeling a continuous function and then experimenting with its behavior.

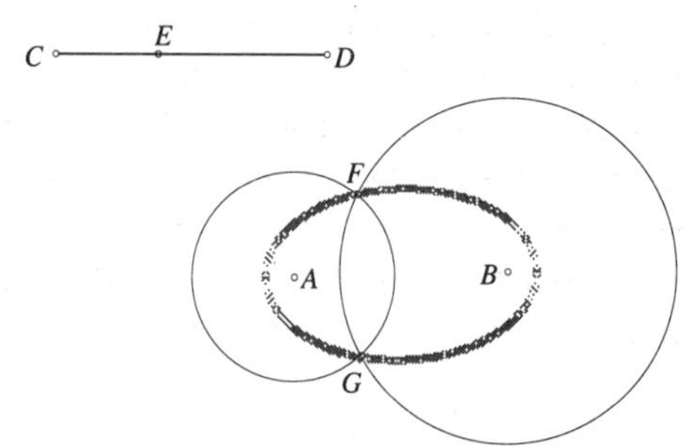

FIGURE 6. $FA + FB = CD$ and $GA + GB = CD$.

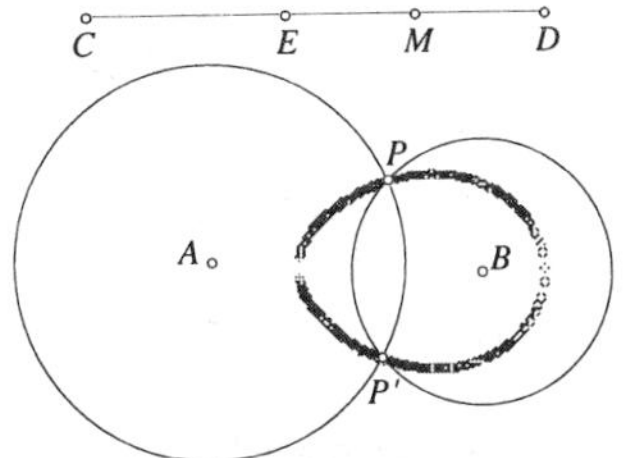

FIGURE 7. $PA + 2PB = CD$.

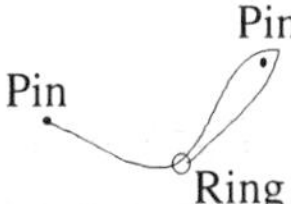

FIGURE 8. A pin-and-string device for constructing the locus of points P such that $PA + 2PB$ is constant.

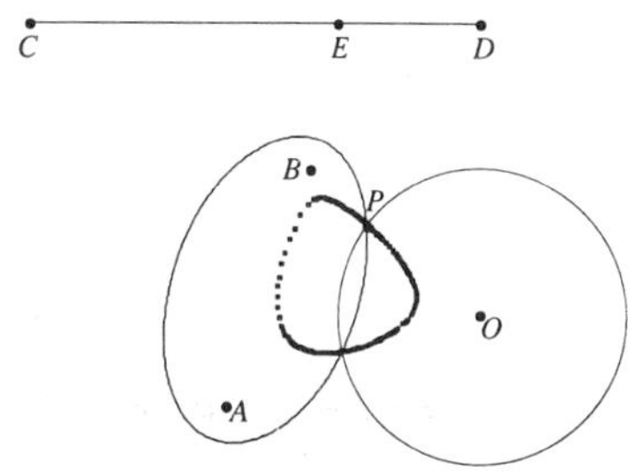

FIGURE 9. $PA + PB + PO = CD$.

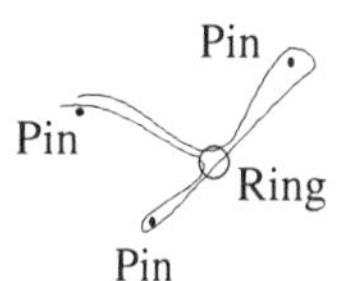

FIGURE 10. A pin-and-string construction for a curve of constant sum of distances from three points.

Dynamic geometry environments allow students to experiment with very general loci in this way (that is, by constructing functions that produce the loci). For example, rather than looking at all the points P so that $PA + PB$ is constant, students can construct functions that produce the set of points P so that $PA + 2PB$ is constant.

In Figure 7, points P (and P') are functions of E that are constructed in the following way: Circles of radius CE and $EM = \frac{1}{2}ED$ are drawn with centers A and B respectively, intersecting at P and P'. As E is dragged along CD, P moves in a way so that $PA + 2PB = CD$ is constant (and similarly for P').

Of course, electronic experiments are not the only kind that give a sense of the dynamics of this situation. One gets a different set of insights from a pin-and-string construction of this curve, which can be made with the aid of a small ring, as illustrated in Figure 8.

Tie the string to one pin, lace it through the ring, around the second pin, and tie it to the ring. Now trace the curve by putting a pencil in the ring. This method extends to allow for curves defined by $PA = kPB$ for any rational k.

The set of possible loci that students can build in dynamic geometry environments depends on the available primitives. In *Cabri-II*, conic sections are built in primitives, so that loci defined as the intersection of conics are possible. For example, "a generalized ellipse with three foci," the locus of points P so that $PA + PB + PO$ is constant, can be gotten from the intersection of an ellipse and a circle.

In Figure 9, $\overline{CD}$ is divided at E. An ellipse with major axis EC is constructed on foci A and B, and a circle with radius ED is constructed with center O. P is the intersection of the ellipse and the circle. As E slides along $\overline{CD}$, the locus that is the generalized ellipse is traced out by P.

Again, this curve can be drawn with a pin-and-string construction, laced as shown in Figure 10.

When one pulls all the "slack" out of the system, one gets a physical sense of what it means to minimize the distances. With some fixed slack—the equivalent of CD from Figure 9—one can place a pencil in the ring and trace the curve for the function $PA + PB + PO = CD$.

The continuous dependence of the locus on the "sliding" point E in the above examples is just the beginning. Once the locus is obtained, students can ask how *it* depends on the parameters in the sketch. In most dynamic geometry environments, it's possible to "encapsulate" loci into objects that can be manipulated via other functions. For example, Figures 11, 12, and 13 show families of ellipses, "quasi-ellipses", and "generalized ellipses" that grow and shrink as a function of the "length of the string."

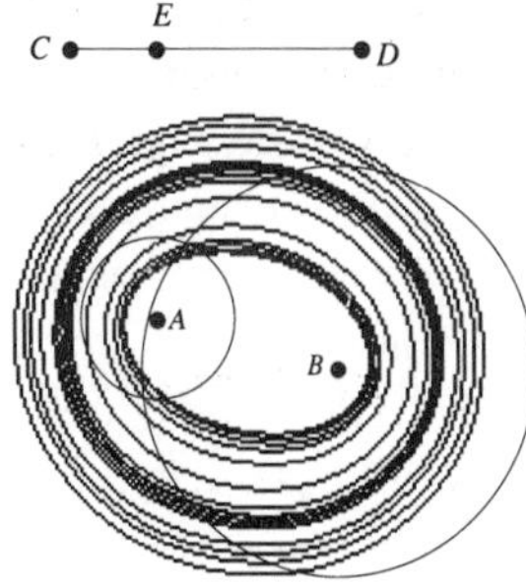

FIGURE 11. A family of ellipses.

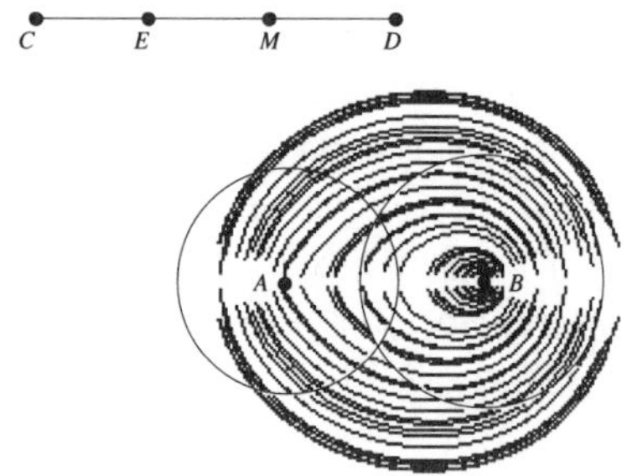

FIGURE 12. A family of "quasi-ellipses."

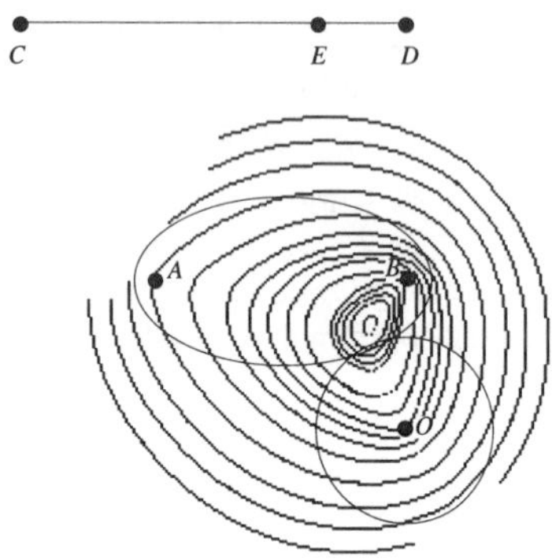

FIGURE 13. A family of "generalized ellipses."

Figure 11 shows a family of ellipses, constructed as in Figure 6, with foci A and B. Each ellipse is the result of a different length of $\overline{CD}$ as A and B remain fixed. As the "string" $\overline{CD}$ gets shorter, the ellipses approach $\overline{AB}$. Figure 12 shows a family of "quasi-ellipses" constructed as in Figure 7. Each egg is the result of a different length of $\overline{CD}$. Figure 13 shows a family of "generalized ellipses" constructed as in Figure 9. Each curve represents the locus of points P that have a constant total distance from P to the three points A, B, and O. That total distance is varied by stretching or shrinking $\overline{CD}$, obtaining a new curve. The point with the minimal total (inside the smallest three-corner egg) is the Fermat point. (The Fermat point (if it exists) is the point P so that the angles $\angle APB$, $\angle APO$, and $\angle OPB$ all have measure 120°.)

As a final example of how dynamic geometry environments can be used to construct and manipulate loci in a continuous way, our favorite example of a "new" geometry problem actually goes back to a thought experiment proposed in 1956 by Wittgenstein [12]. Imagine a rod of fixed length passing through a loose sleeve which is pivoted to the wall (see Figure 14).

As one endpoint of the rod describes a circle, what does the other endpoint describe? Thinking about the locus of the left endpoint of the rod requires one to imagine continuous motion of the system as the right endpoint traverses the circle. Dynamic geometry software allows one to try the experiment first with one's hands and eyes, rather than having to make the entire leap to thought all at once (see Figure 15).

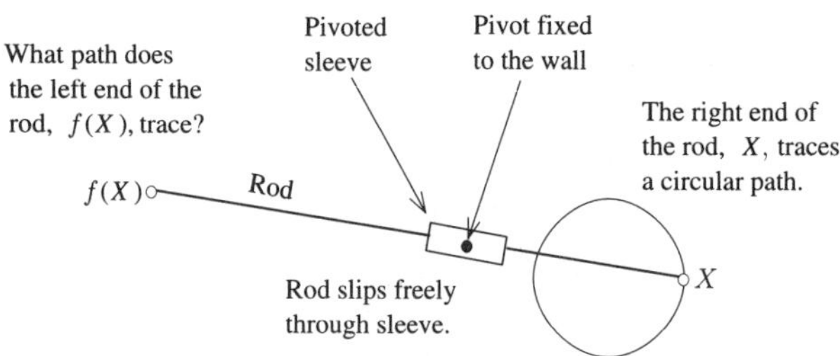

FIGURE 14. Wittgenstein's experiment.

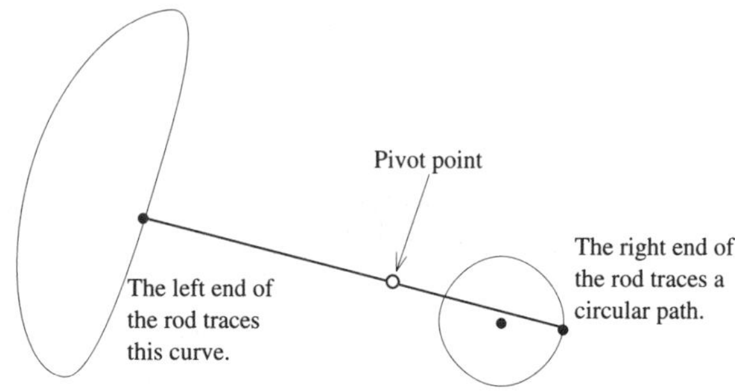

FIGURE 15. The locus.

The actual locus produced is a (continuous almost everywhere) function of several parameters: the length of the rod, the position of the pivot, and the size of the circle. And dynamic geometry software allows students to vary these parameters and to study the effects on the resulting locus. Figure 16 shows several frames of a dynamic experiment that results from varying the rod length. Figure 17 shows how these very different shapes relate, by showing the fam-

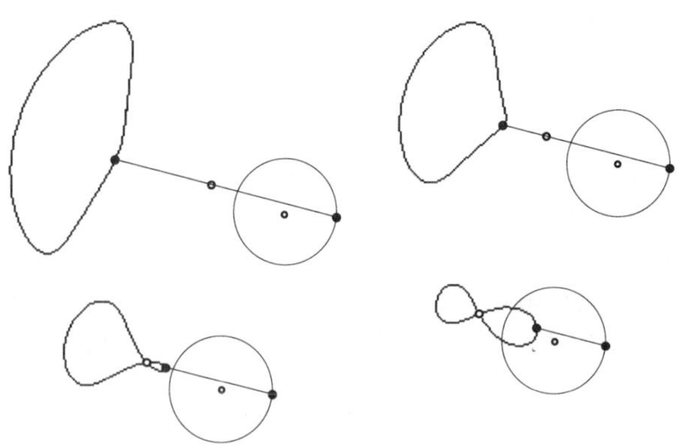

FIGURE 16. Four shapes for the locus of points.

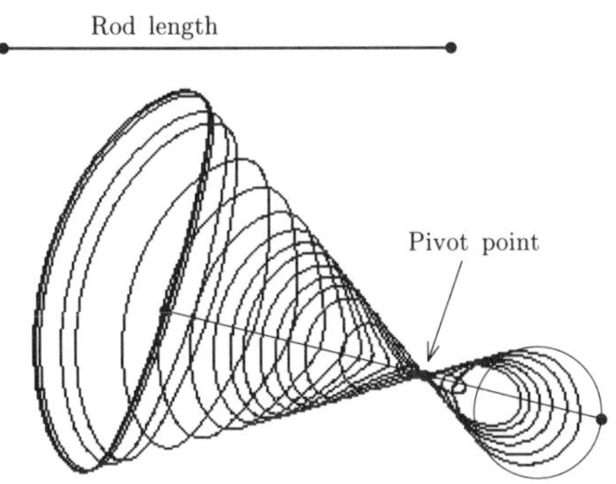

FIGURE 17. The locus as a function of the length of the rod.

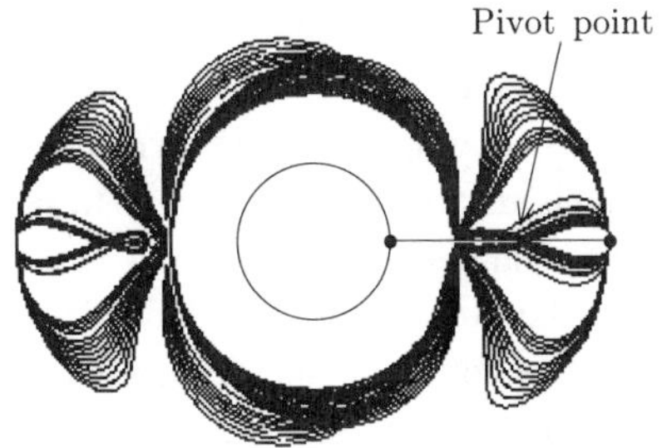

FIGURE 18. The locus as a function of the pivot point.

ily of loci that results from a gradual shrinking of the rod. (This "higher-level" experiment is analogous to asking the role of m in the Cartesian graph of $f(x) = mx + b$. The function value, for each setting of the three variables, is an entire set of points. Indeed, these experiments support the kind of reasoning that can lead to a mathematical analysis and classification of the resulting curves.) Notice that some of these curves result from situations that could not be realized with the mechanical device in Figure 14—for example, situations in which the rod slips out of the sleeve.

Quite a different family of curves is generated when the rod-length and radius are kept fixed and the pivot point is moved. In Figure 18, the pivot point is moved horizontally from roughly one diameter left of the circle, through the center, to one diameter right of the circle.

2. Optimization. The topic of geometric optimization has a long history (see the citations in [1] and [11]). The solution of optimization problems without techniques from calculus has made occasional inroads into the K–14 curriculum, but often only as a collection of *ad-hoc* solutions to isolated problems.[3] Dynamic geometry environments make it possible to develop some coherent principles and habits of mind with which students can approach optimization problems before calculus. What's more important is that these techniques, because they make essential use of reasoning by continuity, lay a strong foundation for the ways of thinking that will be developed in calculus and analysis.

In this section, we describe one such technique—the method of contour lines. The basic philosophy of this approach is described in [9]; dynamic geometry gives new life to this beautiful approach to optimization problems.

In calculus, students learn to find the (possibly local) maximum or minimum of a differentiable $\mathbb{R}$-to-$\mathbb{R}$ function f by deriving a second function f'—one that describes a characteristic (the slope) of all straight-line tangents to f—focusing only on those tangents that happen to be horizontal (for only they indicate extrema), and locating those horizontal tangents. *Optimization: a Geometric Approach* [4], a book designed for high school students, develops a different style of reasoning, through which students see such horizontal straight-line tangents as a special case of a bigger idea that they've already developed and used extensively. This bigger idea involves distinguishing what measurement is being made (e.g., "the 'height' of any function") from the object to which the measurement is being applied (e.g., "the particular function f"). A function g (e.g., the 'height'-measuring function) reports the measurement. When the "level curves" or "contour lines" of g (regardless of their shape) "just touch" the object of measurement, they indicate extreme values of that measurement along f. This notion of contour lines is completely analogous to the contour lines used by cartographers and meteorologists; the notion of "just touching" the "near" or "far" end of the object to be measured fits intuitively with students' understanding of measurement.

This perspective contrasts with what may be the most common way of introducing students to the use of the derivative to find extrema of an $\mathbb{R}$-to-$\mathbb{R}$ curve. Typically, they derive a formula for the slope of the curve and look where that slope is zero to find extrema. In effect, they are considering *all* tangents to the curve (though they may not, in fact, know how to *find* any of them), selecting those that are horizontal, and then examining to see which indicate extrema and which indicate plateaus, or horizontal inflection points.

The contour-line approach considers all horizontal lines and selects those that are tangent to the curve. This is not computable in the same way, and so is not the prevailing perspective, but it provides a much more general insight. Horizontal lines are the "contour lines" for the "height function." Where some curve c crosses them, it is clearly changing "height"—becoming greater or smaller in value. Where c just touches but does not cross the contour lines, it has reached some extreme (max or min) value. There is no exception for horizontal inflection points: they are places where the curve crosses the contour lines.

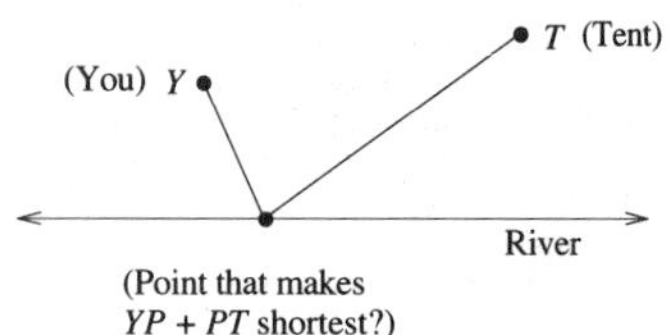

FIGURE 19. The "Burning Tent Problem."

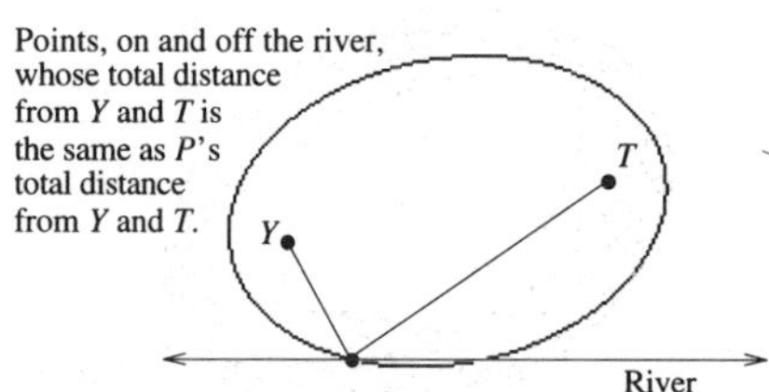

FIGURE 20. A locus of points "as bad as P."

We illustrate this method of contour lines with three examples.

2.1. Tents and Pools. Here's a problem that is getting a great deal of attention in K–12 geometry courses:

> You are on a camping trip, returning from a walk. Standing at Y, you notice that your tent (located at T) is on fire. You need to run to the river, fill your bucket with water, and get to the tent. Where should you land on the shore to minimize the total distance of the trip? (This situation is illustrated in Figure 19.)

There is a simple geometric solution to the problem: reflect Y over the shore to obtain a point Y', and connect Y' to T. Where $\overline{Y'T}$ intersects the shore is the best spot to land.

Although this technique does not spring out of the mathematical blue, it often feels that way to students—a brilliant coup, but how would anyone ever have thought of it?! By taking the time to develop this way of thinking in advance, of course, one can make its application to this problem seem less "out of the blue," but the effort to develop the contour-line approach appears to pay off more richly with mathematical connections, and there is a surprise bonus in using it!

The contour-line approach depends upon students looking for a function in the situation—already a good habit of mind to develop. In this case, we want them to think about the sum of the distances as a function of P, and we want them to think of that sum as a continuously changing quantity as P moves back and forth along the shore.

Suppose you are at a point P on the shore, and you look at the set of points on the *plane* that are equally as "bad" as P. That is, $PY + PT$ is some number, and you look at all the points Q on the plane so that $QY + QT$ is that same number. The locus of points, the sum of whose distances from Y and T is constant, is an ellipse with foci Y and T, and students could construct this ellipse with pins and string.

One sees the continuous variation more readily when one constructs the ellipse in a dynamic geometry software environment. The particular choice for P shown in Figure 20 is clearly not the optimal one, because there are points on the shore that are inside the ellipse (and hence have a smaller sum of distances to Y and T). But the dynamic construction allows one to slide P along the shore and study how the ellipse grows and shrinks. It's very striking to see this family of con-focal ellipses change dynamically, to locate the ellipse that is tangent to the shore, and to verify that it is, in fact, the solution to the problem.

And now the extra bonus. By connecting this solution-by-tangent-ellipse to the geometric solution involving the reflection of Y over the shore, one gets a simple proof of the "tangent property" of the ellipse:

> *A tangent to an ellipse makes congruent angles with segments drawn from the foci to the point of contact* (see Figure 21).

In this problem, we looked at $PY + PT$ as a function of P as P moved along the shore. Visualizing the ellipses suggests that we enlarge the domain of our function to include all points in the plane. If we do this, the ellipses are *contour lines* for our function (now a function from $\mathbb{R}^2$ to $\mathbb{R}$ given by $P \mapsto PY + PT$), curves along which the value of the function (the sum of the distances to Y and T) is constant.

Traveling along *any* line ℓ, the minimum value of $P \mapsto PY + PT$ will be at the place where a contour line (one of the ellipses) is tangent to ℓ (see Figure 22). This way of thinking is very powerful, and can be used to solve a wide range of optimization problems including meandering rivers c as well

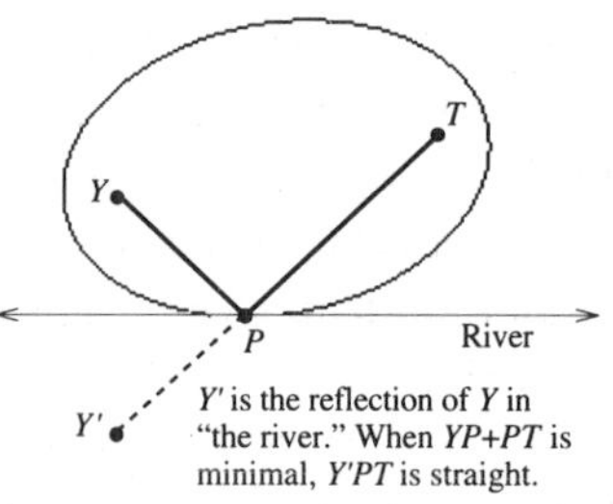

FIGURE 21. The tangent property of the ellipse.

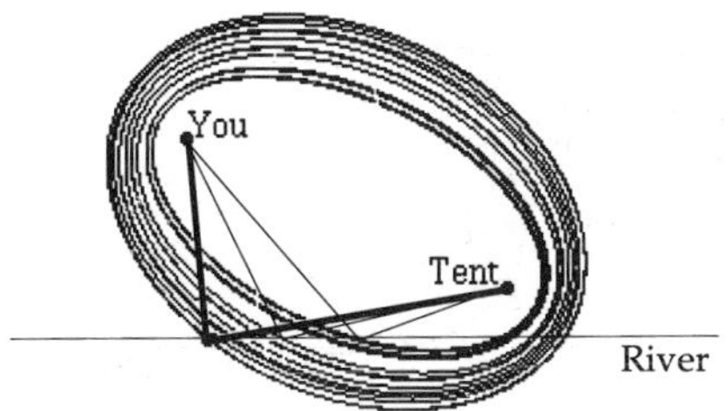

FIGURE 22. Some contour lines for $P \mapsto PY + PT$.

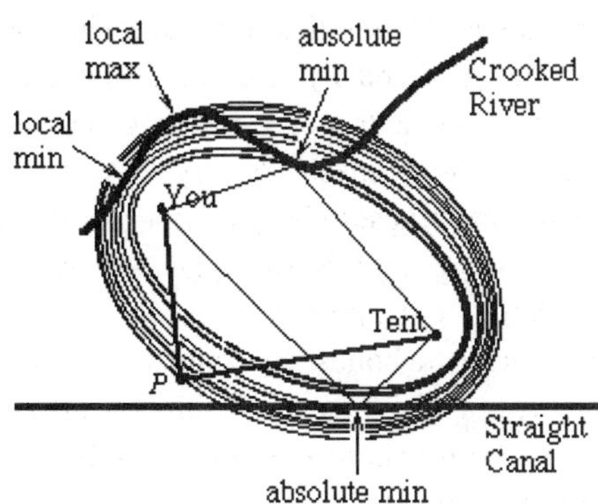

FIGURE 23. The "burning tent problem" with a crooked river.

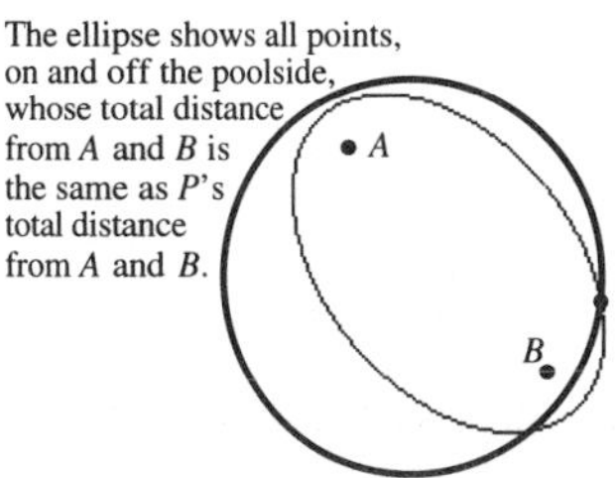

FIGURE 24. P's location isn't optimal.

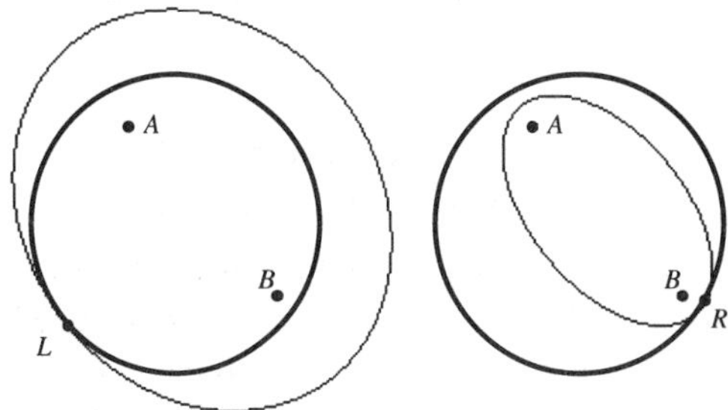

FIGURE 25. Absolute maximum (L) and absolute minimum (R).

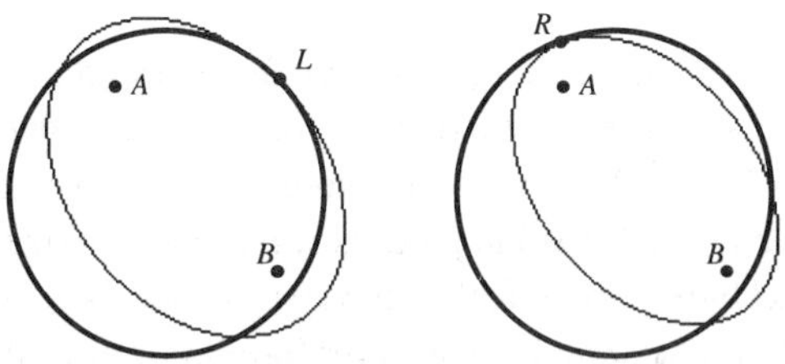

FIGURE 26. Local maximum (L) and local minimum (R).

as straight canals ℓ (see Figure 23). The minimum value of $P \mapsto PY + PT$ will be at the place where a contour line (one of the ellipses) is tangent to the river c, because anywhere that the river *crosses* a contour line, it is *changing* its total distance from Y and T, and therefore not at an extreme value.

As another application of the contour-line method of visualizing functions, consider the swimming pool problem:

> You are in a circular swimming pool at A, you want to swim to the edge of the pool to drop off your sunglasses, and then swim to your friend at B. Where should you land on the edge of the pool to keep the trip to a minimum?

This time we imagine P moving around a circle, so that the function $P \mapsto PA + PB$ is defined on $\mathbb{R}/\mathbb{Z}$. But, if we extend our function's domain to $\mathbb{R}^2$, we can, at each location of P on the circle, construct the contour line of points that have the same "sum of distances" to A and B as P does, again an ellipse (see Figure 24).

Once again, this location for P isn't optimal, because there are points on the circle inside the ellipse. But the behavior of the contour lines for our function is much richer in this case, because the ellipse and the circle can be tangent in various ways. For certain positions of A and B, we can get a non-Cartesian visualization of many of the behaviors that are important in beginning calculus. These are illustrated in Figures 25 and 26.

Here are some questions and problems that students might think about in connection with this problem.

(1) Why does an absolute minimum occur when the ellipse is internally tangent to the circle? Why does an absolute maximum occur when the ellipse is externally tangent to the circle?
(2) Why does a local minimum occur when the ellipse is tangent to the circle as in the right frame of Figure 26? Why does a local maximum occur when the ellipse is tangent to the circle as in the left frame of this figure?
(3) What kind of relationship would the ellipse have to the circle at an inflection point for our function?
(4) What is a geometric description of a point that produces an absolute minimum?

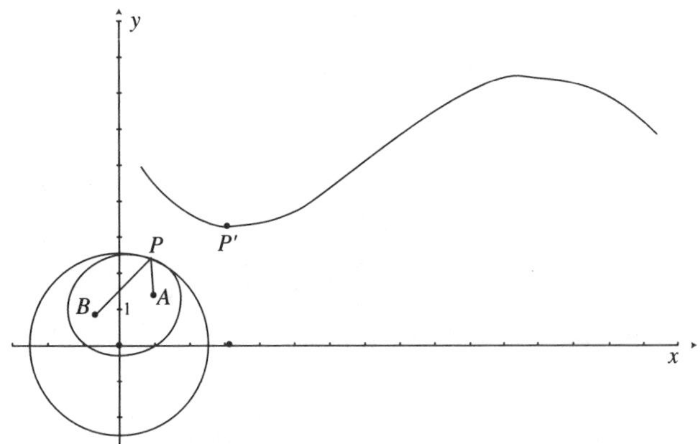

FIGURE 27. One absolute minimum.

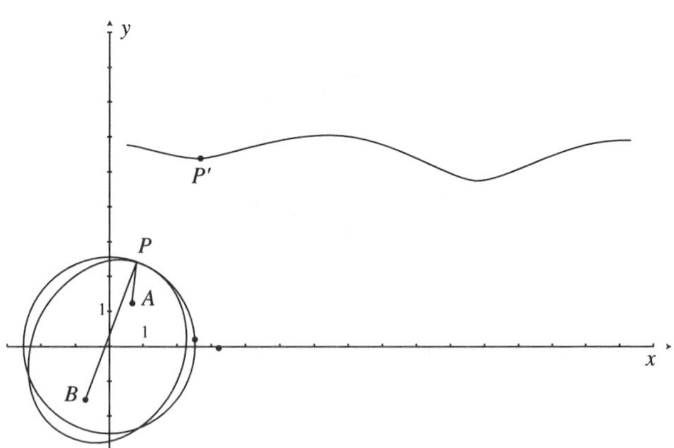

FIGURE 28. Local minimum.

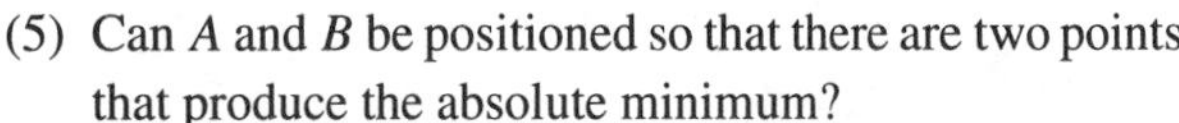

(5) Can A and B be positioned so that there are two points that produce the absolute minimum?
(6) Can A and B be positioned so that there are no points that produce the absolute minimum? So that there are no points that produce a local minimum?

Questions like 5 and 6 ask students to think of the dependence of the function $P \mapsto PA + PB$ on the parameters A and B. In other words, students are being asked to encapsulate $P \mapsto PA + PB$ into an object that can be manipulated (by varying A and B). Dynamic geometry environments that allow for easy construction of Cartesian graphs make this kind of encapsulation more tractable for students.

Looking at our function $P \mapsto PA + PB$ as a function defined on $\mathbb{R}^2$ leads to a consideration of contour lines, and these provide us with a useful way to visualize the behavior of our function. But even greater insights can be obtained by looking at the interplay between this visualization and a more traditional Cartesian one obtained by returning to the point of view that considers the domain of our function to be points on the circle.

In Figures 27 and 28, the independent variable is the length of the counter-clockwise arc between the intersection of the circle with the positive x-axis and P, and the dependent variable is $PA + PB$. P' is the point on the Cartesian graph that corresponds to P. This static picture cannot convey the impact that the dynamic environment has on a person manipulating this construction.

Viewing $P \mapsto PA + PB$ as a function on the circle means that it is periodic. This property of the function is clearly demonstrated by the dynamic geometry experiment: as P passes through the intersection of the circle and the x-axis (in the counter-clockwise direction), P' leaves the right end of the graph to return to the left end.

2.2. The Weighted Tent Problem. When we brought the burning tent problem into Jane Gorman's high school class in Brookline, Massachusetts, one student complained that the problem ignored the crucial fact that the camper can run faster with an empty bucket than with a full one.

Investigating this "weighted tent" can lead to a nice excursion for a calculus class.

If you can run r times as fast from A to P as you can from P to B, it turns out that the best place to land on the river is at the point P so that (in Figure 29)

$$\cos \angle BPS = r \cos \angle APQ$$

This is a geometric version of Snell's Law.

But the problem can be investigated without calculus, using dynamic geometry software. Suppose you can run twice as fast with an empty bucket as with a full one. Then the distance traveled from the river to the tent counts half as much as the distance from you to the river, so you want to find the point P such that P is on the river and $2PA + PB$ is minimized.

But the contourplot for the function $P \mapsto 2PA + PB$ is just a family of "quasi-ellipses" generated in Figure 12 (see Figure 30).

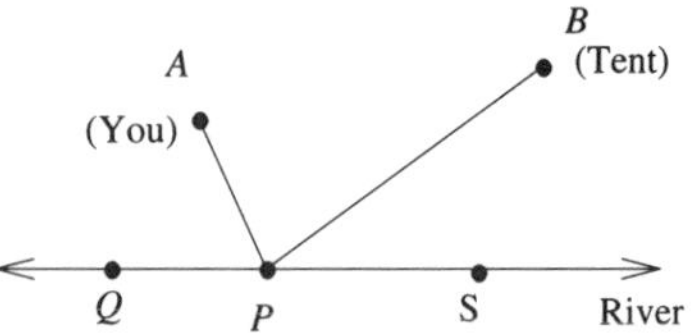

FIGURE 29. The diagram for the weighted burning tent problem is the same, but the interpretation is different.

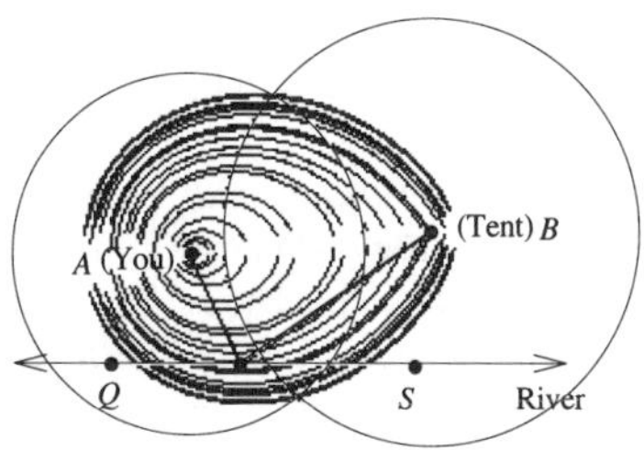

FIGURE 30. Contour plot for $P \mapsto 2PA + PB$.

So, a precise geometric characterization of the solution is the point where the river is tangent to the quasi-ellipse. This can be found experimentally, and precalculus students who know about trigonometry can use the experimental solution to conjecture the analytic one.

2.3. The airport problem. In [4], students investigate the "airport problem":

> *Three cities get together and decide to build an airport. Where should they put it?*

One solution (the "environmental solution") is to locate the airport at a point that minimizes the new road building; all other things being equal, this amounts to finding a point that minimizes the sum of the distances to three fixed points.

Students are again faced with a function from $\mathbb{R}^2$ to $\mathbb{R}$, and once again, contour lines can help students develop a mental image for the situation and see, intuitively, that there is a unique minimum value. The contour lines for the function $P \mapsto PA + PB + PO$ are just the "generalized ellipses" described earlier; the minimum sum occurs at the Fermat point (see Figure 13).

Experimenting with such contour plots can lead to conjectures about the location of the point that minimizes the function, and in [4], students study a completely geometric proof of the fact that the Fermat point minimizes the function. This contourplot also allows students to solve related problems with additional constraints, for example, finding the best spot for an airport that has to lie along an existing road.

Conclusion

Where does reasoning by continuity as described in this paper belong in the curriculum? We see at least two places:

- Thinking about continuous change can be a unifying theme in the study of geometry. In addition to applications like the ones illustrated by Figure 4, dynamic geometry can bring reasoning by continuity to the study of similarity (dilations), congruence (isometries), trigonometry, and coordinate geometry. And, thinking about continuous change provides ways for students to develop proofs and explanations based around notions like invariants and extreme cases (see [3] for an example).
- Topics like geometric optimization can be included in geometry or in precalculus as a context for developing ideas about functions and their behavior. While most precalculus courses consist of technical preparation for calculus, we see topics like the ones described in this paper as conceptual precursors to calculus and analytic thinking. They encourage students to think about continuous change, and they give motivation for the kinds of tools that calculus provides for solving optimization problems.

Acknowledgements. This work was supported by National Science Foundation grants RED-9453864 and MDR-9252952. The opinions expressed here are ours, and do not necessarily reflect the views of the NSF.

Endnotes

1. See also [7] on a role that the dynamic representation of functions may play in preparing students for concepts of analysis.
2. Our repeated mention of the kinesthetic is not a preference for this over other senses, and certainly not over articulated thought. But we have observed [5] that students can have difficulty recognizing the independent variable in a function. They may be able to *say*, for example, that x is the variable in $f(x) = mx + b$, but they *see* nothing obvious to distinguish the four letters in that expression. Computer-graphing experiments can exacerbate the confusion, if students vary m and b while leaving x alone. In the dynamic geometry experiment described here, there is no question where the variable is: it is in the students' hands.
3. Thoughtful and deep treatments of calculus-free optimization methods exist (see, e.g., [8]), but seem not to be part of the mainstream at the undergraduate level, and have had no discernible coherent impact on the K–12 curriculum.

References

1. Courant, R. and Robbins, H. (1941). *What is Mathematics?* New York: Oxford University Press.
2. Cuoco, A. (1995). "Computational Media to Support the Learning and Use of Functions." *Proceedings of the Advanced Nato Workshop: Computational Media to Support Exploratory Learning*. A. diSessa, C. Hoyles, and R. Noss, with L. Edwards (eds.) New York: Springer Verlag.
3. Cuoco, A., Goldenberg, E.P., and Mark, J. (1994). "Technology in Perspective." *Mathematics Teacher*. **87**(6):450.

4. EDC. (1996). *Optimization: A Geometric Approach.* Part of the *Connected Geometry* series. Chicago: Everyday Learning Corporation.
5. Goldenberg, E.P. (1988). "Mathematics, Metaphors, and Human Factors." *Journal of Mathematical Behavior*. 7, pp. 135–173.
6. Goldenberg, E.P. and Cuoco, A. (1996). "What Is Dynamic Geometry?" In *Designing Learning Environments for Developing Understanding of Geometry and Space*. R. Lehrer and D. Chazan (eds.) Hillsdale, NJ: Erlbaum.
7. Goldenberg, E.P., Lewis, P.G., and O'Keefe, J. (1992). "Dynamic representation and the development of an understanding of functions." In G. Harel and E. Dubinsky (eds.). *The Concept of Function: Aspects of Epistemology and Pedagogy*, MAA Notes, Volume 25, 1992. Washington, DC: The Mathematical Association of America.
8. Niven, I. (1981). *Maxima and Minima Without Calculus.* Washington, DC: The Mathematical Association of America.
9. Pólya, G. (1954). *Mathematics and Plausible Reasoning.* Princeton, NJ: Princeton University Press.
10. Scher, D. (1996). "Dynamic Geometry: New Tools Create Fresh Challenges." *Mathematics Teacher,* **89**(4).
11. Tikhomirov, V.M. (1990). *Stories about Maxima and Minima.* Providence, RI: The American Mathematical Society.
12. Wittgenstein, L. (1983). *Remarks on the Foundations of Mathematics*. Cambridge, MA: MIT Press.

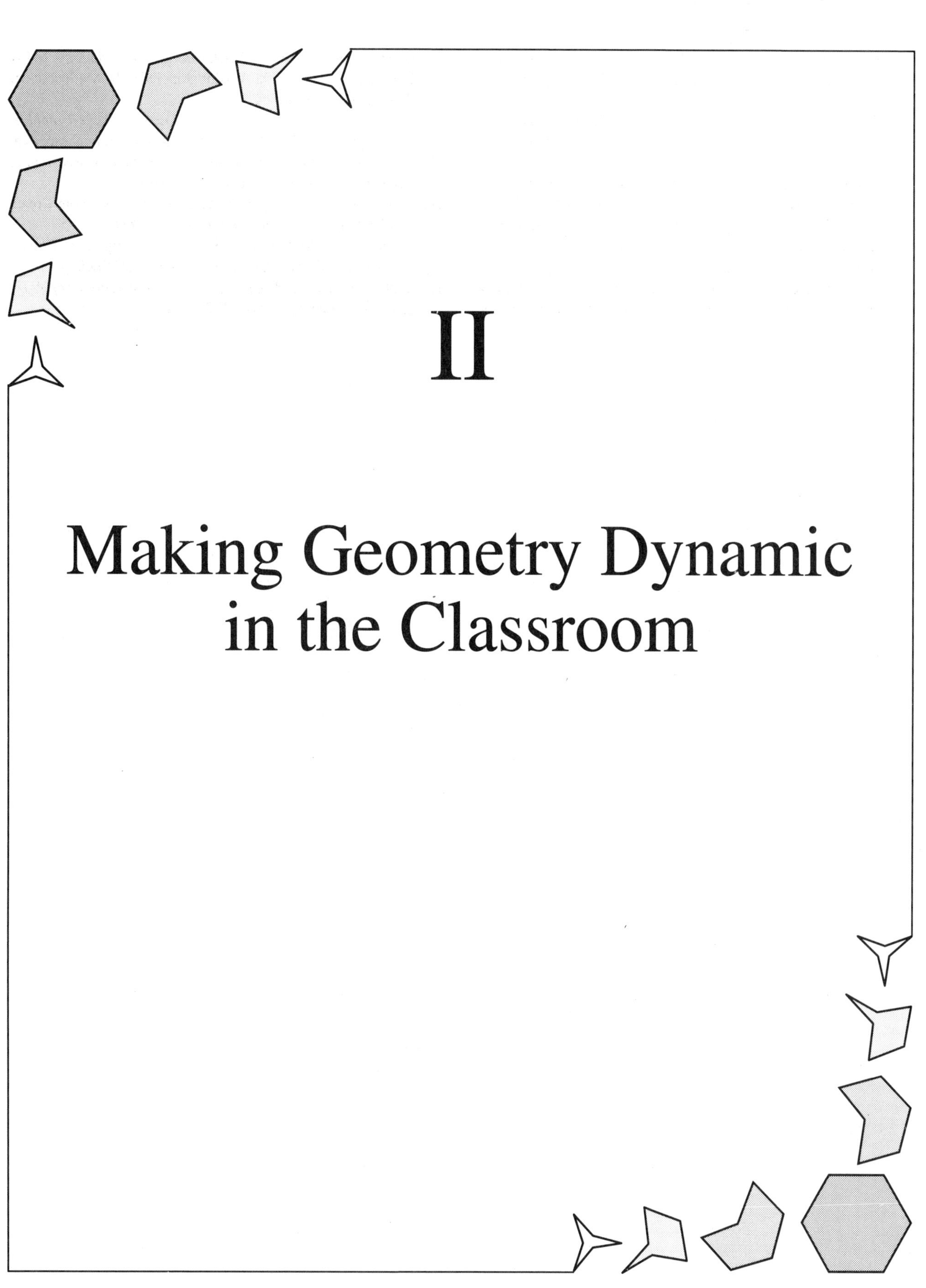

II

Making Geometry Dynamic in the Classroom

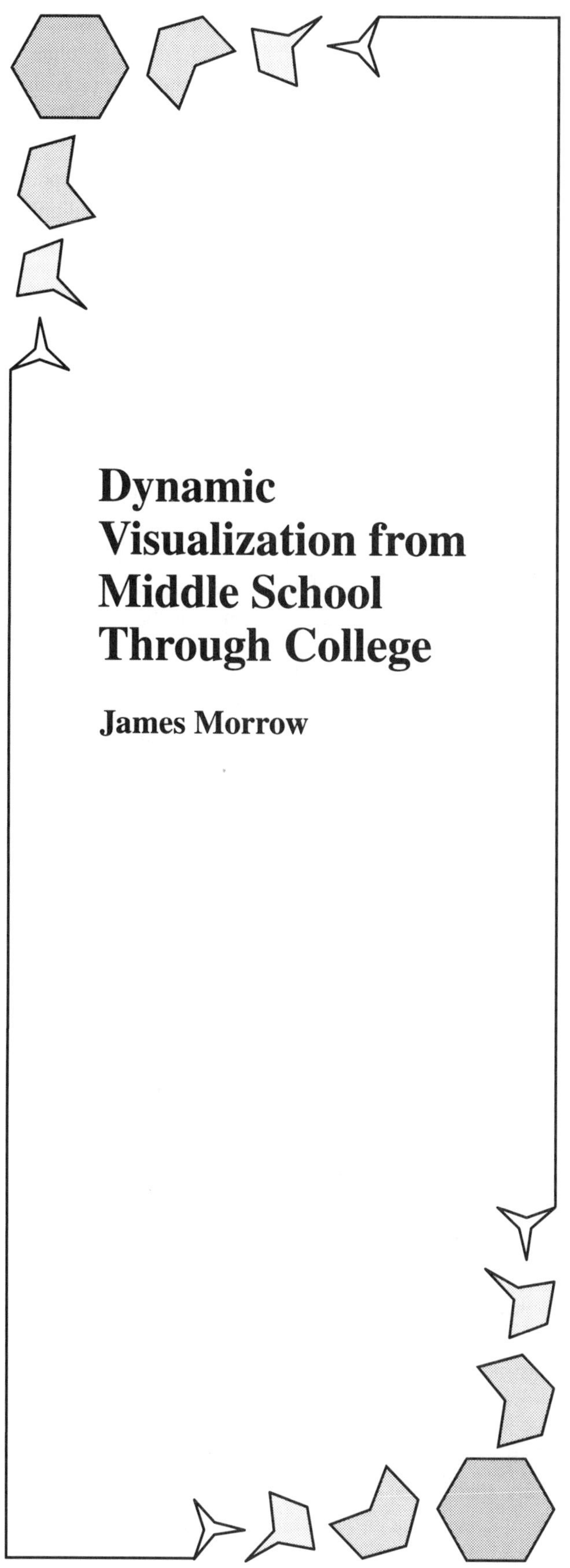

Dynamic Visualization from Middle School Through College

James Morrow

What is Dynamic Visualization?

Imagine a sixth-grader at a computer terminal operating a set of controls to operate a robotic arm. Now imagine her in high school constructing a robotic arm and determining its "reach." Finally, in your mind, follow her to college, where she constructs "scripts" for others to build robotic arms and instructions for animating parts of these arms. The idea of this paper is to make the above scenario, and many others similar to it, a reality.

Dynamic geometry software can be used to aid the process of visualization in all mathematics classes, not just in the study of geometry. Students can construct, revise, and continuously vary geometric sketches. While visualization in itself is a powerful problem-solving tool, the capacity for students to make instantaneous and precise variations to their own visual representations adds a new dynamic dimension whose implications are only beginning to be understood. *Dynamic visualization* refers to the broad use of this dynamic problem-solving tool. In the discussions that follow I will assume the use of *The Geometer's Sketchpad* (hereafter referred to as *Sketchpad*), because that is the tool with which I have most experience, but other dynamic software could also be used.

Before I discuss dynamic visualization, let me state my working assumptions about the process of learning. Very briefly, I assume that, when people learn something, they must construct something new in their mind, connected to what they knew before, and which will actually change the nature of what they knew before. For example, as [8] points out, in citing the work of C. Kieran, "... students... tend to think of the equal sign as a 'do-something' signal." Only later, through reflection on exercises done, do students "... reorganize their understanding to see the equal sign as a relational symbol." Such a reorganization is often not easy to accomplish. To make such a mental construction, learners must face some conflict and some inconsistency in how they think about their experiences. One of the roles of the teacher is to push inconsistencies to the forefront. And dynamic visualization provides an excellent medium in which teachers can do so.

Why Use Dynamic Visualization?

Conjecture Formation and Problem Posing. Dynamic geometry software encourages students to play, to explore, and, with the encouragement of teachers, to form conjectures and pose problems about what they observe. Such software is fun for students *and* for teachers because it allows us to experiment without much penalty. We can try to do something with a visual representation, and, if we don't like what we see, return to the original representation. Since testing ideas is so simple, visual exploration is encouraged. As teachers, we then can encourage our students to form conjectures on the basis of what the students have experimented with. For example, if students have constructed points on a number line for x and $1/x$ and have experimented with positions for those points on the number line, a teacher might ask what happens as x gets larger and larger. In an open atmosphere where student questions and conjectures are rewarded, even when answers may seem simple, everyone will soon be testing ideas, and coming to conclusions that are unexpected by both teachers and students. A key ingredient to making student exploration so rewarding is for the student to be in control of the experimentation. Then they are asking and answering their *own* "why-questions" [4] rather than the teacher's "why-questions."

A fundamental notion which is facilitated by exploration using dynamic geometry software is that of *invariance:* What remains the same in the sketch as another element of the sketch is varied? For example, what happens as a vertex of a triangle is dragged along a line parallel to the side opposite that vertex? Does the triangle's shape, area, perimeter, etc., stay the same? Or, what happens to a triangle's centroid and its circumcenter as one vertex moves in a "random" way? (Cf. [1])

Deepening Understanding of Mathematical Concepts. Dynamic visualization deepens the experiences students have with concepts that have inherently dynamic aspects. For example, the idea of *angle* may be thought of in many ways. One of the most powerful is to view an angle as a rotation, a very dynamic notion. Sketchpad allows angles of rotation to be specified by units of angular measure or selected arbitrary angles that are highlighted by an arc moving from initial side to terminal side of an angle. Contrast such angular representations with the traditional static textbook drawings of angles.

Variable. The idea of using a symbol, say x, to represent an unknown, but knowable, real number in a problem is also an example of a dynamic concept. To solve a problem that involves determining some number, initially known only very indirectly through some properties that it has, by introducing a symbol for it and then manipulating that symbol, is not the most intuitively obvious way to go about solving the problem. Many students do not, of their own choosing, introduce a symbol in this way, because it is counter-intuitive to them. If the symbol can be attached to a point on a number line and the point can be dragged about by hand and that causes other quantities to change (and this can be directly observed), then the concept of *variable* is better understood. A connection between the symbol x and the more intuitive problem-solving method of guessing and checking is also made.

Function. Consider solving a problem by dragging a point x along a number line to see the effects on a related point on that number line. This method leads to the function concept, which is also a dynamic one. Sketchpad allows its user to construct both independent and dependent variables on a single number line, to vary the physical position of x, and to see the position of $f(x)$ change correspondingly. Sketchpad allows students to construct two points, representing, say, x and $2x + 1/x$, so that one can be manually varied while the other varies according to the prescribed relation. In this way, mental constructions are encouraged by a parallel physical construction. In the words of [3], "... there is a sense in which even the dynamic pictures you create in your mind do not quite convey the extraordinarily concrete sense of functional relation one gets when a function behavior is manipulated and perceived dynamically, mouse-in-hand, on the computer. ... But when you *hold the variable in your hand*—starting and stopping, running it backward, ... you can *feel* the relationship between variable and image ... "

Derivative. A derivative of a function at a point may be thought of as a limit of chord slopes. Many have used graphing devices very effectively to "zoom in" on a point to visualize local linearity. Sketchpad can complement such an approach by responding to a student's dragging of the point $a+h$ towards the point a with the corresponding varying of the point $(a+h, f(a+h))$, the chord joining $(a, f(a))$ to the point $(a+h, f(a+h))$, and the slope of that chord.

Connect Visual Thinking to Ordered Logical Thinking. In Sketchpad, the process of constructing geometric objects, from blank sketch to final product, can be recovered and automatically recorded in sentences as a sequence of steps. This can be done by copying a completed sketch to a script, or recording a script as the construction takes place. Not only can the script be applied to create variations or replications of the sketch, but a record of the construction process, step by step, indicates how each geometric object has been constructed and on what other objects it depends.

Examples

I hope that as you look at the sketches in this section you can imagine some of the possibilities for truly dynamic visualization.

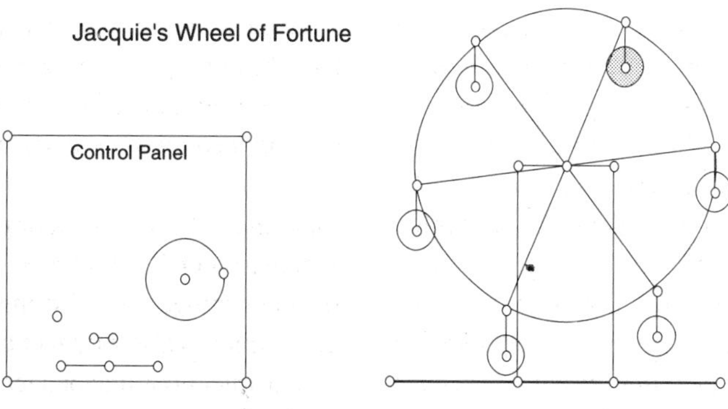

FIGURE 1

Middle School Examples. The middle school examples here are of sketches that have been constructed for the students. These sketches are linked to mathematical questions that some middle school students might not be ready for. Such questions may be deferred until a later grade when the necessary skills have been developed to answer them (or given to those middle school students who are ready to benefit from a challenge). The expectation is that students will manipulate and change the sketches, but won't necessarily construct anything new. See [10] for some exercises for middle school students that involve teacher-guided constructions by students.

Ferris Wheel: Jacquie's Wheel of Fortune. In Figure 1, students use the control panel to operate the Ferris wheel (make the wheel rotate, change the size of the cars, move the whole structure, and change the width and height of the supports). The wheel is one of many possibilities to illustrate the linkage of a machine to its controls through a simple mathematical relationship. It can be used to encourage experimentation and "what if" questions. For example, "What if the 'cars' didn't hang but were attached rigidly to the wheel?" (You could construct the Ferris wheel this way originally, and have students see this as a problem and suggest how to fix it up.) Students have asked, "How fast are the cars moving?"

Some related exercises: You could make a list of other simple machines (pulleys, seesaws, etc.) from which students could choose to make constructions of their own. Questions about the dynamic concepts of linear and angular speed can be dealt with on an empirical basis using the wheel, the measurement and animation tools of *Sketchpad*, and a timing device. Comparison of the loci of points on the wheel and on the cars and elsewhere can be explored using the *Sketchpad* Trace Locus feature.

Spinner Games. Spinner games involve rotating a ray or line (the spinner) so that it lies in a region of a square

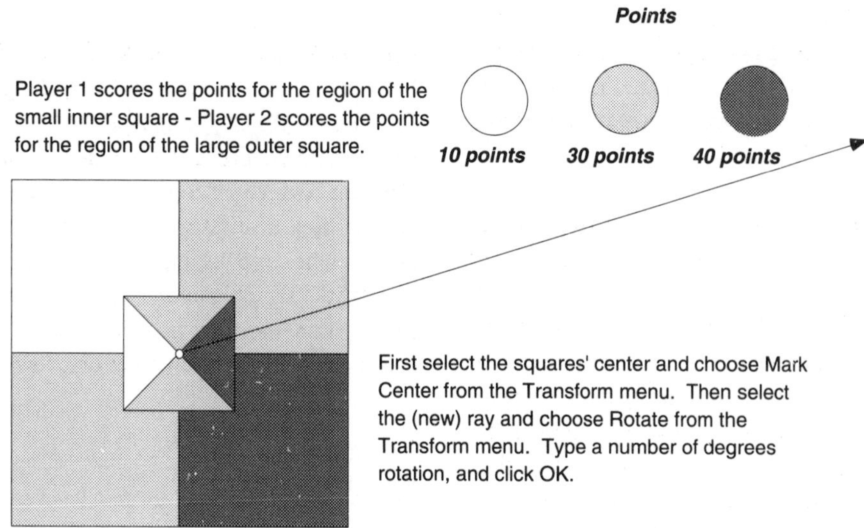

FIGURE 2

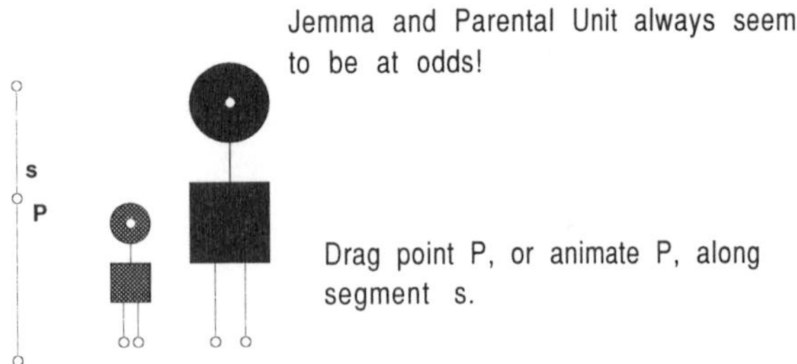

Jemma and Parental Unit always seem to be at odds!
s
P
Drag point P, or animate P, along segment s.

FIGURE 3

advantageous to the player rotating the spinner. The players get to control the number of degrees of their own spin, providing motivation to develop an intuitive sense of degree measure. Students can play a given version, vary the points associated with each region, or construct new regions to make an entirely new game.

The version of the game shown is designed to help develop intuition about the measure of angles in multiples of 45°, while other games with triangular shaped "game boards" have a geometry that encourages an intuitive understanding of multiples of 30°.

As a starting exercise for the games, students should rotate the spinner with the free-hand rotation tool to get a kinesthetic sense of the rotation. Then as the game is played, students control the rotation of the ray by specifying the number of degrees in the rotation, thus connecting the kinesthetic sense of rotation to angular measure.

Scaling, Scale Models, and Similarity. To gain familiarity with scaling and similarity, simple geometric shapes are transformed by dilations. Students can see the effects of a dilation change as the center of the dilation and the dilation ratio change, see how perimeter and area are affected, and guess at the dilation ratio for similar figures.

In Figure 3, animation or dragging operates the diagram so that the smaller figure, Jemma, jumps up (goes down) as the larger figure, Parental Unit, goes down (jumps up). Students are asked to draw lines joining pairs of corresponding points and to note anything remarkable. After students locate the center of dilation, they should take measurements of the line segments that go from the center of dilation to each of the corresponding points. In this way, the general concept of dilation may be introduced.

In related sketches, students are encouraged to develop an intuitive sense of a dilation ratio by playing the game of guessing how much Jemma must be stretched to fit the proportional Parental Unit.

For students who wonder how to simultaneously make one figure go up as the other goes down, the hidden elements can be shown, and a series of more and more difficult animation exercises can be given. These exercises illustrate the use of translation, reflection, and dilation to produce the desired coordination of up and down motion. The prototype is dilated to form Jemma, and a reflection of a translation of the prototype is dilated to form Jemma's Parental Unit.

A Robotic Arm. The machine in Figure 4 has five controls to operate the robotic arm. It's a fun way to start a study of robotic arms. Students can move the test object to different positions in the plane and try their hands at operating the controls to grasp the object without "bumping it." This experience provides good intuition for more systematic and extensive activities, such as the FAIM (Faculty Advancement

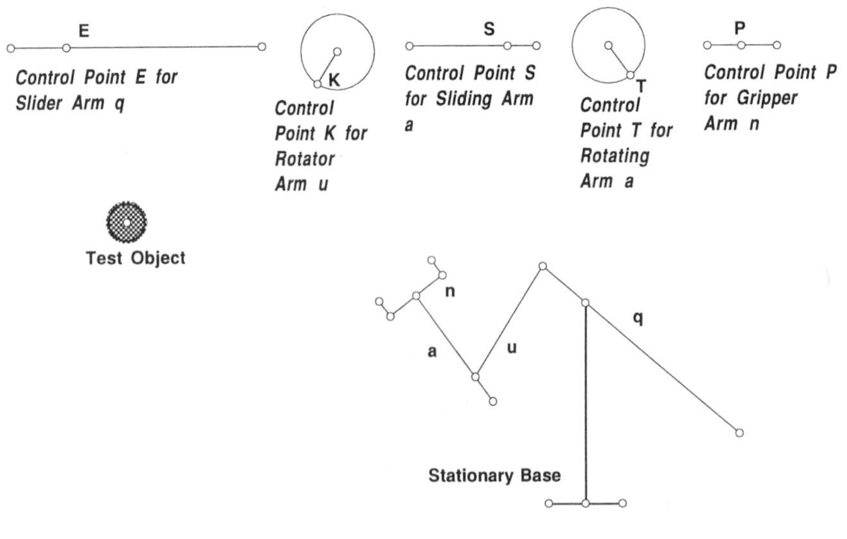

FIGURE 4

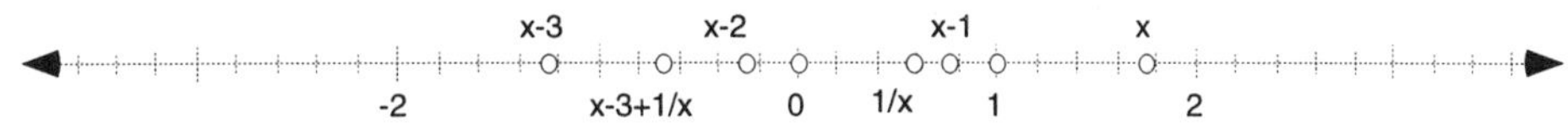

FIGURE 5

In Mathematics, from COMAP) module on robotic arms [2]. Although this module was written for undergraduates, it can be modified for use by advanced middle school students. The collection [6] has scripts for the basic elements of the arm: slider, rotator, grasper, and rotator-slider combination, with which students can build different versions of robotic arms.

Secondary School Examples. These examples, in contrast to those for middle school, have been, or at least could have been, constructed by students.

Algebraic Expressions. The idea of this unit (taken from [5]) is to construct algebraic expressions on a number line using geometric transformations: translate, reflect, dilate, and invert. Then linear, quadratic, and some linear fractional equations can be "solved" by dragging a point, representing the independent variable in the equation, on the number line. The first three transformations are accessed from the *Sketchpad* Transform menu and the fourth by playing a script. Algebraic identities can be tested and verified experimentally, and linear and linear fractional functions can be explored. Here is an outline of the unit:

Step 0. Students are given problems involving geometric patterns, numerical patterns, or some "real world" problem. They then construct algebraic expressions to represent those problems. The arithmetic operations in those algebraic expressions thus are connected to something more concrete. An example of such a problem is: "Each page of the *Student Gazette* can have an area of 93.5 square inches, the same as an 8.5 inch by 11 inch page. If the side margins are each 1 inch and the top and bottom margins are 0.5 inches and 0.25 inches, respectively, what shape page will use the least amount of paper?" The two-week unit, based on the desire to solve such problems, follows the steps below.

Step 1. Students learn the meaning of the transformations: translate, reflect, dilate, and invert, perhaps restricted to transformations of a point on a number line.

Step 2. Students learn the use of vector, mirror, center of dilation, and circle of inversion in *Sketchpad* to accomplish translations, reflections, dilations, and inversions.

Step 3. Students connect geometric transformations to arithmetic operations: Translations of points on a coordinate line accomplish addition and subtraction; reflections in a mirror line perpendicular to the number line through 0 accomplish negation; dilations accomplish multiplication and division by constants; and inversion in a unit circle centered at 0 accomplishes algebraic inversions.

Step 4. Students construct a variable point and related variable points on a number line to represent algebraic expressions geometrically and approximate solutions to equations. Figure 5 is an example of what Step 4 looks like for an approximate solution to the equation $x - 3 + 1/x = 0$. (Note that $x - 3 + 1/x$ is not yet equal to 0 in the diagram; as x is dragged to the right, $x - 3 + 1/x$ gets closer to 0.)

This unit develops the sense of variable by having students construct a point on a line to represent geometrically that variable, and then "drag" the point arbitrarily along the line. The unit also develops the sense of how simple variable expressions are constructed by geometric transformations. It provides the experience of witnessing those expressions vary as the independent variable is dragged along the same line. The functions are constructed geometrically step by step, which forces a hierarchical order of arithmetic operations in constructing an algebraic expression.

Although Figure 5 is somewhat cluttered, it shows all the steps involved in constructing the point that represents $x-3+1/x$. Students may suggest ways to avoid the clutter of points on the line, such as hiding all the points except $x, x - 3 + 1/x$, 0, and 1, or perhaps by putting $x - 3 + 1/x$ on a different line. The dynamic coordination of $x-3+1/x$ with x on a single number line provides some scaffolding for students to make an appropriate interpretation of a conventional two-dimensional graph of the function f given by $f(x) = x-3+1/x$. A next stage might be to use different parallel lines for x and $x - 3 + 1/x$. Later, the conventional graph of f could be constructed in the following way: Construct a point on the line through x, perpendicular to the number line, and a directed distance $x - 3 + 1/x$ from x; then trace the locus of that new point as x is dragged along the number line.

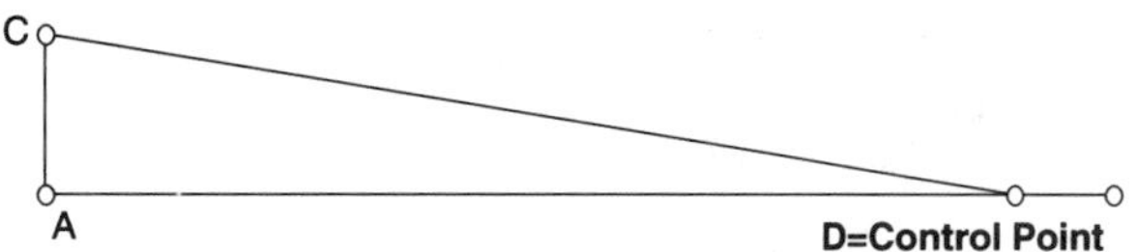

FIGURE 6

Applied Trigonometry. Typical applications of trigonometry to find approximate distances and angles may be done using the measurement features of *Sketchpad*. Scaling will usually be necessary.

> *You need to construct a wheelchair ramp to gain access to a home that is three feet above street level. To be safe, the ramp shouldn't be too steep, and it shouldn't cost too much in building materials. How long a ramp should you build?*

A dynamic construction to study this problem might look like Figure 6.

In the diagram, a straight ramp is assumed and cost is assumed to be proportional to the length of the ramp. By moving D, the control point, one may see that there is a tradeoff between a numerical variable—cost, and the categorical variable—ease of access.

Exploring Trigonometry with The Geometer's Sketchpad [7] contains more examples of the use of *Sketchpad* in trigonometry problems.

Upper Secondary/College Examples.

Optimization. Sketchpad can be used very effectively to find approximate solutions to optimization problems that arise in geometric contexts.

> *If three sides of a trapezoid are each 6 inches long, how long must the fourth side be if the trapezoid's area is a maximum?*

Figure 7 shows a construction that allows the student to analyze the trapezoid problem:

This problem offers the challenge of constructing both a constant constraint and a variable. The student must make a sketch so that the line segments EF, FG, and GH are equal to each other, EH is parallel to FG, and yet EH is allowed to vary in length. Once an appropriate sketch has been constructed, finding an optimum shape for the trapezoid is an experimental activity. A solid understanding of the problem is required (or developed) in order to make the construction of the dynamic geometric model.

Imagining the Limit of Secant Lines: Constructing the Idea of Derivative at a Point. This activity uses *Sketchpad* to develop a "dynamic" concept of derivative at a point with a continuously varying secant line and continuously varying secant slope.

Figure 8 illustrates the exploration. The control point in the diagram is labeled $a+h$. As the point labeled $a+h$ is dragged toward the point labeled a, all of the related points, segments, labels, and measurements move in a coordinated way, showing the point $(a+h,\ f(a+h))$ approach the point $(a,\ f(a))$; the secant line approaches the tangent line to the graph at the point $(a,\ f(a))$; and the slope of the secant line approaches a fixed number.

The idea is for this exercise to precede the definition of derivative at a point, and to put it into the context of students' prior understanding of tangent to a circle (and perhaps develop that idea a bit). After students have explored enough to make and test their conjectures about what will happen to $[f(a+h)-f(a)]/h$, the students are asked to construct

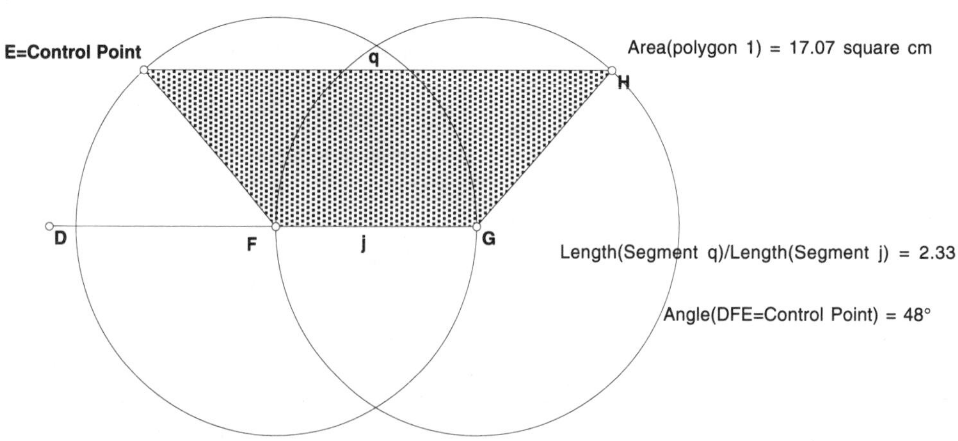

FIGURE 7

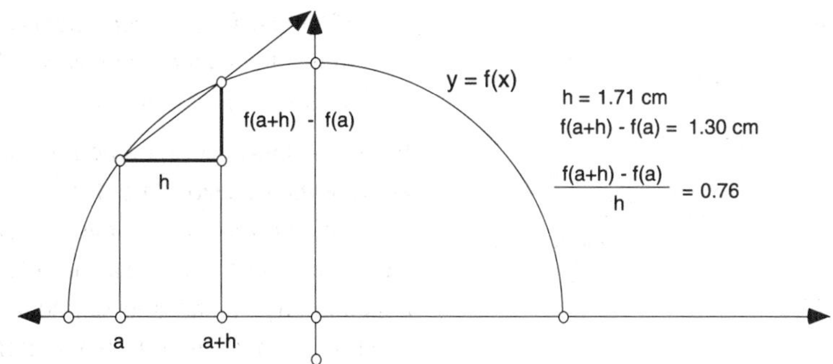

FIGURE 8

the radial line at the point $(a, f(a))$ and to form a conjecture about its slope.

In looking at the sketch above, it may be striking that it doesn't appear much different from the usual textbook or classroom diagram (except that it is a little "clunkier"!). It is important to imagine the effect of being in control of the limit process as you drag point $a + h$ towards point a. No longer is the limit conveyed by a static picture on a page or a finite sequence of numerical difference quotients, but is the result of physically moving a point.

Non-Trivial Pursuit. This example involves the classical chase problem that often appears in a course in differential equations.

> *A dog spots a rabbit moving on a straight line at a constant speed. The dog heads straight for the rabbit, moving at three times the rabbit's speed. If the rabbit continues, oblivious to the dog, at the same speed and on the same line, and the dog's direction is always towards the rabbit at three times its speed, describe the motion of the dog.*

Figure 9 shows the trajectories of the dog and the rabbit, and is produced by a recursive process. The "step size" noted in the sketch represents the actual physical length of the dog's step.

This construction, which shows a discrete geometric model solution, could be considered preparation for a more analytic approach to the problem, providing intuition for a more symbolic construction. One such symbolic construction, in [9], is quite laborious, and I'm not sure the analytic solution, interesting though it is to some of us, provides more understanding of the physical situation. Of course, construction of the differential equations modeling the pursuit problem and seeing the graphical solution using appropriate software are very valuable experiences. Alternatively, the script that produced the sketch in Figure 9 could be considered a solution in itself.

Concluding Remarks

Merely looking at a visual representation gives a pale shadow of the experience of controlling and changing that

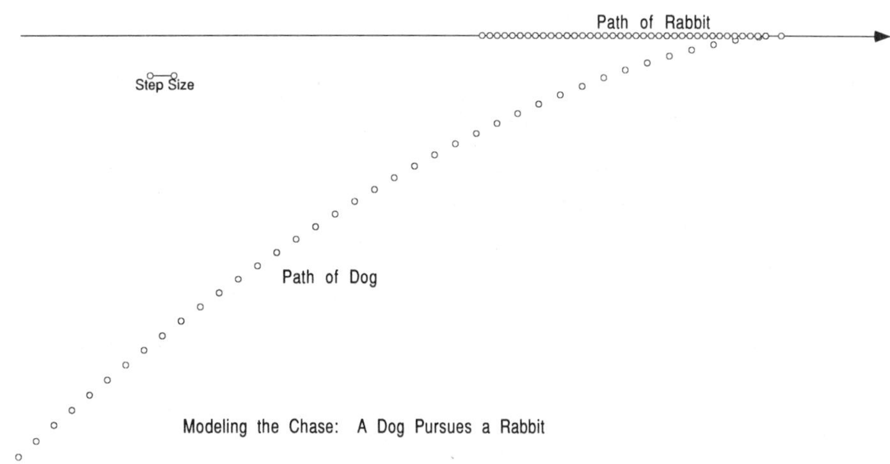

FIGURE 9

visualization. Similarly, viewing and manipulating a sketch is not nearly so powerful an experience as constructing a sketch. The process of constructing sketches in itself poses a challenge that pushes for a deeper conceptual understanding of key concepts. I encourage you and your students to construct sketches to see for yourselves!

One of the great strengths of mathematics is to imagine all possibilities within a universe determined by some constraint: we say, for example, "For each epsilon greater than 0, . . . ," "Given any line in the plane, . . . ," and ". . . , where G_i is any group of order p_i . . . " Yet, the static symbol representing an arbitrary instance out of a universe of possibilities may not convey to an inexperienced user more than a single object subject to formal manipulation. Software that allows the user to physically construct an object subject to a constraint, and then to physically vary the object, makes the experience in some sense closer to a powerful mathematical reality.

References

1. Cuoco, Albert A., Goldenberg, E. Paul, and Mark, J., "A Potpourri," in *Technology Tips, Mathematics Teacher,* 87 (1994), p. 567.
2. Dettmers, R., Doraiswamy, I., Gorini, C., and Toy, C. *Where Can a Robot Arm Reach?,* COMAP, Inc., Lexington, MA, 1994.
3. Goldenberg, P., Lewis, P., and O'Keefe, J., "Dynamic Representation and the Development of a Process Understanding of Function," in *The Concept of Function, Aspects of Epistemology and Pedagogy, MAA Notes, Volume 25,* 1992, pp. 235–260.
4. Henderson, David W., "I Learn Mathematics from My Students—Multiculturalism in Action," *For The Learning of Mathematics*, Vol. 16, June 1996, pp. 34–40.
5. Morrow, J. *Algebra Meets Geometry and Technology,* a two-week workshop at the SummerMath Program, Mount Holyoke College, available from the author at 50 College Street, South Hadley, Massachusetts 01075-1441.
6. Morrow, J. *Scripts for Robotic Arm Parts,* available from the author as above.
7. Shaffer, D. *Exploring Trigonometry with The Geometer's Sketchpad,* Key Curriculum Press, Berkeley, 1994.
8. Silver, E.A., Kilpatrick, J., and Schlesinger, B. *Thinking Through Mathematics,* College Entrance Examination Board, New York, 1990, p. 7.
9. Simmons, G.F. *Differential Equations with Applications and Historical Notes,* McGraw-Hill, New York, 1972, pp. 55–56.
10. Taylor, L., "Exploring Geometry with The Geometer's Sketchpad," *Arithmetic Teacher,* 40 (1992), pp. 187–191.

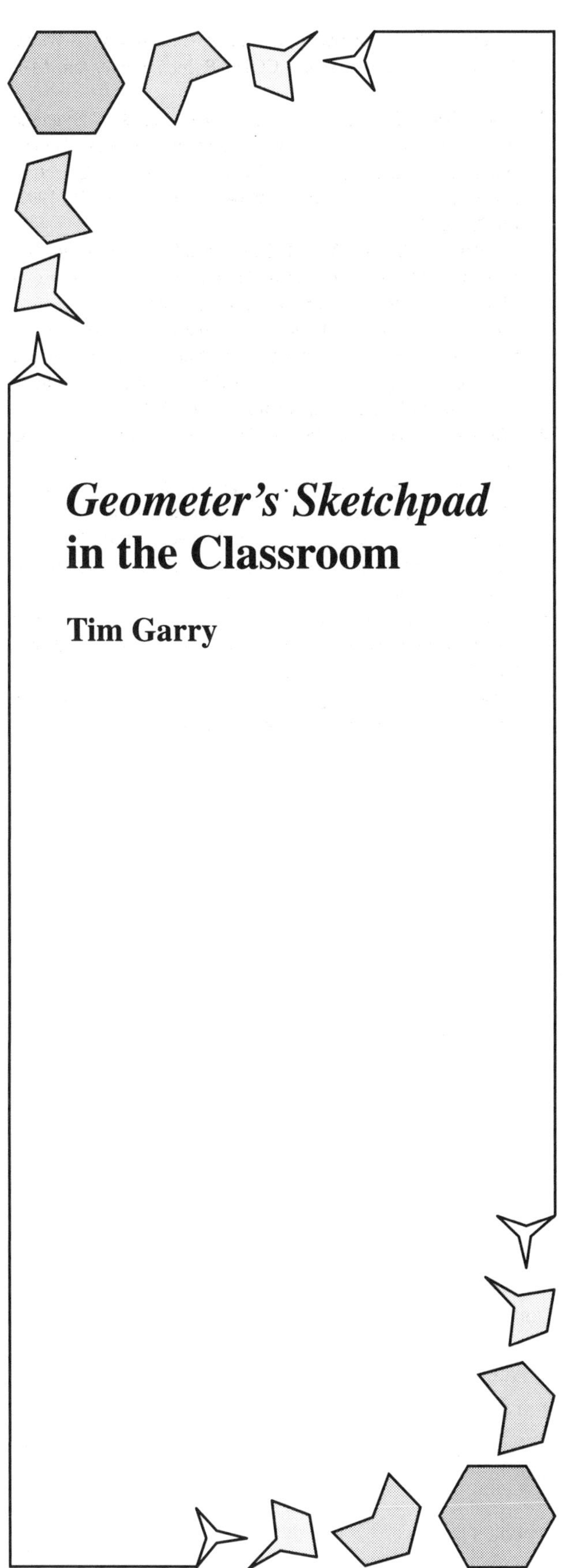

Geometer's Sketchpad in the Classroom

Tim Garry

In 1984 Judah L. Schwartz (M.I.T. and Harvard) and Michal Yerushalmy (University of Haifa) wrote a paper entitled *Using Microcomputers to Restore Invention to the Learning of Mathematics.* In that paper they submitted the following opinion:

> There is something odd about the way we teach mathematics. We teach it as if assuming our students will themselves never have occasion to make new mathematics. We do not teach language that way . . . the nature of mathematics instruction is such that when a teacher assigns a theorem to prove, the student ordinarily assumes that the theorem is true and that a proof can be found. This constitutes a kind of satire on the nature of mathematical thinking and the way new mathematics is made. The central activity in the making of new mathematics lies in making and testing conjectures.

The notion of guiding students to "make new mathematics" became far more practical and accessible to teachers with the development of *The Geometric Supposer* software application (Sunburst Communications) by Schwartz and Yerushalmy back in the early 1980s. It allowed students to perform a wide range of compass and straightedge constructions with dynamic measurement capabilities. *The Geometric Supposer* had profound effects on the teaching style of many math teachers. It gave them a teaching tool that was more stimulating, open-ended, and flexible than a textbook. Students could test their own mathematical ideas and conjectures in a visual, efficient, and dynamic manner and—in the process—be more fully engaged in their own learning. The National Council of Teachers of Mathematics (NCTM) in North America published an instructive booklet in 1989 entitled *How to Use Conjecturing and Microcomputers to Teach Geometry* (Chazan and Houde). It was the direct result of educators' experiences with *The Geometric Supposer.* The power and versatility of explorative-based geometry software made a major leap forward with the introduction of *The Geometer's Sketchpad* (Key Curriculum Press) in 1991. Other products—such as *Cabri Geometry* (Texas Instruments)—have further established dynamic geometry software as an enriching and practical tool for the study of mathematics—and not only in the area of geometry. In this paper I have outlined a few examples from my experience with *Sketchpad* in teaching secondary mathematics. Although the examples are limited in number and scope, I do believe they give some insight into how *Sketchpad* can enable students and teachers to be active participants in "mathematical thinking."

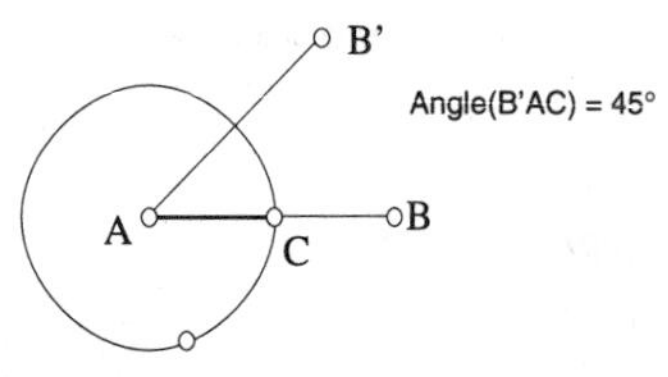

INVESTIGATING THE RATIO OF PERIMETER TO DIAMETER FOR A REGULAR POLYGON

If n = # of sides of a regular polygon and x = 360/n, then a point must be rotated x degrees n times around circle to construct the n vertices of the regular n-gon.
1. Select pt. A ; Mark Center [pt. A center of rotation]
2. Shift select pt. B and segment AB
3. Rotate x degrees to create angle(B'AC)
4. Shift select pts. B', A and C [in that order]
5. Mark Angle [angle(B'AC) is marked angle]
6. Open Reg. Polygon Script
7. Shift select pts. P, Q, B', A and C [in that order]
8. Play Script ; Depth of Recursion = n
9. Select all the vertices of the regular polygon
10. Construct Polygon Interior ; Measure Perimeter
11. Calculate ratio Perimeter/Diameter

Diameter (2r) = 8.54 cm

Perimeter = 10.29 inches

Perimeter/Diameter = 1.21

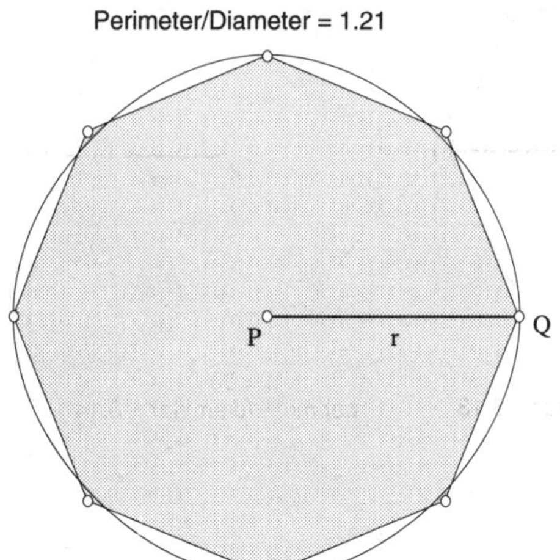

FIGURE 1

Regular Polygon Template

This is a recursive script (contains a loop) that constructs a reg. polygon with n sides with n determined by angle(DEF). n is equal to 360 divided by angle(DEF). For example, if angle(DEF) = 72, then the script constructs a reg. 5-gon (360/72 = 5). The depth of recursion specified when playing the script is equal to the number of sides.

Given:
1. Point A
2. Point B
3. Point D
4. Point E
5. Point F

Steps:
1. Let [1] = Circle with center at Point A passing through Point B (hidden).
2. Let [C] = Image of Point B rotated by angle D-E-F about center Point A.
3. Let [j] = Segment between Point [C] and Point B.
4. Recurse on A, [C], D, E and F.

FIGURE 2

Before dynamic geometry software appeared, I once attempted to provide students with a meaningful introduction to the constant π by constructing ever larger regular polygons and computing the ratio of perimeter to diameter each time. Perhaps it was a good idea in theory, but in practice it took too much time, and students lost interest because they were passive spectators. With *Sketchpad*, and applying no more effort than my low-tech lesson required, I prepared a sketch template (Figure 1) and a recursive script that constructs a regular polygon for any desired number of sides (Figure 2). After selecting the "givens" in the template—the points that define the radius (circumscribed circle) and the angle to rotate one vertex to the next about the center—one can "play" the script at three different speeds. The result is a sketch containing a regular polygon with the number of sides determined by the user-defined angle of rotation. The sketch template contains instructions so that students with some experience using *Sketchpad* (and an understanding of perimeter, circumference, and diameter) can get directly involved. Figure 1 shows the result if one were to follow the instructions in the template for constructing a regular octagon. Depending upon various factors (such as time and equipment), the investigation can be a group enterprise with one computer projecting its display via an overhead projector and LCD panel or, preferably, with students working at computer stations. Students are instructed to compile data on the ratio of perimeter to diameter for various regular polygons of their choice (Figure 3) and challenged to draw their own conclusion(s) from this data. The script could certainly be applied to other geometric investigations. Some natural

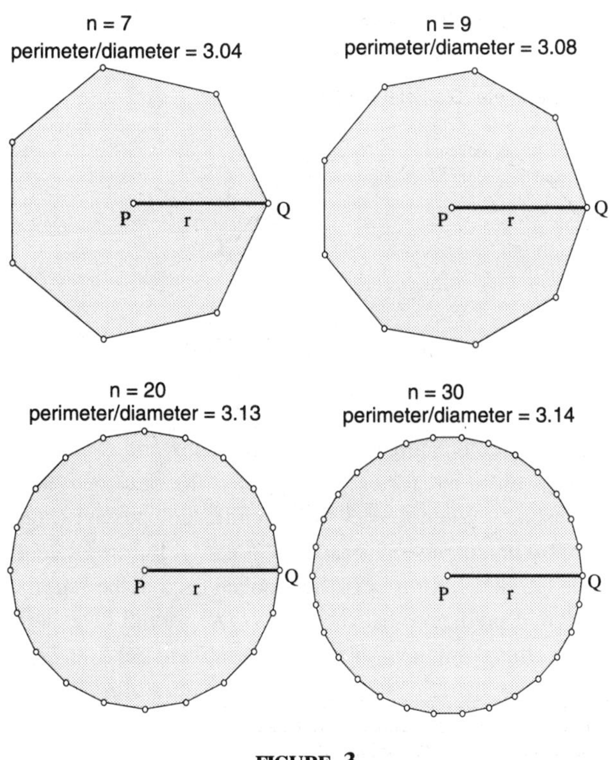

FIGURE 3

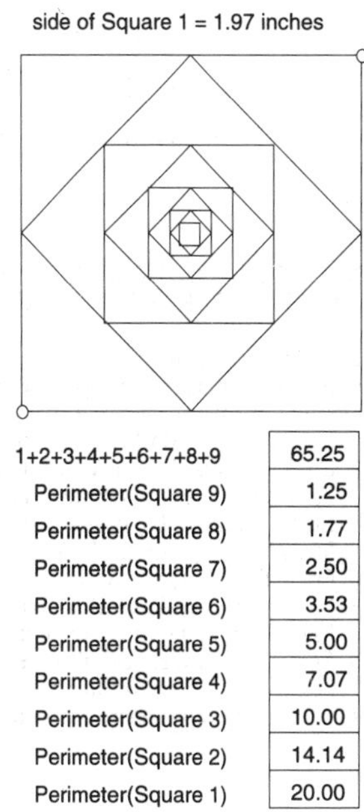

1+2+3+4+5+6+7+8+9	65.25
Perimeter(Square 9)	1.25
Perimeter(Square 8)	1.77
Perimeter(Square 7)	2.50
Perimeter(Square 6)	3.53
Perimeter(Square 5)	5.00
Perimeter(Square 4)	7.07
Perimeter(Square 3)	10.00
Perimeter(Square 2)	14.14
Perimeter(Square 1)	20.00

FIGURE 4

extensions include comparing results using inscribed circles, investigating the properties of regular polygons and measuring the circumference and diameter of "real" circles (manually), or for circles constructed using *Sketchpad*.

On an algebra 2 assignment students were asked to calculate the sum of an infinite geometric series consisting of the perimeters of squares in which each successive square was constructed using the midpoints of the previous square's sides as vertices (Figure 4). A recursive script produces a diagram with as many inscribed squares as can be seen (or memory allows). Students can choose to write their own script, be given one, or construct the diagram step by step. The connection between the geometric diagram and the algebraic problem of finding the sum of the infinite series prompted several interesting student conjectures. After measuring some perimeters and calculating a few sums, the students agreed that the series converged, and eventually they were able to estimate the limit of the sum of the perimeters. The discussion continued with questions on how this limit depended on the size of the first square. Some students started to explore patterns created by using a shape other than a square.

After practice with *Sketchpad* and composing conjectures, a worthwhile exercise is to pose an open-ended question to students. The task is to collect data, search for generalizations and patterns, test conjectures, and ask more questions, in other words, to think mathematically. Students document their conjectures and analysis with written work and *Sketchpad* documents. Results can be shared, collected, or presented electronically. Students enjoy presenting their results to classmates, especially if they have spiced up their *Sketchpad* documents with animations and sound. A question which I offered to a 10th grade geometry class early in the year can open the door to some intriguing further exploration. The question was "Given a circle, how many circles congruent to it can be placed around it such that each circle is tangent to the given circle and to the circle on either side of it?" A follow-on question that hopefully a student will suggest is, "Does the size of the given circle make a difference?" Figure 5 shows one student's "answer." Students can be asked to provide a deductive proof based on their construction or to pose more questions for further investigation.

One very interesting extension of this previous example is to investigate some general properties of "circles around circles." When students realize that the situation of six circles "fitting" exactly around a circle of the same size is a special case of a more general problem, a more general question may be: Given a circle with radius R, how many congruent circles (of *any* radius r) can be placed so that each of these circles is tangent to the given circle and tangent to the circle on either side of it? Are there constraints on the number of circles? Most students come up with the fact that the minimum number of circles is always three, and, after probing

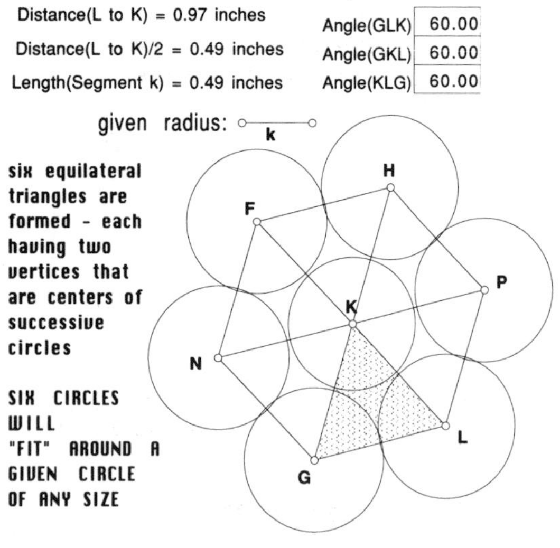

FIGURE 5

with a few experimental sketches on *Sketchpad* (and perhaps some algebra and trigonometry), students will also realize that there is no limit to the number of circles (r goes to zero). Are there constraints on r? Again, a few sketches lead many students to conjecture that, depending on the value of R (radius of given "inside" circle), r (radius of "outside" circles) is limited to certain values. Gaps or overlapping occurs for some values of r (i.e., the number of "outside" circles, n, must be an integer).

A student in a pre-calculus class decided to take a different approach. Given n, what values can R and r take on? Furthermore, is there a relationship between R and r? The student produced sketches like the ones in Figures 6 and 7 for $n = 3, 4$, and 5. By dynamically changing the radii of both the "inside" and "outside" circles the student

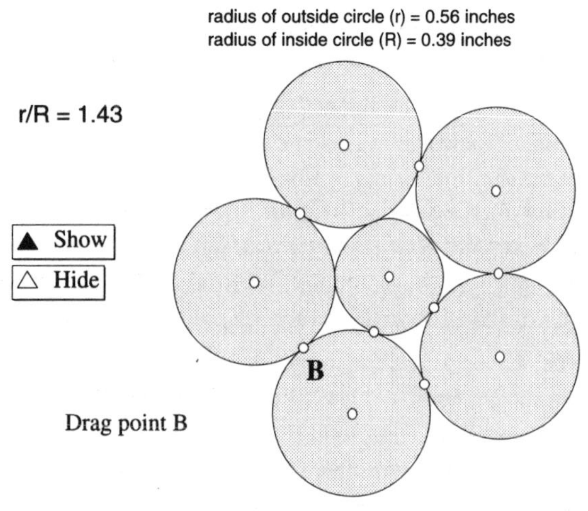

FIGURE 6

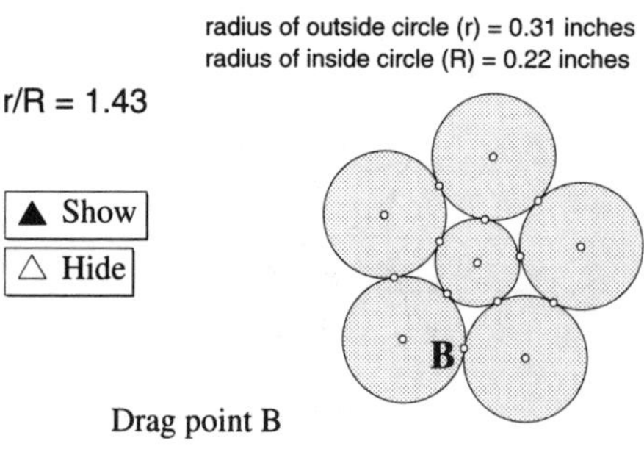

FIGURE 7

provided a convincing demonstration that r and R can take on any values but for each value of n the ratio of r to R is constant. When $n = 3$ (minimum), the constant (r/R) is approximately 6.464 (exactly $3 + 2\sqrt{3}$). Before students made the sketch for $n = 6$, they looked back at the "special case" and predicted that the ratio (r/R) should be exactly one. At this point, it wasn't enough simply to use *Sketchpad* to compute the ratio (r/R) for various values of n. Since n (# of outside circles) determined the ratio (r/R), then it must be possible to write r/R as a function of n and, therefore, produce a general result for any n (integer greater than two). Students went off in groups of two or three to investigate. One group started by taking an earlier sketch for $n = 5$ (Figure 6) and activating the Show button to reveal all of the construction lines (Figure 8). One conjecture that arose is whether the segment connecting the centers of two tangent circles will always pass through the point of tangency. Some dynamic movement of the sketch quickly reminded them of this geometric property. They continued by focusing on a portion of the sketch and inserted appropriate labels (Figures 9 and 10). Then out came pencil and paper, and, by applying some trigonometry and algebra, the students eventually convinced themselves that the ratio r/R is a function of n, and, likewise, that n is a function of the ratio $r/(R+r)$.

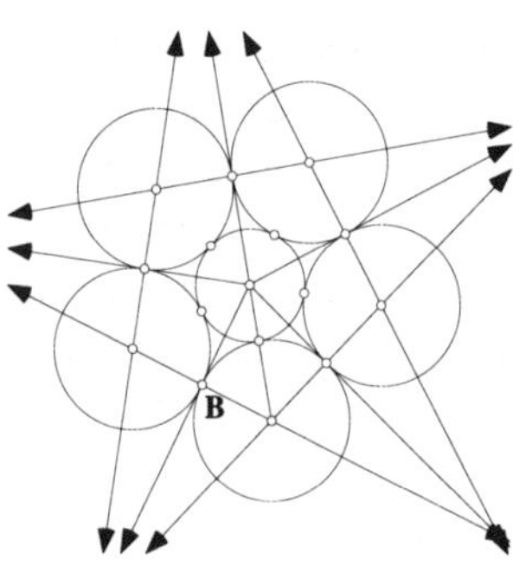

FIGURE 8

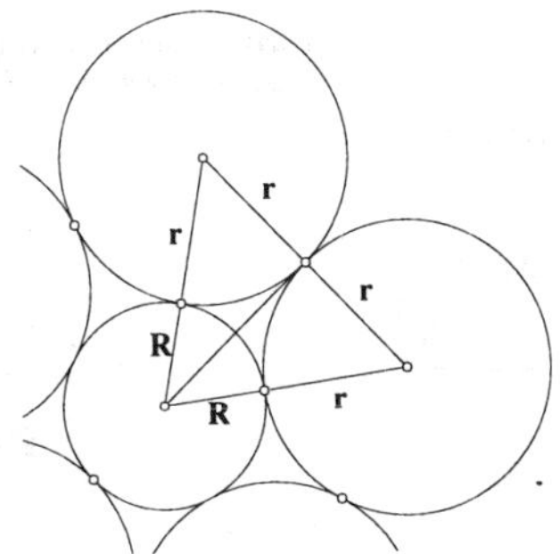

FIGURE 9

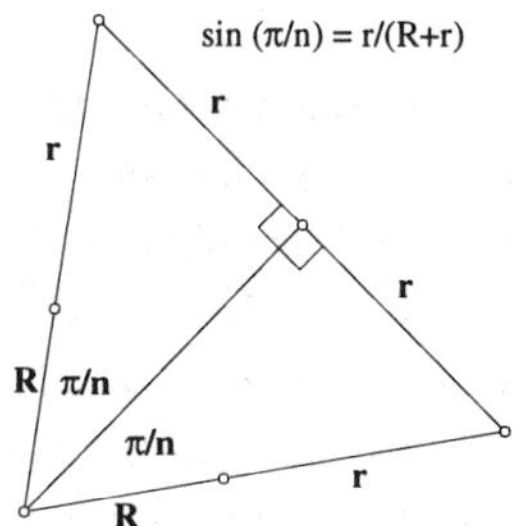

FIGURE 10

Symbolically speaking,

$$\frac{r}{R} = \frac{\sin(\pi/n)}{1-\sin(\pi/n)} \quad \text{and} \quad n = \frac{\pi}{\arcsin(\frac{r}{R+r})}.$$

All the students were encouraged to demonstrate how to prove these results. My students were intrigued by how such a modest question led to some interesting results. They appreciated *Sketchpad* as a tool to help them gain insight into their questions. I appreciated that they combined the tools of dynamic geometry with algebraic tools to satisfy their questions.

Sketchpad's use as a demonstration tool can provide a teacher with the means to present a concept in a dynamic, efficient, and meaningful way. The concept of a locus of points can be demonstrated very effectively, for example, by constructing a parabola tracer (Figure 11). The construction of the tracer itself is instructive for students. The exercise is to construct a point that is equidistant from a point (focus) and a line (directrix), and in such a way that the condition of equidistance from the point and line remains if the point is dragged. The Trace Locus feature of *Sketchpad* can then be employed to trace out the locus of points that are equidistant from the point and line, producing a parabola (Figure 12). A wide variety of tracers can be constructed—including other conic sections and trigonometric functions. Any student will gain a better understanding of the sine function by seeing it "unwrap" as a

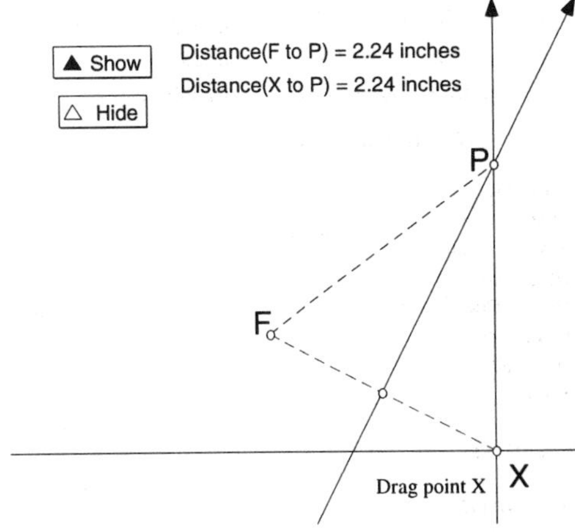

FIGURE 11

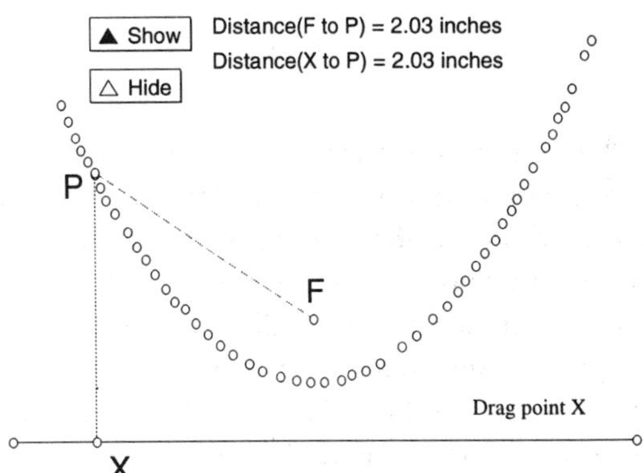

FIGURE 12

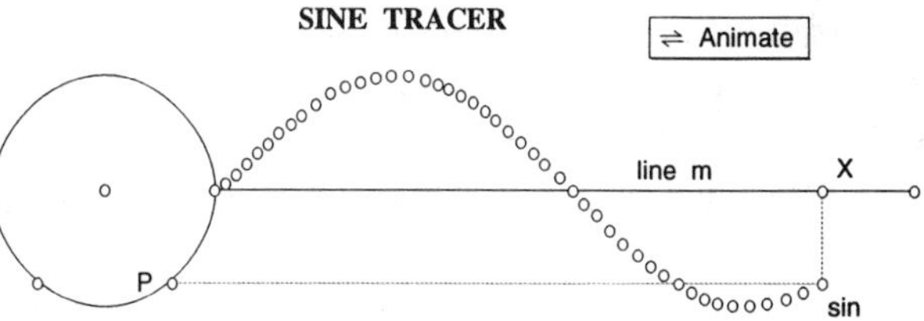

Point P is animated around the circle and, simultaneously, the point X is animated along the line m. The point labeled "sin" is the intersection between the line through P parallel to m and the line through point X perpendicular to m.

FIGURE 13

point revolves around a circle. Students can benefit further by constructing their own tracers. The sine tracer shown in Figure 13 must be seen in "action" to be appreciated. The tracer can be modified to trace the cosine and tangent functions.

My last example is the product of an open-ended investigation and illustrates that *Sketchpad* can also be utilized as a classroom experimental tool. This investigation is concerned with a two-dimensional random walk. Although random walks can be found in a range of mathematical literature,

I must give credit to Bill Finzer of Key Curriculum Press for providing the idea for this class activity.

The activity is initiated by simply asking the question: What happens when someone takes ten steps of constant length, each in a random direction? If time permits, it is best to first have students physically model the "problem." Using a person's foot as a constant unit and some type of "spinner" which produces a random direction when spun, students "walk" a specified number of random steps (heel to toe). The distance from the starting point to the final position is measured in terms of a number of constant units (i.e., the length of a student's foot). The experiment my students did in class consisted of ten random steps.

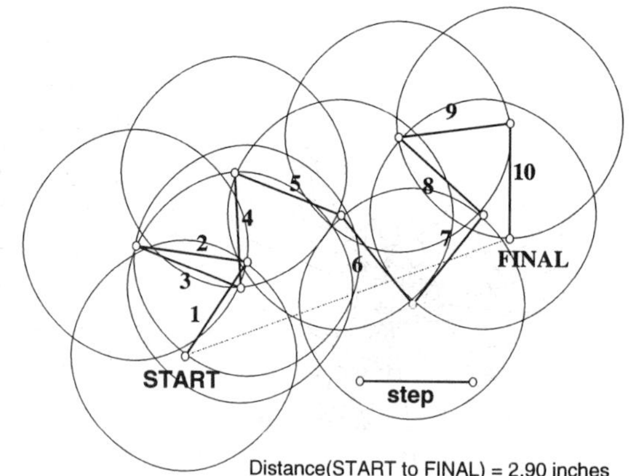

FIGURE 14

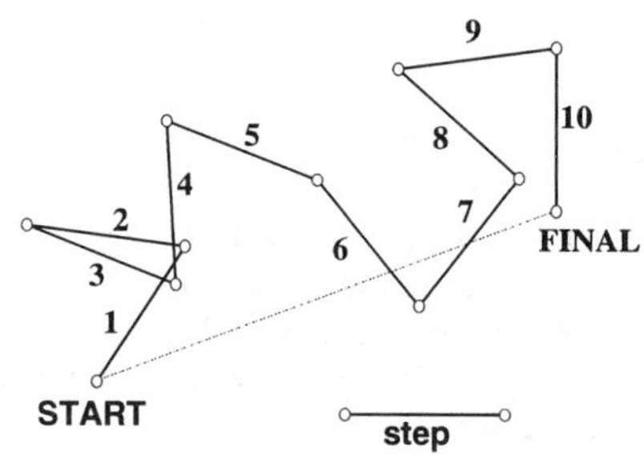

FIGURE 15

I prefer not to give students a specific question to be answered, but rather let them pose a question or set of questions of their own. If carried out in a mathematics class that has studied some probability and/or statistics, students often want to find an average distance. They are curious to see whether a person is most likely to end up a certain distance from the starting point (ask them what the minimum and maximum will be). Data from the students' physical experiments can be studied, but students should agree that the set of data gathered by physically stepping and measuring is not large enough to give a reliable average (expected value) of the distance. In order to carry out the experiment for, say, more than a hundred trials, the students need an efficient modelling tool.

At this point, reveal to students that, when *Sketchpad* is instructed to construct a point on a circle, it places it at a random location on the circle. Soon students are creating on *Sketchpad* the equivalent of a "spinner." They also soon realize that the circle with the random point on it can also be used for producing "steps" of constant length. This is done by creating a sketch (or, better, writing a recursive script) which constructs a series of ten congruent circles with the center of each successive circle being the random point constructed on the previous circle (Figure 14). The length of each "step" is the constant radius. To complete each trial, measure the distance from the starting point to the final random point constructed on the tenth circle (automated in a script). The final sketch is far less busy if the circles for each step are hidden (Figure 15). Students now have a handy tool for conducting numerous trials and collecting the data. Students who modeled the random walk with a script could gather data for any number of walks with any number of steps (given memory limitations). In Figure 16 three mini-sketches show the resulting chaotic, unpredictable pattern of performing ten random walks of ten steps each. The sketches show a line segment for each step. A few students, working outside class, obtained average distances for ten-step walks after 500 trials (2.264), 750 trials (2.432) and 1000 trials (2.504). But what is the significance of the average distance? What is the relationship between the number of steps and the average distance? This was an exercise that generated more questions than could be answered. Students do not often experience that in a math classroom. However, such an experience should not be unexpected when "thinking mathematically" with such a versatile and powerful tool.

Although the examples presented in this paper are limited in scope, I do believe that they do illustrate the power and versatility of dynamic geometry software as a learning and teaching tool. In the hands of students and teachers, a software application such as *The Geometer's Sketchpad* can open the door to classroom experiences in which students engage themselves in more genuine mathematical thinking. The interactive nature and dynamic qualities of the software lead students to propose their own conjectures and efficiently test them. The feedback that students can get from utilizing dynamic geometry software is efficient and exciting. Students can gain a better understanding and visual grasp of the mathematics they are investigating. And teachers have a tool that offers limitless possibilities to guide students to be more fully engaged in their own learning and mathematical thinking.

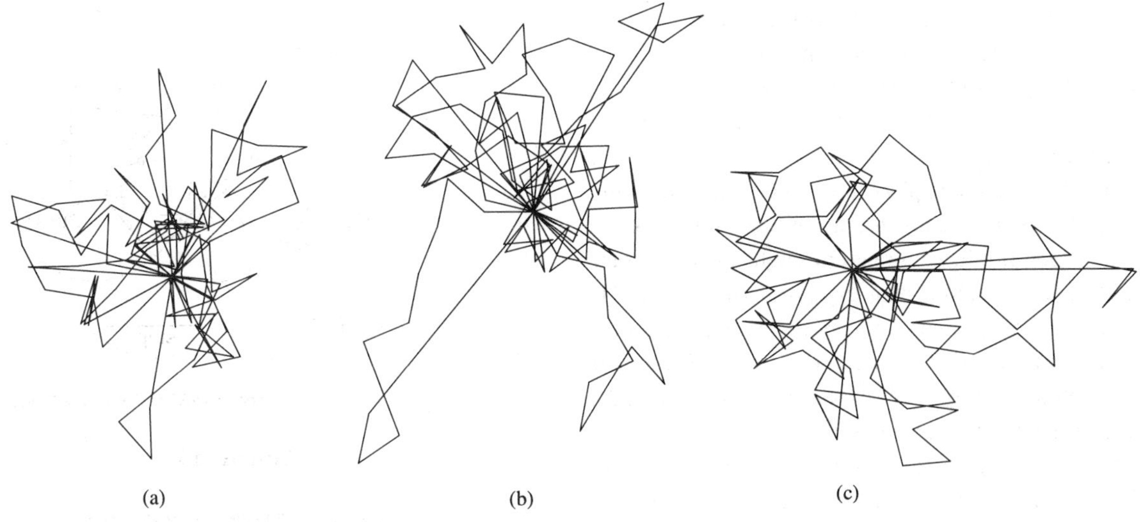

(a) (b) (c)

FIGURE 16

References

1. Chazan, D. and Houde, R. (1989). *How to use Conjecturing and Microcomputers to Teach Geometry*, National Council of Teachers of Mathematics, Reston, VA.
2. Jackiw, N. (1992–95). *The Geometer's Sketchpad*, Key Curriculum Press, Berkeley, CA.
3. Schwartz, J. and Yerushalmy, M. (1986). "Using Microcomuters to Restore Invention to the Learning of Mathematics." *Contributors to Thinking* (D. Perkins and R. Nickerson, eds.), Lawrence Erlbaum Associates, Hillsdale, NJ, pp. 293–298.

Addendum
A Two-Dimensional Random Walk

The problem presented in this paper investigates how far (on the average) someone will be from their starting point after taking ten steps of equal length with the direction of each step being random. An "answer" for a one-dimensional random walk is significantly easier to obtain. The final location of a one-dimensional random walk of ten steps can only be located at one of the following locations on the axis (0 is the starting point or origin):

$$-10 \quad -8 \quad -6 \quad -4 \quad -2 \quad 0 \quad 2 \quad 4 \quad 6 \quad 8 \quad 10$$

It is not possible to end up at an "odd" location (true for any walk with an even numbers of steps). The probabilities for "landing" on each of the locations can be computed by first finding the number of outcomes for each position—there is one (${}_{10}C_0$) way to "land" on 10 or -10; ten (${}_{10}C_1$) ways to "land" on 8 or -8; forty-five (${}_{10}10_2$) ways to "land" on 6 or -6; etc. The total number of walks possible is 2^{10}. One can then compute the expected value for the final distance from the origin to be approximately 2.4609375 units (or steps).

In a similar fashion, one can approach the two-dimensional random walk. This is made more manageable by restricting the "random" directions of each step to the four compass point (N, S, E, and W). Whereas in the one-dimensional walk, the outcomes for each location are values from Pascal's Triangle (binomial coefficients), the outcomes for this restricted two-dimensional walk can be envisaged as a Pascal's Pyramid. The faces of the pyramid are Pascal's Triangle with the edges of each row being a row of the triangle. The internal rows of each level are just appropriate multiples of the edge. It is then possible to generate outcomes for each point on the "grid". Again, one can use these results to directly compute the expected value for the final distance from the origin. Using Euclidean distances rather than "taxicab" distances (i.e., the length of the path to the location along grid lines), this expected value of the distance is approximately 2.800234706.

The unrestricted two-dimensional walk is much more difficult to tame. The following constitutes an attempt to do so and reveals a significant jump in complexity. It also shows the usefulness of having a tool like *Geometer's Sketchpad* by which to model the given two-dimensional random walk having a constant length but a random direction for each step. Repeated trials can give a decent approximation to, what I will call here, the expected value of the displacement (i.e., final distance from the origin).

The definition of the displacement of the process after N steps can be given as:

$$D_N = \left[(X_N - X_O)^2 + (Y_N - Y_O)^2\right]^{1/2},$$

where X and Y are random variables representing the horizontal and vertical position of the process, respectively. Assuming that the process begins at the origin gives:

$$D_N = \left[X_N^2 + Y_N^2\right]^{1/2}.$$

If at each step the process moves to a random point on a unit circle centered on the current position of the process, the X coordinate after N steps is the sum of all the cosines of the N randomly chosen angles. The Y coordinate is the sum of all the sines.

$$D_N = \left[\left(\sum_{k=1}^{N} \cos\theta_k\right)^2 + \left(\sum_{k=1}^{N} \sin\theta_k\right)^2\right]^{1/2}$$

$$= \left[\sum_{k=1}^{N}(\cos^2\theta_k + \sin^2\theta_k)\right.$$

$$\left. + 2\sum_{k=1}^{N}\sum_{j=k+1}^{N}(\cos\theta_k\cos\theta_j + \sin\theta_k\sin\theta_j)\right]^{1/2}$$

$$= \left[N + \sum_{k=1}^{N}(\alpha_{N,k}\cos\theta_k + \beta_{N,k}\sin\theta_k)\right]^{1/2},$$

$$\alpha_{N.k} = 2\sum_{j=k+1}^{N}\cos\theta_j, \quad \beta_{N,k} = 2\sum_{j=k+1}^{N}\sin\theta_j.$$

The angles are independent and identically distributed (uniform on $[0, 2\pi]$), so the expected value of the displacement is:

$$\mathrm{E}D_N = \int_{X_N=0}^{2\pi}\int_{X_{N-1}=0}^{2\pi}\cdots\int_{X_1=0}^{2\pi}\left[N + \sum_{k=1}^{N}(\alpha_{N,k}\cos x_k\right.$$

$$\left. + \beta_{N,k}\sin x_k)\right]^{1/2}\prod_{k=1}^{N} f_\theta(x_k)dx_1\,dx_2\ldots dx_N$$

$$= (2\pi)^{-N}\int_{X_N=0}^{2\pi}\int_{X_{N-1}=0}^{2\pi}\cdots\int_{X_1=0}^{2\pi}\left[N+\right.$$

$$\left.\sum_{k=1}^{N}(\alpha_{N,k}\cos x_k + \beta_{N,k}\sin x_k)\right]^{1/2} dx_1\,dx_2\ldots dx_N.$$

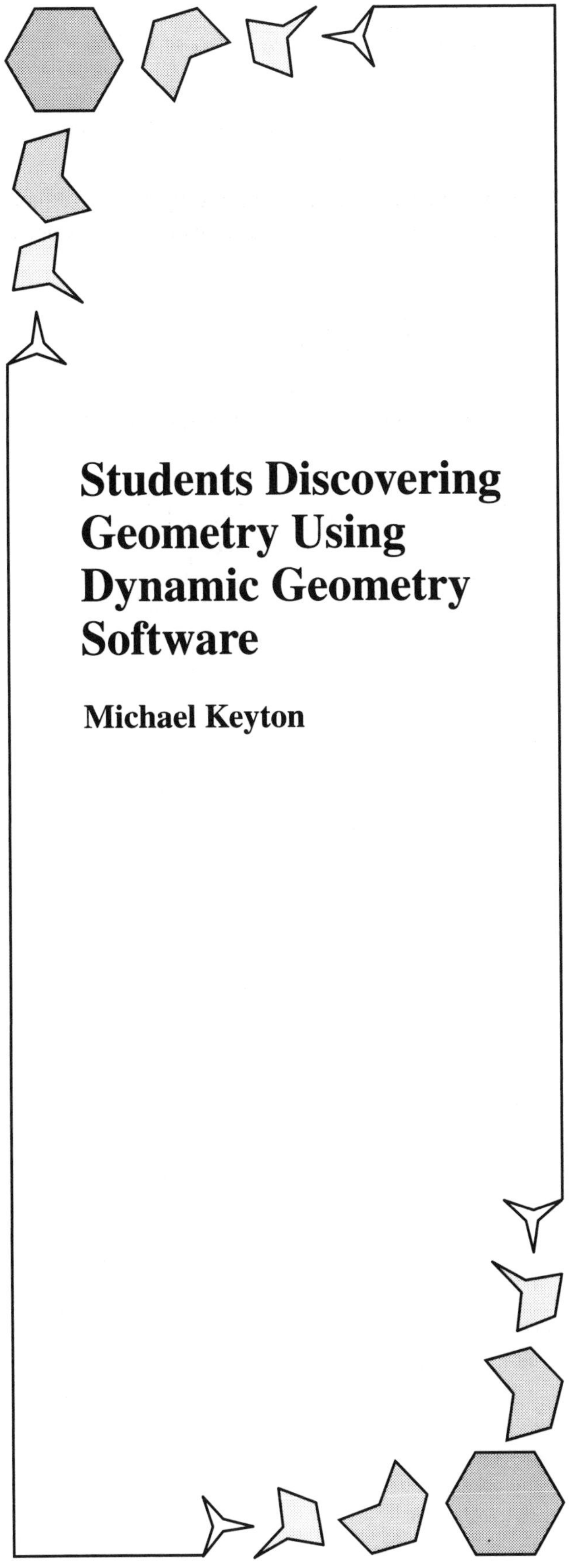

Students Discovering Geometry Using Dynamic Geometry Software

Michael Keyton

> The best way to learn anything is to discover it for yourself.
> Let them learn guessing. Let them learn proving.
> Do not give away your whole secret at once—
> let the students guess before you tell it—
> let them find out for themselves as much as is feasible.
>
> —George Pólya [2]

Many teachers believe this approach to learning by Pólya is best; yet, often in the past it has been difficult to turn students loose on their own, especially in geometry where constructions can be unwieldy. Bertrand Russell also realized the value of generating belief before attempting proof. He wrote:

> What is best in mathematics deserves not merely to be learnt as a task, but to be assimilated as a part of daily thought, and brought again and again before the mind, with ever-renewed encouragement. ...
>
> In geometry, instead of the tedious apparatus of fallacious proofs for obvious truisms which constitutes the beginning of Euclid, the learner should be allowed at first to assume the truth of everything obvious, and should be instructed in the demonstrations of theorems which are at once startling and easily verifiable by actual drawing, such as those in which it is shown that three or more lines meet in a point. In this way belief is generated; it is seen that reasoning may lead to startling conclusions, which nevertheless the facts will verify; and thus the instinctive distrust of whatever is abstract or rational is gradually overcome. Where theorems are difficult, they should be first taught as exercises in geometrical drawing, until the figure has become thoroughly familiar; it will then be an agreeable advance to be taught the logical connections of the various lines or circles that occur. It is desirable also that the figure illustrating a theorem should be drawn in all possible cases and shapes, that so the abstract demonstrations should form but a small part of the instruction, and should be given when, by familiarity with concrete illustrations, they have come to be felt as the natural embodiment of visible fact.
>
> Bertrand Russell [3]

I think Pólya and Russell would enthusiastically embrace the opportunity to use interactive dynamic geometry software such as *The Geometer's Sketchpad* or *Cabri Geometry II* which permits students to discover for themselves and

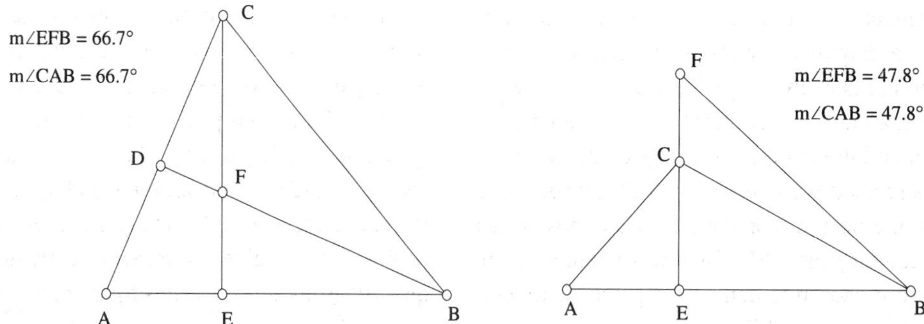

FIGURE 1. The angle formed by two altitudes of a triangle is congruent to the third angle of the triangle.

to generate belief in a geometrical fact before attempting a proof.

For the last three years, my Honors Geometry class (grade 9) has been more exciting and more productive as a result of using *Sketchpad* (the first two years) and *Cabri II* (the last year). Though this class did not use a formal textbook, it covered the material from a classical elementary geometry course. One of the major difficulties and struggles was getting students to draw adequate diagrams for problems. The class constantly verified that well-known, unstated postulate of elementary geometry classes: all triangles drawn by students are either isosceles or right. However, with interactive geometry software and its ability to alter a figure dynamically to produce virtually hundreds of examples quickly, the students were able to make significant discoveries beyond anything I had seen during the eighteen previous years. As a consequence, I have had to rethink the presentation of the material for future classes.

The number of conjectures and new ideas that resulted from the students using *Sketchpad* was not anticipated. Almost daily it seemed there was a new discovery. I had planned a course of study that would develop the processes of explaining why a feature of a drawing is true, which, with adjustments, would lead to formal proof. I did not consider a particular form of proof important, but during the first two months devoted much discussion daily to the topic of how to explain solutions well, until a natural format was accepted. (I admit to a degree of student manipulation here; for by carefully constructing questions, I obtained a format agreeable to me. The students believed that the process was sensible rather than arbitrary.)

By the end of the first quarter, the process of discovering conjectures, looking for counterexamples, and constructing proofs became the normal activity. At this time, I expected all students to know how to sequence both a direct and an indirect proof, and how to construct a proof by contrapositive and by *mutatis mutandis.* I expected them to understand the distinctions between a definition, a postulate and a theorem, to know and understand additional terms such as "lemma" and "corollary," and to realize the distinction and relationship between a theorem and its converse.

Once these aspects had been exposed, discussed, and used, each student began to explore with *Sketchpad*, both at home and at school. On almost the first day, one student returned with the conjecture that the angle formed by altitudes from two angles of a triangle was congruent to the third angle. The remarkable aspect of his investigation was that he also found all three cases easily. (See Figure 1, which shows 2 cases. The third case, not shown, occurs in a right triangle.) For many problems in which multiple cases must be considered, the proof must be altered to account for each one. In previous years, finding the multiple cases was one of the more difficult aspects of proving a theorem completely. Even Euclid had difficulty with this.[1] (One of the reasons I do not like using most textbooks is that authors seem compelled to tell a student when there are multiple cases for a question. Why not put this information in a hint section of the book and allow a good student to discover this extra difficulty?)

In previous years, usually on at least one problem each day I omitted the conclusion; later I would also omit one or more of the hypotheses. This allowed me to discuss with students the concept of improving a theorem by either weakening a hypothesis or strengthening the conclusion. As the class using *Sketchpad* progressed, I found that I could expand this sort of exploration since the drawings were now manageable. By the middle of the third quarter, enough material was developed, enough proofs and counterexamples generated, and sufficient ideas covered to give the students several weeks to explore quadrilaterals on their own.

For this investigation, they were given standard definitions of eight basic quadrilaterals: cyclic quadrilateral, trapezoid, isosceles trapezoid, parallelogram, rectangle, kite, rhombus, and square. The definitions were changed slightly from standard ones, for I preferred inclusion whenever possible. Thus a parallelogram is a trapezoid, a rectangle is an isosceles trapezoid, etc. They were also given definitions of a few parts (e.g., a diagonal joins two opposite vertices, a median joins midpoints of two opposite sides). They were allowed to name new parts: if an object had a standard name, it was then given and used; otherwise they could choose a word that best described the phenomenon observed. On a few occasions in the past, a word was so well chosen that I used it in subsequent years. My favorite student creation was "quord," a word chosen to denote a segment with endpoints on two sides of a quadrilateral (since it involves a quadrilateral and somewhat resembles a chord of a circle). Other favorites were "diacenter" (intersection of the diagonals) and the "medcenter" (intersection of the medians). I found that students especially enjoyed discovering and naming objects. Sometimes they elected to name an object or a theorem after themselves, but most of the time they tried to find a name that connected the object linguistically to at least one of its characteristics.

In previous years I had obtained an average of about four different theorems per student per day with about eight different theorems per class per day. At the end of the three-week period, students had produced about 125 theorems. Corollaries that were special cases were not counted; for instance, one of the first theorems about a rectangle should be that it is a parallelogram, so any theorem for a parallelogram that transfers to a rectangle is excluded from their count. In the first year with the use of *Sketchpad,* the number of theorems increased to almost 20 per day for the class, with more than 300 theorems produced for the whole investigation.

In the process of the investigation, several quadrilaterals were studied in addition to the eight basic ones previously listed. A standard theorem states that the quadrilateral formed by connecting consecutive midpoints of a quadrilateral is always a parallelogram (students called these connecting segments "midments".) This was apparently discovered in the early 18th century by Pierre Varignon. Under certain conditions, the varignon quadrilateral can be a rectangle, a rhombus, or a square. The students were asked to investigate the various converses of that theorem, that is, if the varignon quadrilateral is a rectangle (rhombus, square), then ?????. This investigation produced two new quadrilaterals: one with congruent diagonals (an "isodiagonal" quadrilateral) and one with perpendicular diagonals (an "orthodiagonal" quadrilateral). We made a rule that any new name for a part of a quadrilateral or new type of quadrilateral must have at least one theorem to support the definition; otherwise there is no mathematics, only taxonomy.

I did not give students these new quadrilaterals, nor did I supply any proofs for their conjectures, though I did work on their conjectures for my own interest. Even when I knew the resolution of one of their conjectures, I did not give it to them. I preferred to have them feel that the discoveries were entirely theirs. My role was to collect their results, organize them, and write up pages for their "textbook." Six of the quadrilaterals they discovered and investigated were: (1) bisectogram, (2) perpbisogram, (3) altagram, (4) exvarignon, (5) midvexogram, and (6) tangentogram. In the last section of this article I give a summary of the definitions of these quadrilaterals and the results for them that the students found and (with a few exceptions) proved.

In this class I found that students who struggled with proof began to understand why proof is an essential part of the process. Those who were excellent at mimicking and thinking in small structures found they needed to reevaluate their position, to learn first to outline a proof. I also found that more students became involved in the discovery process, and that transferred to a desire to prove the results. A few students who in the early stages of the course were very good at proving theorems discovered that they lacked the imagination to find new problems, while others who struggled with proof could look at a few examples and quickly discern relationships. More students became involved in the course, and everyone discovered that there is considerably more to mathematics than just proving existing results.

At the end of the year, many conjectures remained unproved. Some of the students continued their explorations in succeeding years, and shared their results with me. One of my greatest pleasures was to have a student continue to explore mathematics and communicate results to me.

Student Discoveries. A "bisectogram" is a quadrilateral formed by the bisectors of the angles of a quadrilateral. (See Figure 2). When the bisectogram degenerates to a point or a line segment, we will say it does not exist. Students proved the following theorems:

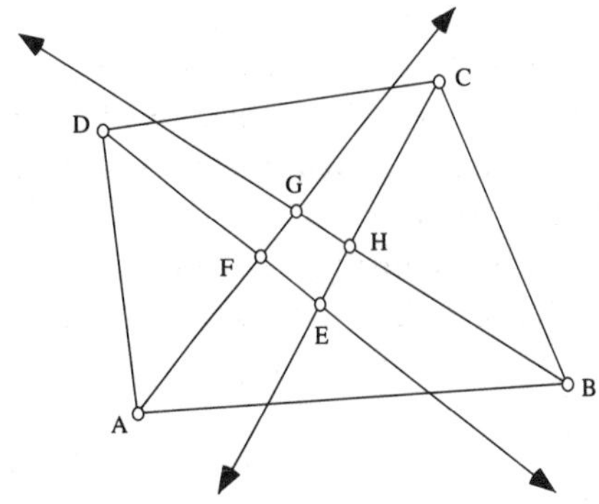

FIGURE 2. Bisectogram $EFGH$ of $ABCD$.

B1: If a diagonal of a quadrilateral bisects the angles at its endpoints (kite, rhombus, square), then its bisectogram does not exist.
B2: If the bisectogram of a quadrilateral exists, it is a cyclic quadrilateral.
B3: The bisectogram of a parallelogram is a rectangle.
B4: The bisectogram of a rectangle is a square.
B5: The bisectogram of an isosceles trapezoid is a cyclic kite (exactly two angles are right angles).

A "perpbisogram" is a quadrilateral formed by the perpendicular bisectors of the sides of a quadrilateral. (See Figure 3). Students proved the following theorems:

P1: If a quadrilateral is cyclic, then its perpbisogram does not exist. (This takes care of the rectangle, isosceles trapezoid, square, and cyclic kite.)
P2: The perpbisogram of a parallelogram is a parallelogram.
P3: The diacenter of a parallelogram is the diacenter of its perpbisogram.
P4.1: The perpbisogram of a trapezoid is a trapezoid.
P4.2: The angles of a trapezoid are congruent to the angles of its perpbisogram.
P5: The perpbisogram of a rhombus is a rhombus.
P6: The diagonals of a rhombus are collinear with the diagonals of its perpbisogram.
P7: The perpbisogram of a non-cyclic kite is a kite.

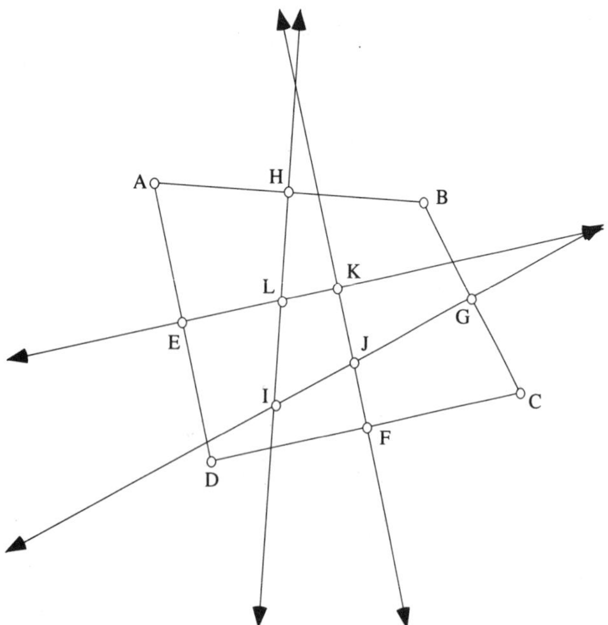

FIGURE 3. Perpbisogram $IJKL$ of $ABCD$.

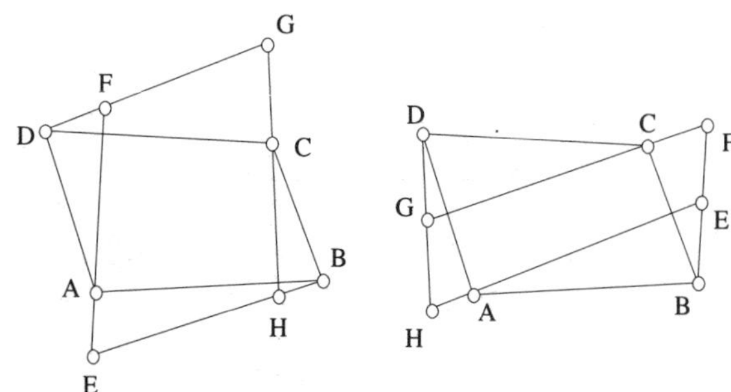

FIGURE 4. The altagrams $EFGH$ of quadrilateral $ABCD$. Left: Dextro-altagram $EFGH$. Right: Levo-altagram of $ABCD$.

P8: The main diagonal (line of symmetry) of a kite is the same for its perpbisogram.
P10: The angles of a kite are supplementary to the corresponding angles of its perpbisogram. (There is a reversal of orientation of the perpbisogram that can occur as the kite is dragged.)
P11: The perpbisogram is non-convex if its quadrilateral is non-convex.

An "altagram" is a quadrilateral formed by the altitudes from the vertices. Since each vertex belongs to two sides of a quadrilateral, there are two "opposite" sides for each vertex. This means that for a given quadrilateral, there are sometimes two altagrams, and under some conditions none. To distinguish between the two altagrams, I supplied the modifiers "dextro" for the altagram formed by going clockwise around the quadrilateral and "levo" for the one formed going counterclockwise (see Figure 4). Students discovered and proved the following theorems:

A1: The angles of the altagrams of a parallelogram are congruent to those of the parallelogram.
A2: The altagrams of a kite are congruent, and they intersect the kite at the vertices of the congruent angles. (They called these angles of a kite the iso-angles; the others were called the trans-angles. The symmetry diagonal was called the trans-diagonal, and the other diagonal was called the vers-diagonal.)
A3: The altagrams of a rhombus are congruent parallelograms.
A4: An altagram of a rectangle is the rectangle.
A5: The altagrams of an isosceles trapezoid are congruent and homothetic (ratio $= -1$) with center on the median between the parallel sides of the isosceles trapezoid. (This median received different names: transdicular, perpitude, transmedian, and altaversal.)
A6: The altagrams of a trapezoid are trapezoids.

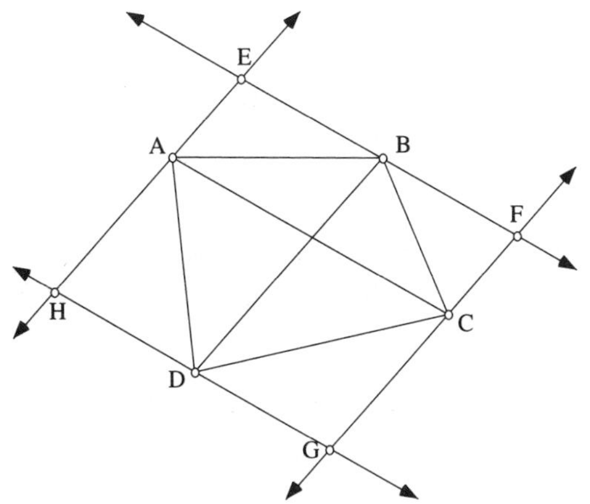

FIGURE 5. Exvarignon $EFGH$ of $ABCD$.

A7.1: The altagrams of a cyclic quadrilateral are cyclic.

A7.2: The circumcircles of the altagrams of a cyclic quadrilateral are congruent.

An "exvarignon" quadrilateral is formed by lines parallel to a diagonal and passing through endpoints of the other diagonal. (See Figure 5.) The following theorems were discovered and proved:

X1: For any quadrilateral, the exvarignon quadrilateral is a parallelogram.

X2: If the diagonals of a quadrilateral are perpendicular, then the exvarignon quadrilateral is a rectangle. (This takes care of the orthodiagonal quadrilateral, which includes the kite, rhombus, and square.)

X3: If the diagonals of a quadrilateral are congruent, then the exvarignon quadrilateral is a rectangle. This takes care of the isodiagonal quadrilateral, which includes the isosceles trapezoid, rectangle, and square.

X4.1: For a quadrilateral, its exvarignon quadrilateral and varignon quadrilateral are homothetic (ratio $= 2$). (The first year, this theorem was stated as: the sides of the varignon quadrilateral are parallel to the sides of the exvarignon quadrilateral. The next year, I started including homothetic figures as a basic element of similarity.)

X4.2: For a quadrilateral, the diacenter of its varignon quadrilateral is the midpoint of the diacenter of its exvarignon quadrilateral and their homothetic center.

A "midvexogram" is a quadrilateral formed by the intersection of quords from a vertex to a midpoint of the opposite side (one year this segment was called a midvex).

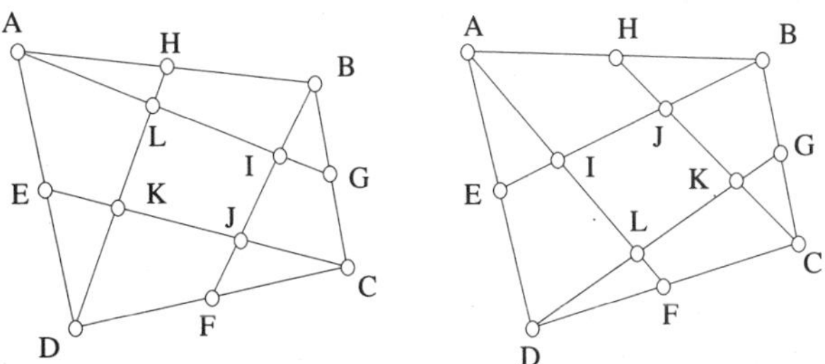

FIGURE 6. Left: Dextro-midvexogram $IJKL$ of $ABCD$. Right: Levo-midvexogram $IJKL$ of $ABCD$.

As before, since each vertex of $ABCD$ has two "opposite" sides, there are two of these quadrilaterals called the dextro-midvexogram and the levo-midvexogram (See Figure 6). These quadrilaterals received more attention than all the others combined. Students proved the following theorems:

M1: For a trapezoid, lines through corresponding vertices of the midvexograms are parallel to the parallel sides of the trapezoid.

M2: If either of the midvexograms is a parallelogram, then the other midvexogram is a parallelogram and the quadrilateral is a parallelogram.

M3: For an isosceles trapezoid (or rectangle), the midvexograms are congruent; in fact, the reflection of one over the transmedian (median from the midpoints of the parallel sides) is the other midvexogram. (The discovery and proof of this theorem provided the proofs for a large number of other theorems, many of which had already been proved.)

M4: For a kite (or rhombus), the midvexograms are congruent; in fact, the reflection of one over the transdiagonal (symmetry diagonal) is the other midvexogram. (The discovery and proof of this theorem also solved a number of other problems. With M3, it connected kites with isosceles trapezoids.)

M5: For a parallelogram, the area of a midvexogram is one-fifth the area of the parallelogram. (In general, they are not congruent.)

M6: For a square, the midvexograms are squares.

The students also produced a conjecture that they could not prove and two unsolved problems.

M7: (Unproved by the students) The area of a quadrilateral is at least five times the area of a midvexogram.

M8: (Unsolved) Under what conditions will be area of the midvexograms be equal (equiareal)?

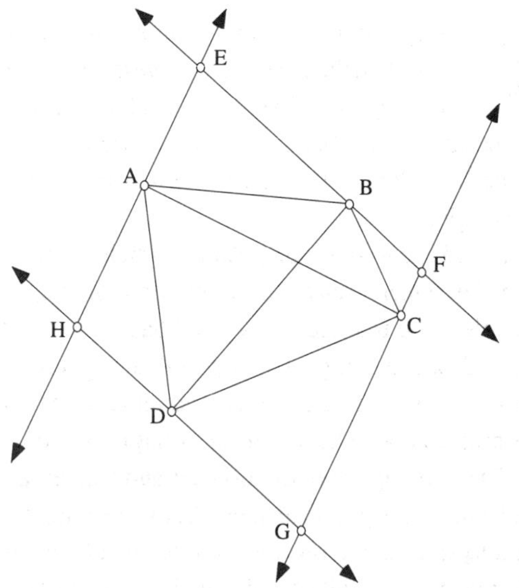

FIGURE 7. Tangentogram $EFGH$ of $ABCD$.

They easily constructed examples with unequal areas. They strongly believed that, if the areas of the midvexograms are equal, then the quadrilateral is a parallelogram.

M9: (Unsolved) Under what conditions will the midvexograms be congruent?

A "tangentogram" is a quadrilateral formed by the perpendiculars to the diagonals at their endpoints. These lines correspond somewhat to tangents of a circle, perpendicular to a radius at the point of intersection with the circle (see Figure 7). The theorems for this quadrilateral are similar to those of the varignon quadrilateral.

T1: The tangentogram of a quadrilateral is a parallelogram.

T2: The tangentogram of an orthodiagonal quadrilateral is a rectangle.

T3: The diagonals of the tangentogram of a cyclic quadrilateral intersect at the center of the circumcircle.

T4: The tangentogram of a quadrilateral with congruent diagonals is a rhombus.

T5: The tangentogram of a square is a square.

References

1. Heath, Sir Thomas L., *Euclid: The Thirteen Books of The Elements,* Dover Publications, Inc., 1956.
2. Pólya, George, *Mathematical Discovery,* John Wiley & Sons, 1962.
3. Russell, Bertrand, "The Study of Mathematics," *Mysticism and Logic,* Doubleday & Co., Inc., 1957.

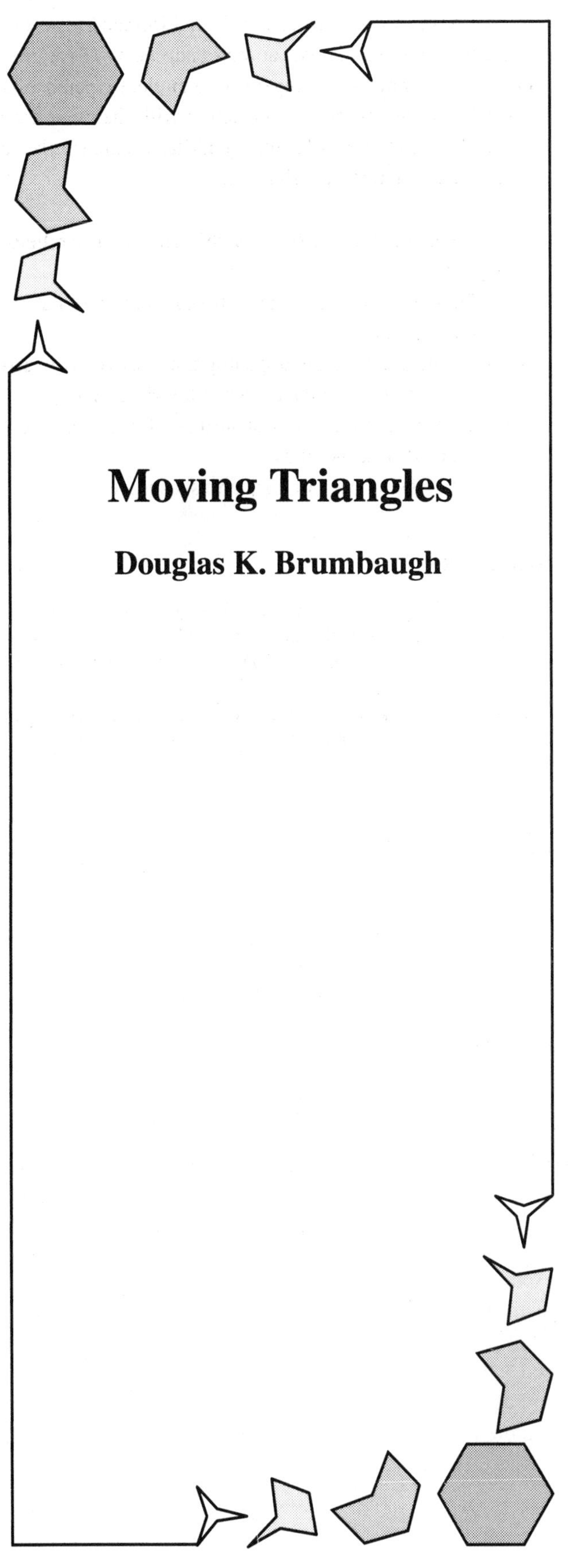

Moving Triangles

Douglas K. Brumbaugh

Recently I taught a middle-school class that dealt with the idea that the area of a triangle is base times height, divided by two. These students had encountered the concept many times before. They still were not convinced that the formula was true, let alone that it would work for *any* triangle.

I used *Geometer's Sketchpad* to help convince them. When they entered class, they saw an animated sketch of a triangle. The triangle had a fixed base CD and the third vertex, E, was moving back and forth along a line segment parallel to CD. The area, base, and altitude measurements were projected with the animated triangle (see Figure 1).

The entering students "ooh"ed and "ah"ed at the projected image. They soon realized they were being confronted with information that involved a dynamically changing collection of triangles, all of which had the same area, altitude, and base. As they settled in, comments like "Look at the screen." Neat!" were common.

After the bell rang, I said nothing to start the class. The students' comments began changing to things like "The triangle keeps changing," "But the area stays the same," and "It has to stay the same because the base and height don't change." Not all students were willing to accept the idea that the area had to be the same. They insisted that the triangles were different.

I was still silent, letting the discussion seek its own course. One boy went to the screen and showed those who were non-believers in his discussion group that the distance between the two endpoints of the base stayed the same as the third vertex moved. He said "See, these two points are not moving, and the distance between them won't change."

He returned to his seat, and, while I still said nothing, a girl then went to the board and pointed to the altitude that was moving along the screen. She clarified for her group that the line segment that held "E" was parallel to base CD and that the altitude EF showed the distance between those parallels. She moved her hands along with EF and then, holding them at that distance, turned to her group saying, "The distance between E and F stays the same."

At this point I entered the discussions for the first time. I asked for clarification of what they were talking about and what the two individuals who had gone to the screen meant. Members from the class told me that for the area to change, either the base or altitude had to change. I said "So you mean that if I change this height, the area will change?" While I was talking, I altered the height, and they could see the area change as I moved the segment holding "E". I continued by saying "And you mean that if I move D the area will change?" and, at the same time, altered the length of the base, CD.

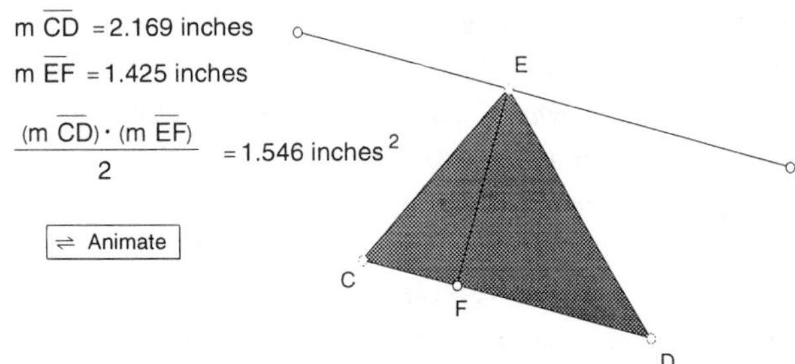

FIGURE 1. Screen students saw as they entered class. E was moving back and forth along AB. Altitude EF moved with E, remaining constant. The lengths of segments CE and DE changed as E moved.

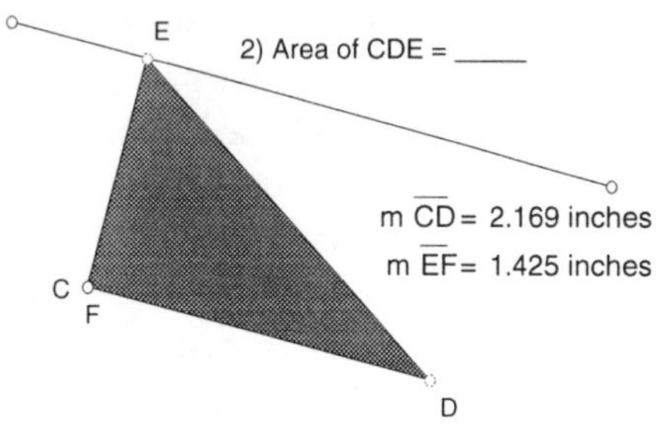

FIGURE 2

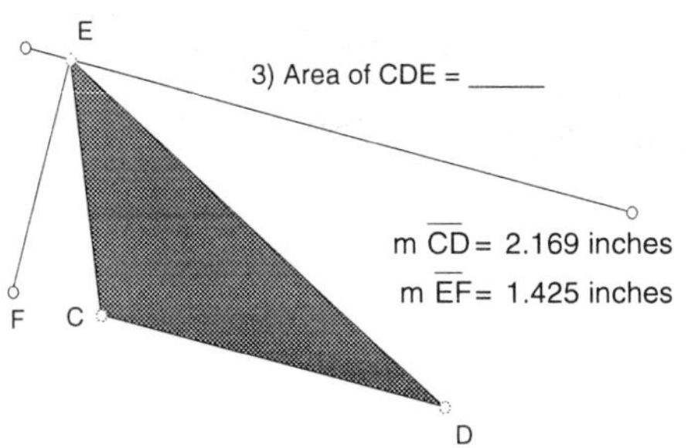

FIGURE 3

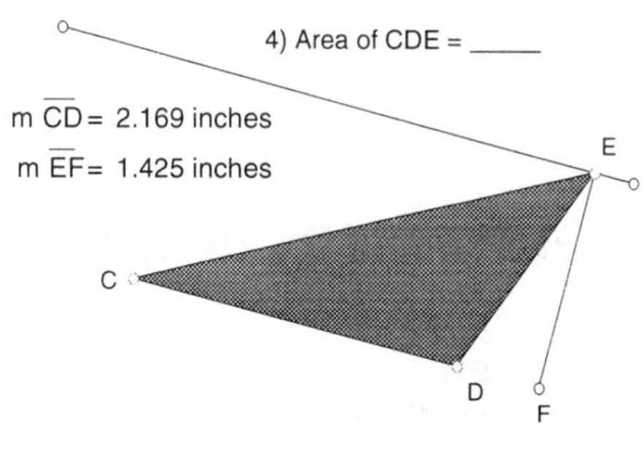

FIGURE 4

The whole lesson was completed in less than 20 minutes. The students seemed to be fully aware that changing either the base or altitude of a triangle would impact the area. In the process of discussing what they had seen, the students also remarked that the measure of the base and height did not have to be counting numbers. They were quite willing to approach a set of problems to show that they could easily determine the area or a triangle if they were given the base and height.

It was now homework time. Their homework consisted of finding the area of ten triangles. Three of the problems, all taken from screen prints, are given in Figures 2, 3, and 4.

The same base and height appeared in each of the ten problems, but the triangles looked different.

The class worked for a few minutes, and it was clear they were finished. I chastised them for not doing their work. They said they were done. I said that could not be since they had so many problems to do. They responded that, since the base and height were the same in each problem, they only needed to work out the first one and then copy the answer for the other nine. It appeared as if they had a much stronger grasp of the concept of the area of a triangle.

Subsequent checks of the students' ability to compute the area of given triangles have shown good results. The student reaction to questions dealing with finding the area of triangles, no matter the configuration, seems to be one of confidence. They indicate that such problems are trivially simple.

Beyond the impact of the lesson on the area of triangles, the demonstration brought comments from students who asked "Why don't we do this kind of stuff more?" As I hear that question, it becomes a mandate to which we must respond, no matter what area of mathematics we teach. We must begin to build technology into our lessons. We can cover more ideas, build strong and deeper conceptualizations in the students, and motivate them to want to learn, to become involved, and explore mathematics in their world.

With the software currently available, we as teachers can begin to explore the world of mathematics in new ways. In the process, we will learn different techniques, and maybe even get a different perspective of many of the branches of mathematics we have studied in the past. As we do that, it could well be that we will get excited about what we learn and enthusiastically tell our classes about our investigations and conjectures. If they want the results, they will have to explore on their own. Let the fun begin!

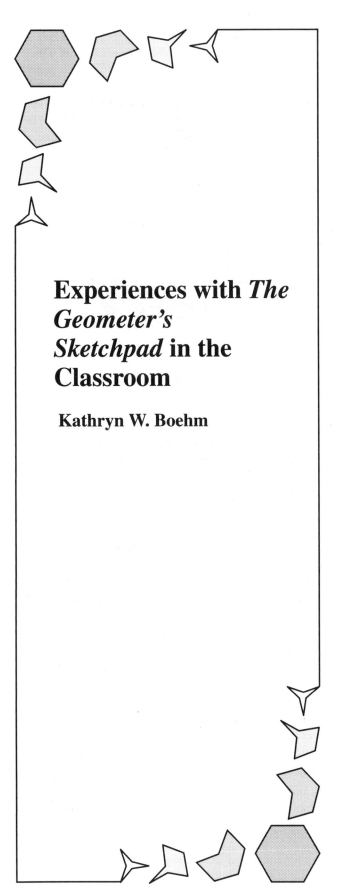

Experiences with *The Geometer's Sketchpad* in the Classroom

Kathryn W. Boehm

In the Spring of 1992, I attended the NCTM Annual Meeting in Nashville, Tennessee, where I saw *The Geometer's Sketchpad* for the first time. I saw that this tool would allow independent student exploration and dynamic teacher demonstrations, so I asked for a demo copy and was quite surprised when the representative handed me a box containing the complete software package. When I returned to school, I immediately began to use the software to model homework problems and to demonstrate concepts in my classes, particularly in my Advanced Geometry class. We used *Sketchpad* to explore transformations, and the sketches helped my students to visualize more clearly the effects of rotation, translation, dilation, and reflection. One homework problem from the text ***Geometry,*** by Jurgensen, Brown, and Jurgensen, sticks in my mind:

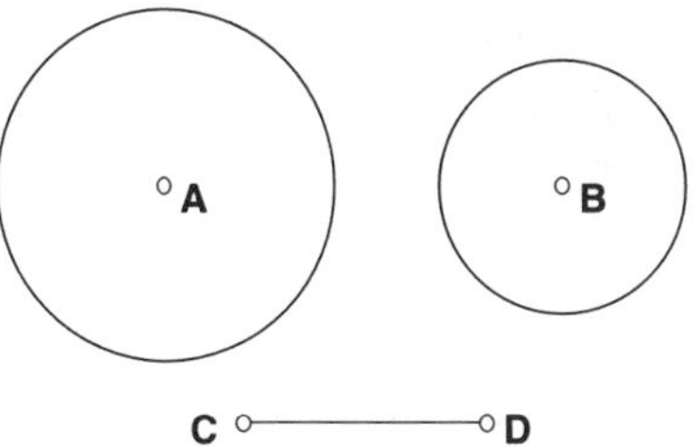

> *Given circle A and circle B and $\overline{CD}$, construct a segment $\overline{XY}$, parallel to and congruent to $\overline{CD}$ and having X on circle A and Y on circle B. (Hint: Translate circle A on a path parallel to and congruent to $\overline{CD}$.) (p. 524)*

Although we discussed the solution in class and drew it on the board, many of my students were visibly uncomfortable with the problem. I herded them to our computer lab, and we looked at the same construction on *Sketchpad.* The software allowed us to vary the size of the circles and to look at variations in the problem. The reason I recall this incident so vividly is that my students said, "Thank you for showing us that," after class was over. I suppose any teacher recalls being thanked! The experiences that I had during that demonstration period convinced me that we needed to purchase the software. (Fortunately, I was able to convince my Head of School as well.)

Sketchpad facilitates the presentation of topics which are not readily explained using static drawings. My students often have difficulty understanding properties of the diagonals of special quadrilaterals. They want the diagonals of a parallelogram to be congruent, but *Sketchpad* makes it easy to demonstrate the contrary. Many textbooks discuss the trapezoid midsegment theorem, then suggest shortening the upper base until the trapezoid becomes a triangle. This allows students to see the validity of the triangle midsegment theorem. Using *Sketchpad,* one can construct a trapezoid, measure all relevant lengths, then drag the upper base until the trapezoid is a triangle. Students can see that the geometric relationships are preserved, and students are able to conjecture a theorem about the triangle midsegment on their own. Classroom demonstrations, whether planned or spontaneous, are often made more effective by using this dynamic software.

I spent most of the summer of 1992 learning to use *Sketchpad* by going through the manual, guided tours, activities, and the exercises in *Exploring Geometry with the Geometer's Sketchpad* (Key Curriculum Press, 1992). Many of the investigations involve topics such as parallel lines, which my students have studied in middle school, or topics about which they have ready, correct intuition. During that summer, however, I came across an investigation on Napoleon's triangle in *Exploring Geometry*. I had never heard of this theorem before, and so I decided to give my Advanced Geometry class an assignment on it. They were all able to conjecture the theorem:

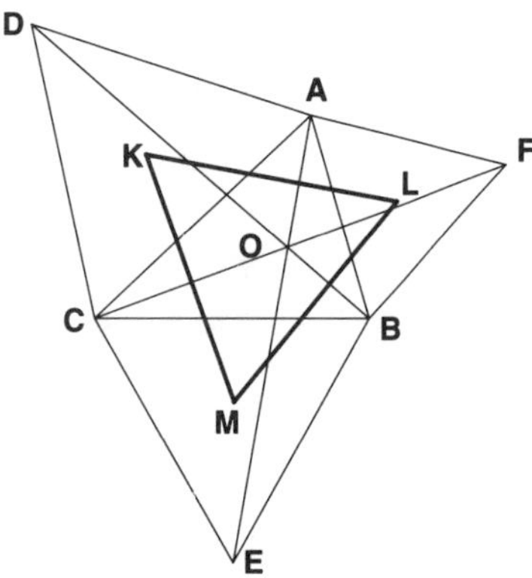

Given any $\triangle ABC$, construct equilateral triangles on each of the three sides. The centroids of the three equilateral triangles, when connected, form another equilateral triangle, called the outer Napoleon triangle.

The interesting part of the activity for me, however, came when my students connected points A, B, and C with points E, D, and F respectively. I knew that these segments are always congruent, but my students commented, "$\overline{AE}$, $\overline{BD}$, and $\overline{CF}$ look like they are concurrent. What's that point called?" My more astute students had already constructed the orthocenter, incenter, circumcenter, and centroid of $\triangle ABC$ and determined that point O is a different point. I confess that, at the time, I did not know the answer to their question, and it was another year and a half before I learned that O is the Fermat point, the point at which the sum of the distances to the three vertices of $\triangle ABC$ is a minimum. I also found a proof of the theorem itself, and discovered that it would be an ideal topic to serve as a theme for the year. In order to prove that $\overline{AE}$, $\overline{BD}$, and $\overline{CF}$ are congruent, students need only to know how to use congruent triangles, so this part of the exercise can be discussed early in the year. The concurrency proof requires the inscribed angle theorem, and can be included in a chapter on circles. The proof that $\triangle KLM$ is equilateral involves similar triangles and 30°-60°-90° right triangles. If the geometry course includes transformations, the vertices K, L, and M of the outer Napoleon triangle can be reflected through $\overline{AC}$, $\overline{AB}$, and $\overline{BC}$ respectively, resulting in the inner Napoleon triangle. Not only is this inner triangle equilateral, but the sum of the areas of the inner Napoleon triangle and the original triangle equals the area of the outer Napoleon triangle.

Any teacher using *Sketchpad* or other dynamic geometry software needs to be prepared for students to ask unexpected questions. I have had to answer, "I don't know; let me get back to you," many times. I have used *Sketchpad* to conjecture general results and to suggest methods of proof. Investigating my students' questions, especially those concerning Napoleon's triangle, has given me many hours of enjoyment and intrigue and has enhanced my own knowledge of geometry. There are many results in classical geometry that would be fruitful areas of investigation for students and teachers alike. For example, Pascal's theorem, Brianchon's theorem, and Simson's theorem all lend themselves readily to *Sketchpad* investigation. When the students see that the teacher is also involved in investigating and learning, the entire process of "doing math" becomes more authentic to them and they sense that we are all learners working together.

I do, however, have two major concerns about the proper use of dynamic geometry software in the classroom, and they are, perhaps, interrelated. First, my advanced students are overwhelmingly positive about their experience with the software, but my regular students are generally lukewarm. One student in my 1993–94 advanced class said, "I found working on the computers easier because of the speed with which we could draw accurate figures, giving us more time to work on the problem, and allowing this to be the focus of our work." At times students in the regular class seem to enjoy the diversion of working on the computers, but they frequently claim to learn more from teacher-centered demonstrations or exposition: "I learn better when you just teach it to me." I am puzzled by the extreme dichotomy between these two groups, and can think of two potential explanations. Perhaps using *Sketchpad* is uncomfortable for weaker students because they are expected to experiment and discover on their own. Advanced students possess more confidence that they will arrive at a correct conjecture than do regular students. Another possibility is that students who are less adept at math may need more time to master the technical aspects of working with the program. Such students are intimidated by the technology and find it difficult to focus on the geometric concepts as a result. This leads to my second concern: Given the time constraints of an academic year, it is difficult to allow for extensive computer instruction on the use of software. I am often reluctant to "give up" a class period to do an exploration in the computer room. I have to be creative in designing *Sketchpad* assignments which fit with the established curriculum, so that a significant lesson may be covered in a class session with the software.

The above concerns notwithstanding, *Sketchpad* has added another dimension to my teaching of geometry. Not only does it provide variety of presentation, but it presents material in a way which is qualitatively different from that of a textbook. It is not "technology for technology's sake," but rather a unique tool which could enhance any geometry curriculum. It has revitalized my approach to teaching geometry and, thus, has helped my students as well.

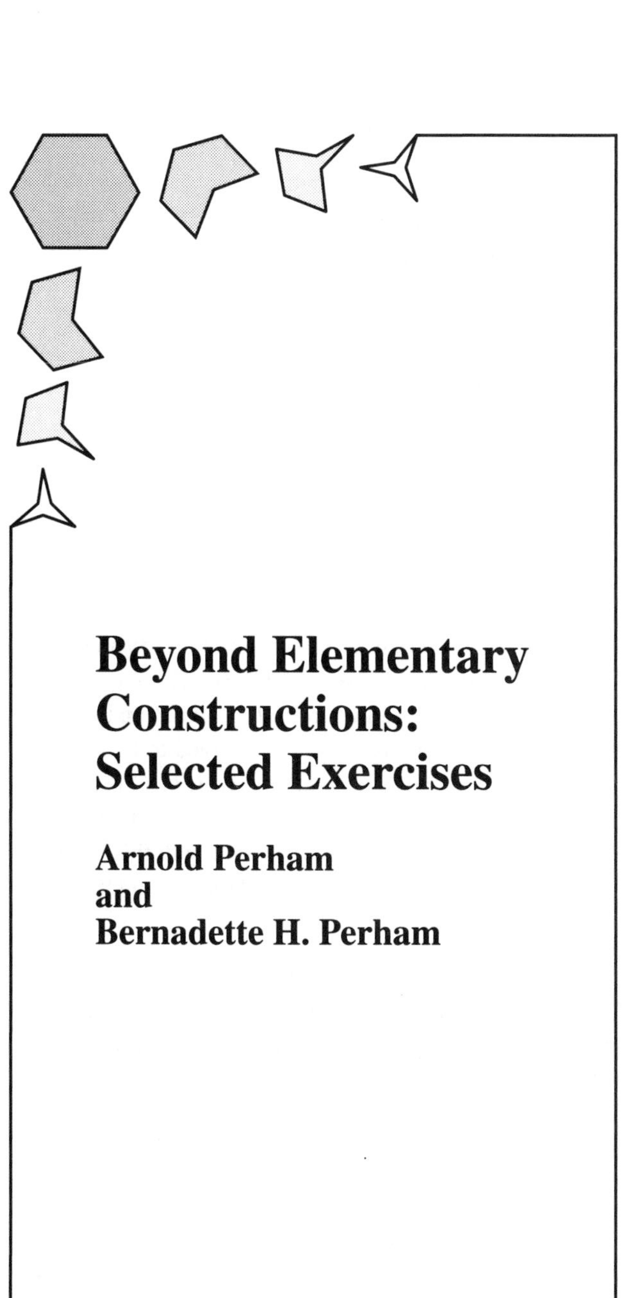

Beyond Elementary Constructions: Selected Exercises

Arnold Perham
and
Bernadette H. Perham

In Lewis Carroll's *Through the Looking Glass*, the Red Queen speaks a line that resonates with teachers, "Now, here, you see, it takes all the running you can do, to keep in the same place [2, p. 147]." With the Queen's caution in mind, it is appropriate to ask the question: "Given the demands on our time, is the educational payoff big enough to warrant introducing geometry construction software into the syllabus?" This question is difficult to answer unless we experience a thorough demonstration of the software or use it ourselves.

In assessing the pedagogical value of software, the experience, vision, and knowledge of people in the mathematics community can help inform our judgment. To this end, it is important to be aware that internationally respected educators like Colette Laborde (FRA) [4], Tommy Dreyfus (ISR) [3], and Benoit Mandelbrot (USA) [5], among many others, urge an experimental, investigative approach in learning geometry. In this rebirth of experimental mathematics, the computer and construction software (such as, CABRI, used for our exercises [1]) supplement the traditional compass and straightedge.

There is value in having students work with paper and pencil. A student who successfully prepares a paper-and-pencil construction exhibits ownership of the concepts in a way different from diagrams produced by the computer. This observation is reflected in the language students use referring to their constructions, e.g., "on *my* drawing," or "*my* three lines don't intersect in a point, should they?" On the other hand, drawings on paper by nature are static, whereas computer-drawn diagrams can be freely altered. It is the dynamic property of the software that permits students to experiment and form conjectures as the elements of the construction change in measure, but invariant relationships remain unchanged. In the process, students develop a sense of "doing mathematics."

Construction software makes it possible to visually present geometry in a way not rendered by any other medium.

Selected Constructions and Investigations

Exercise 1.

Construction. Create the centroid, circumcenter, and the orthocenter of an arbitrary triangle. Draw a line segment from the orthocenter to the circumcenter.

Investigations.

(i) Where is the centroid located with respect to the line segment?
Does this seem to hold true for every triangle?

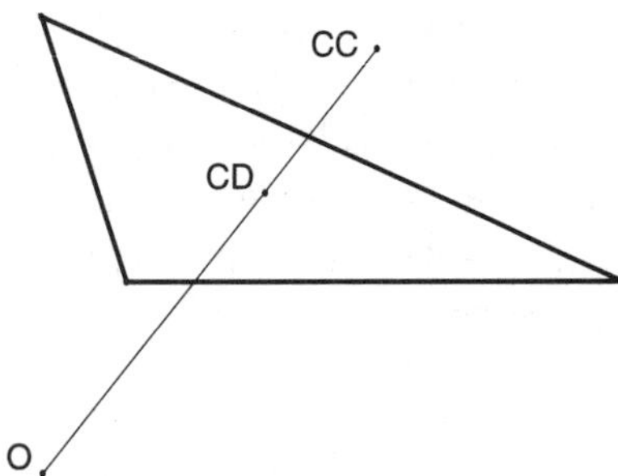

FIGURE 1. CD: centroid, CC: circumcenter, O: orthocenter.

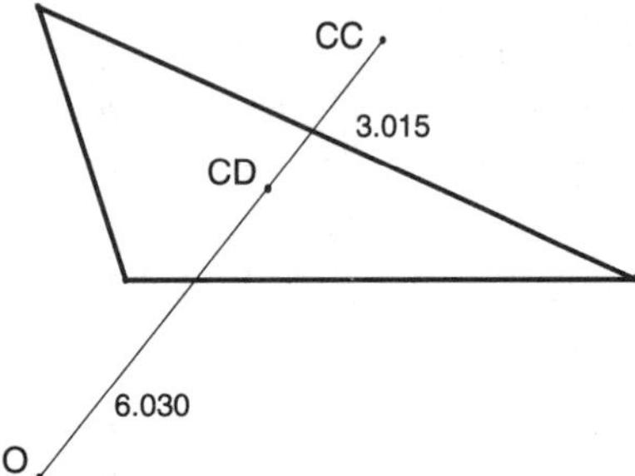

FIGURE 2. O: orthocenter, CD: centroid, CC: circumcenter.

After observing the collinearity of the three points (they lie on the Euler line), the student can pursue other questions (see Figure 1). Is the centroid always between the circumcenter and the orthocenter? For what types of triangles will these points lie inside the triangle?

(ii) What are the distances from the centroid to the circumcenter and from the centroid to the orthocenter? What is the ratio of these distances, respectively? Does it seem that the ratio of these distances is the same for every triangle (within the limits of round-off error)?

Figure 2 illustrates how the software automatically calculates the length of each segment of interest. For several arbitrary triangles, the student can record these lengths and calculate the ratio, verifying that the centroid trisects the line segment formed by the circumcenter and the orthocenter.

Exercise 2.

Construction. Inscribe a triangle in a circle. Select a point on the circle and draw perpendiculars from the point to each side of the triangle. Draw a line through two of the three feet of the perpendiculars.

Investigation. Where is the foot of the third perpendicular with respect to the line segment? What conjecture, if any, can you make?

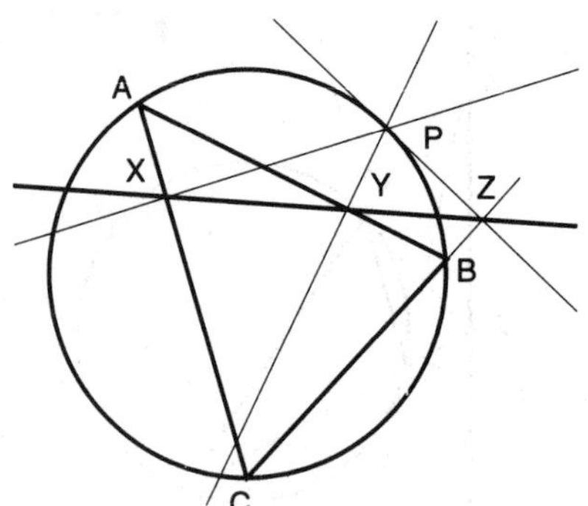

FIGURE 3

In this exercise, it is easier to draw the circle first and then position the required vertices A, B, C on the circle of the inscribed triangle (see Figure 3). The feet X, Y, Z of the perpendiculars drawn from the point P on the circle to the sides of the triangle can be observed to be collinear even when any one of the points P, A, B, or C is dragged to new positions on the circle. The English mathematician Robert Simson (1607–1768) is credited with this observation; the line joining X, Y, Z is called the Simson line.

Exercise 3.

Construction. Inscribe a triangle in a circle. Draw a tangent at each vertex such that it intersects the extended side of the triangle opposite the vertex.

Investigation. What do you observe about these points of intersection?

The inscribed triangle ABC is constructed as in Exercise 2. The tangents to the circle at points A, B, and C, produce collinear points of intersection X, Y, and Z (see Figure 4). As in Example 2, by dragging any vertex of the triangle, the collinearity of points can be seen to be an invariant property.

Exercise 4.

Construction. Draw the bisectors of the external angles of a triangle. Mark the point of intersection of each bisector with the extended side opposite the angle.

Investigation. What do you observe about the points of intersection?

Figure 5 shows the construction of triangle ABC in which the external angle bisector at A meets the extended opposite side at X, the external angle bisector at B meets the

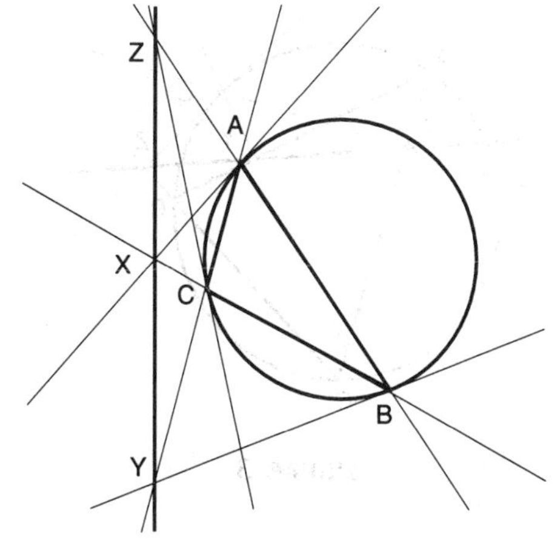

FIGURE 4

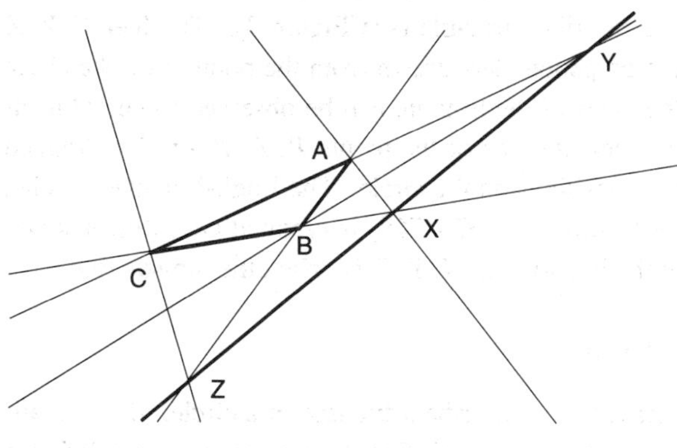

FIGURE 5

extended opposite side at Y, and the external angle bisector at C meets the extended opposite side at Z. When ABC is dragged to a new shape, the collinearity of points X, Y, and Z remains an invariant.

Exercise 5.

Construction. Draw a triangle. Draw lines from each vertex to its opposite side such that the three lines are concurrent.

Investigation. Each side of the triangle is divided into two segments by one of the lines. The six segments, whose measures can be computed, can be partitioned into two sets of three segments such that each set is made up of nonconsecutive segments. What can you say about the product of the line segments in each set (within the limits of round-off error)?

In Figure 6, AX, BY, and CZ are the three lines intersecting at P. These lines intersect the opposite sides at X, Y, and Z. The purpose of the construction is to have the student discover that

$$AZ * BX * CY = ZB * XC * YA$$

for any triangle ABC. This theorem is due to the Italian mathematician Giovanni Ceva (1647–1736). A good follow-up exercise is to discover that, if points X, Y, Z are chosen on sides BC, AC, and AB, respectively, so that the equation above is true, then the three segments AX, BY, CZ (called Cevians) are concurrent.

Exercise 6.

Construction. Create a triangle and its altitudes such that the orthocenter P lies within the triangle.

Investigation. Each altitude is divided internally into two parts by the orthocenter P. For each altitude, measure the length of the segment from P to the foot of the altitude and also the length of the altitude; take the ratio of the 2 lengths. What can you say about the sum of the three ratios (within the limits of round-off error)?

The construction is shown in Figure 7, where each of the segments of interest is labeled with its measure. By dragging on a vertex, the shape of the triangle will change, but the sum of the ratios will be constant for every position of the orthocenter that lies within the triangle, that is,

$$PX/AX + PY/BY + PZ/CZ = 1$$

Exercise 7.

Construction. In Figure 8, A, P, and B are the vertices of a triangle, and C and D are points on (extended) side AB so that PC is a bisector of angle APB and PD is the external bisector of angle P. Construct the circle whose diameter is CD.

Investigation. As P moves along the circle, examine the relationships between the three ratios $\frac{AC}{CB}$, $\frac{AP}{PB}$, and $\frac{AD}{BD}$. Make a conjecture.

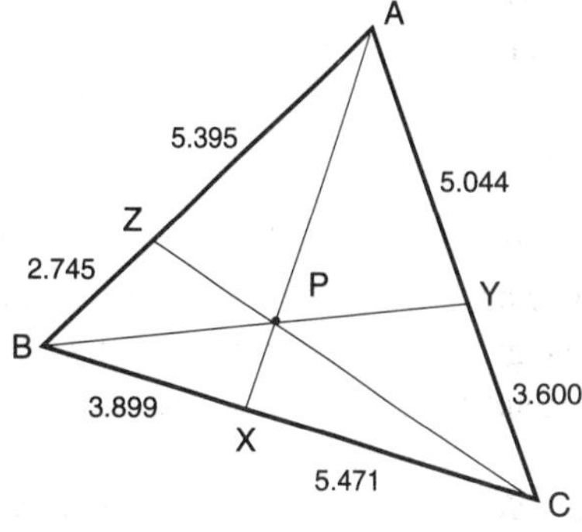

FIGURE 6. Note: $5.395 * 3.899 * 3.600 = 75.726378$ and $2.745 * 5.471 * 5.044 = 75.75026238$.

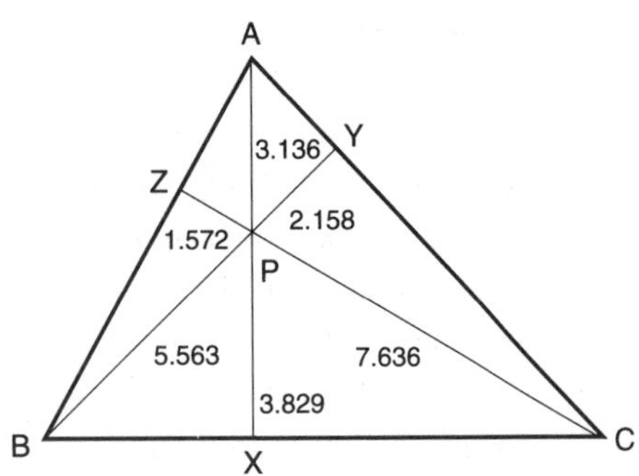

FIGURE 7. Note: $3.829/(3.829 + 3.136) + 2.158/(2.158 + 5.563) + 1.572/(1.572 + 7.636) = .99997$.

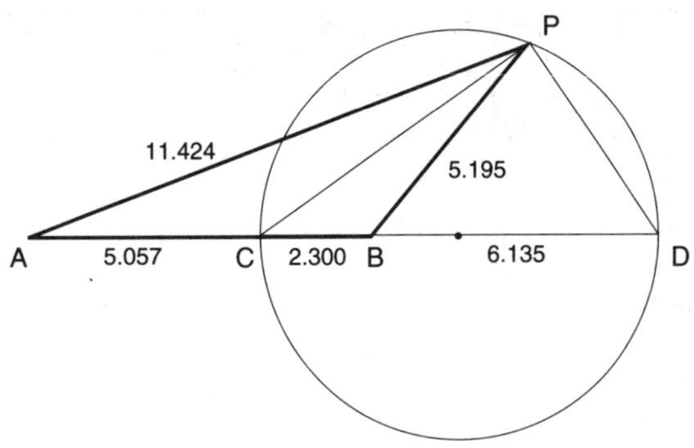

FIGURE 8. Note: $5.057/2.3 = 2.199$, $11.424/5.195 = 2.199$, and $(5.057 + 2.3 + 6.135)/6.135 = 2.199$.

After completing the essential construction, each segment of interest is labeled with its measure, and the three ratios are calculated. When P is dragged along the arc of the circle, the three ratios are seen to be equal, even though the shape of the triangle changes.

Conclusion

Geometry construction software gives students a tool for drawing basic geometric objects, e.g., point, line, segment, circle, and triangle, with ease and accuracy. The dynamic nature of the software, together with its measurement feature, encourages students to explore the relationships that exist between these objects. Even though secondary-school students have limited experience in geometry, they can formulate and test interesting and significant conjectures derived from the basic objects.

In these exercises we have extended the students' conjecturing into areas not typically explored in high school geometry. Perhaps these exercises will suggest other fruitful topics more typical in the secondary curriculum. For example, if the radius of a circle is doubled, tripled, or quadrupled, how is its circumference and area affected? How is the measure of an exterior angle of any triangle related to the nonadjacent interior angles? If the midpoints of two sides of a triangle are connected, how is the measure of this segment related to the measure of the third side? The answers to these questions become obvious if both the dynamic and measurement features of the software are employed. These investigations which prompt conjecturing can lead in each case to a satisfactory proof.

References

1. Baulac, Y., Belleman, F., and Laborde, J.-M. (1992). *CABRI, The Interactive Geometry Notebook*, Macintosh Version 1.7. Pacific Grove, California: Brooks/Cole Publishing Company.
2. Carroll, L. (1960). *Alice's Adventures in Wonderland & Through the Looking Glass*, A Signet Classic, The New American Library, Inc., New York, NY.
3. Dreyfus, T. (1994). Imagery and Reasoning in Mathematics and Mathematics Education, *Selected Lectures from the 7th International Congress on Mathematical Education*, Les Presses de l'Université Laval, Sainte-Foy, 107–122.
4. Laborde, C. (1994). Teaching Geometry: Permanences and Revolutions, *Proceedings of the 7th International Congress on Mathematical Education*, Les Presses de l'Université Laval, Sainte Foy, 47–75.
5. Mandelbrot, B. (1992). *Fractals and the Rebirth of Experimental Mathematics*, in: Fractals for the Classroom, H.-O. Peitgen, H. Jürgens, and Dietmar Saupe, Springer-Verlag, New York, NY, 1–16.

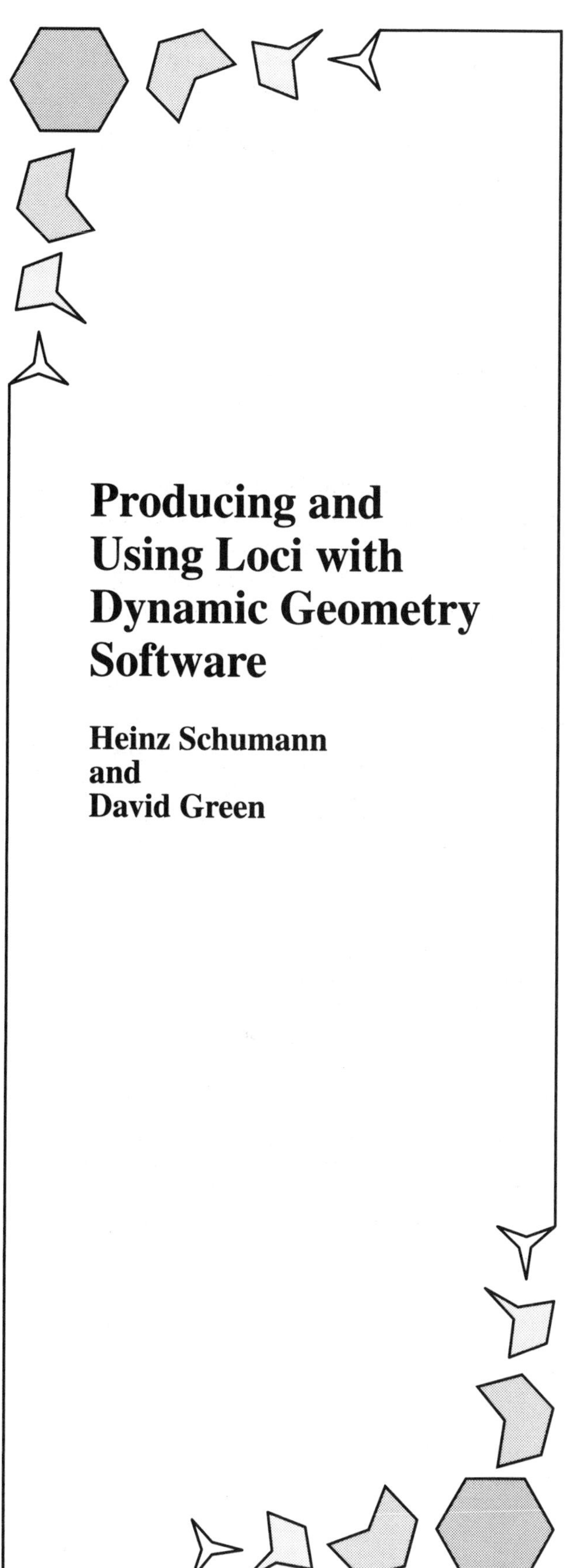

Producing and Using Loci with Dynamic Geometry Software

Heinz Schumann
and
David Green

The Moving Finger writes,
and having writ, moves on.

Edward Fitzgerald

"The moving finger" in this article appears in the dynamic geometry program *Cabri-géomètre*. It guides a point along a path and, in so doing, produces another sequence of points, the locus of some chosen point. Until the availability of interactive dynamic geometry software, the possibility of envisioning a locus in animation as it was traced out by a moving point depended on students' and teachers' powers of visual imagination. Now, with the software, students can explore the production of loci interactively.

Description of the method of producing a locus with Cabri-géomètre I

- Make a construction. In our example in Figure 1, a triangle and its medians are constructed.
- The so-called basic points which determine the construction are free to move. In Figure 1, these are the three vertices of the triangle.
- Connect a basic point or other basic object to a specially selected path. In Figure 1, B is the basic point, and the path is the circumcircle of the triangle.
- Move the basic point along the chosen path. This causes the constructed configuration to vary. In Figure 1, the icon of a grabbing hand indicates the cursor dragging point B.
- As the basic point is dragged along the path, we follow the movement of one selected point whose construction depends on the basic point. In Figure 2, this selected point is A, the centroid of the triangle where the medians intersect.

We ask the question: *What is the locus of point A as point B is moved on its guide-path? Cabri* gives a trace of **discrete** points on the screen, which are images of A as it moves. The user must associate this discrete trace with the continuous shape of the locus. In Figure 2, the locus of A is a circle.

This process produces a mapping from points of the circumcircle of the triangle onto the circular path followed by the centroid of the triangle. Normally, the concept of function comes into geometry in the context of volume, area, or length measurements, but the context of loci offers a new possibility of introducing functional dependencies. (The locus-path is a function of the guide-path.)

Interactive production of loci can be used effectively in these aspects of mathematics teaching:

— in the heuristic phase of construction task problem-solving;

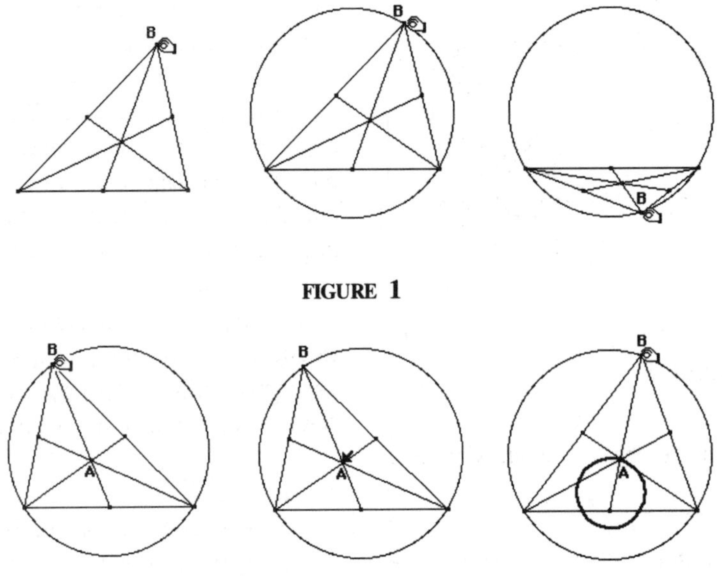

FIGURE 1

FIGURE 2

— in the experimental verification of construction results;
— in investigations of the position and shape of the image of a transformed original shape;
— in the construction of conic sections and algebraic curves of order > 2;
— in investigations of the shapes of loci generated by the movement of special points in a triangle or in a more complex configuration.

Loci as an Aid to Finding Solutions for Constructions

Using loci is particularly valuable for solving a certain class of geometry problems known as insertion problems.

Example 1. A square is to be inscribed in a triangle so that one side of the triangle contains two vertices of the square and each of the other two sides of the triangle contains one vertex of the square.

Solution. First we construct a square with three of its vertices in the required positions, i.e., leaving out one condition (Figure 3(a)). We then vary the size of the square by dragging a vertex that lies on a side of the triangle. Along what path does the fourth corner of the square move? Its locus, together with an experimental solution, are shown in Figure 3(b). From this, the construction method is discovered.

Example 2. In this example three parallel lines are given. We seek to construct an equilateral triangle so that each line contains one vertex of the triangle.

Solution. First we construct an equilateral triangle ABC with basic point B on one of the given lines and basic point C on a second of the given lines while ignoring the condition that A is to be on the third given line. We move B and generate the locus of A (Figure 4). With this experiment, we see that whatever the independent positions of the three

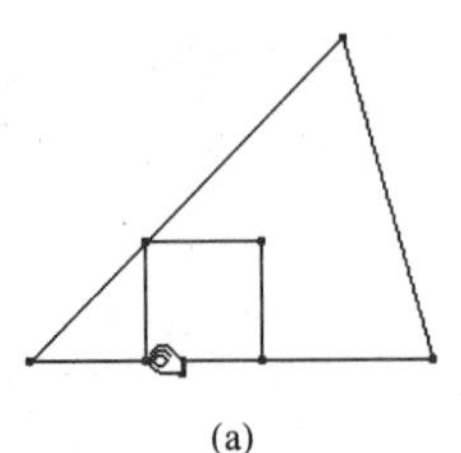

(a)

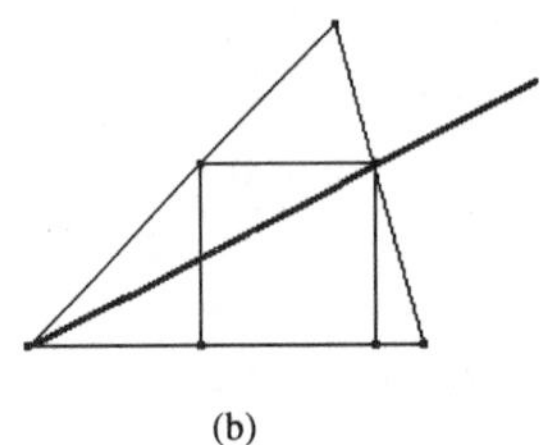

(b)

FIGURE 3

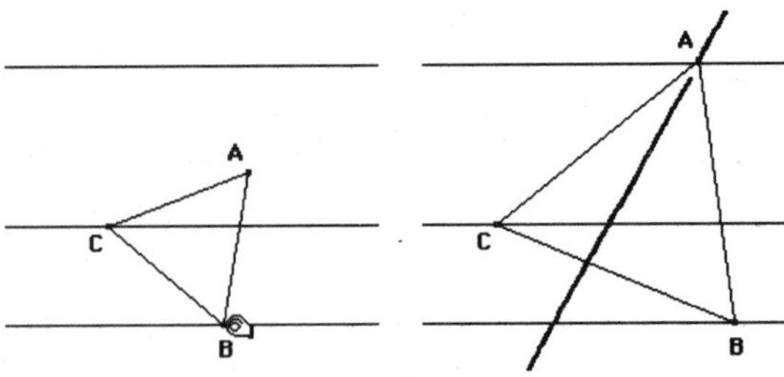

FIGURE 4

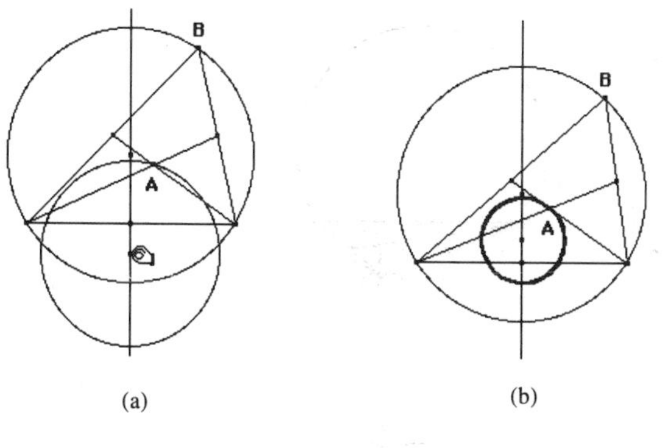

FIGURE 5

parallel lines, the locus of A makes the same angle with these lines, namely 60°.

Example 3 [A harder example]. We return to our introductory locus in Figure 2. There, the centroid A of a triangle traced out a circle when a vertex B of the triangle was moved around the circumcircle of the triangle. Where does the center of this centroid-circle lie? What is its radius?

Solution. We recognise that the centroid-circle is symmetrical to the perpendicular bisector of the fixed side of the triangle. We construct a circle through the centroid A with moveable center lying on this perpendicular bisector (Figure 5(a)). We keep moving the center until this circle covers the centroid-circle (Figure 5(b)). By measuring we find that the radius of the centroid-circle equals one-third the radius of the circumcircle and that its center is one-third the distance along the perpendicular line from the fixed side of the triangle to the circumcenter of the triangle.

Experimental Checking of Construction Results

Suppose we have constructed a figure using line segment, line, or circle as basic objects, as is often the case. If the same figure can also be produced point by point as a locus, an opportunity to check the result arises: Does the object constructed as a whole correspond with its locus? We give two examples for such a test, useful for checking whether mistakes have been made in a construction. This testing can give information about further properties of a constructed object and can help to reinforce geometric knowledge. From a logical viewpoint, such a test does not, of course, suffice to prove the correctness of a construction.

Example 4. Two straight lines meet at a vertex to produce an angle. The angle bisector is constructed (Figure 6(a)). The locus facility of the software can then be used to check that the angle bisector has the following properties:

- It is an axis of reflective symmetry for the two-line configuration; i.e., a reflection in it reproduces one of the given lines on top of the other.
- It consists of the midpoints of all pairs of points (one on each of the two lines) which are equidistant from the vertex (Figure 6(b)).
- Its points have the same perpendicular distance from the two straight lines (Figure 6c).

Example 5 [A harder example]. We wish to know: what is the image of a small circle with center M and radius MN under circle inversion in the large circle with center O that contains it. (Figure 7(a)). As a first attempt we find the image M' of the center M and the image P' of a variable point P on the small circle and infer that the circle's image is another circle having center M' and passing through P'. Using the **Locus** command to create the image circle, we see that our assumption was wrong! Under inversion, the image of the center is not the center of the image (Figure 7(b)).

Investigating Image-Figures as Loci

In the following examples, the images of circles are generated as loci. In each, a single basic point P on the given

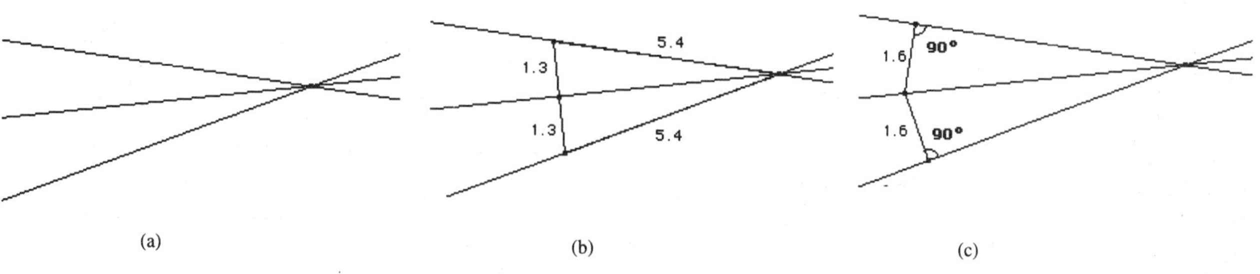

FIGURE 6

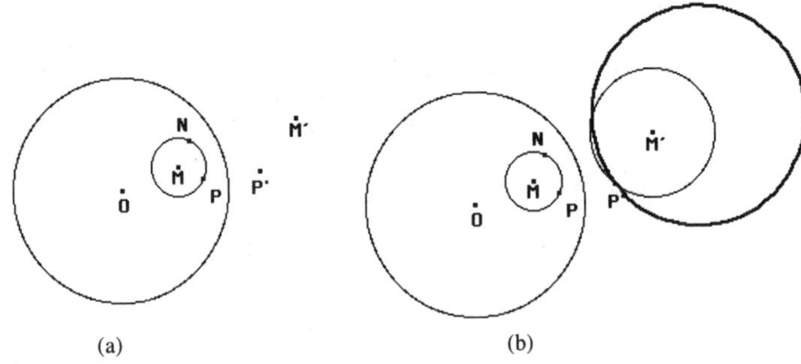

FIGURE 7

circle is transformed to image point P' according to some rule, and then, as P is dragged around the circle, the locus of its images is produced.

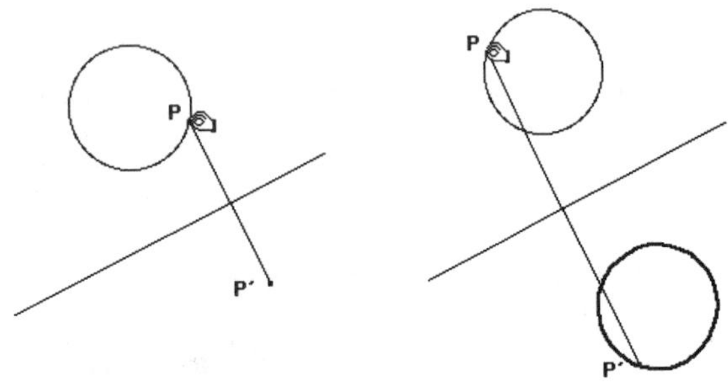

FIGURE 8

Example 6. In Figure 8, the circle image is produced by reflecting P in a line.

Example 7. In Figure 9, a "screen pantograph" is used for enlarging a circle. The point Z is the center of the enlargement.

Example 8. In Figure 10, an ellipse is produced as the image of a circle under an axis-affinity transformation. In the figure, g is the fixed axis, and P maps onto P' by a one-way stretch with stretch factor -2.

Example 9. Figure 11 shows the images of a circle under a central projection with center Z, vanishing line v, and axis s. The nature of the circle's image depends upon whether the circle misses, touches, or intersects the vanishing line v, producing, respectively, an ellipse (Figure 11(a)), a parabola (Figure 11b), or hyperbola (Figure 11(c)).

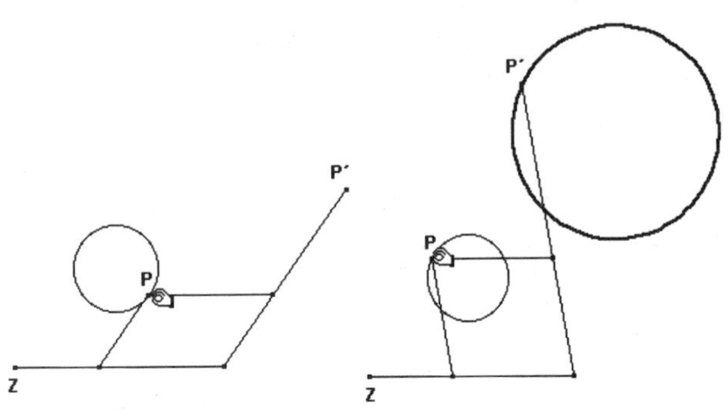

FIGURE 9

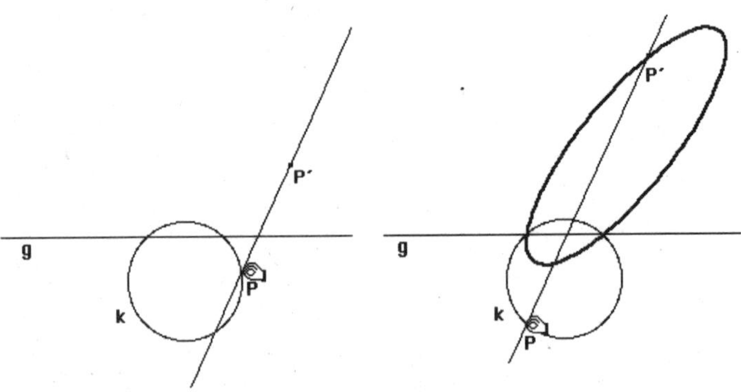

FIGURE 10

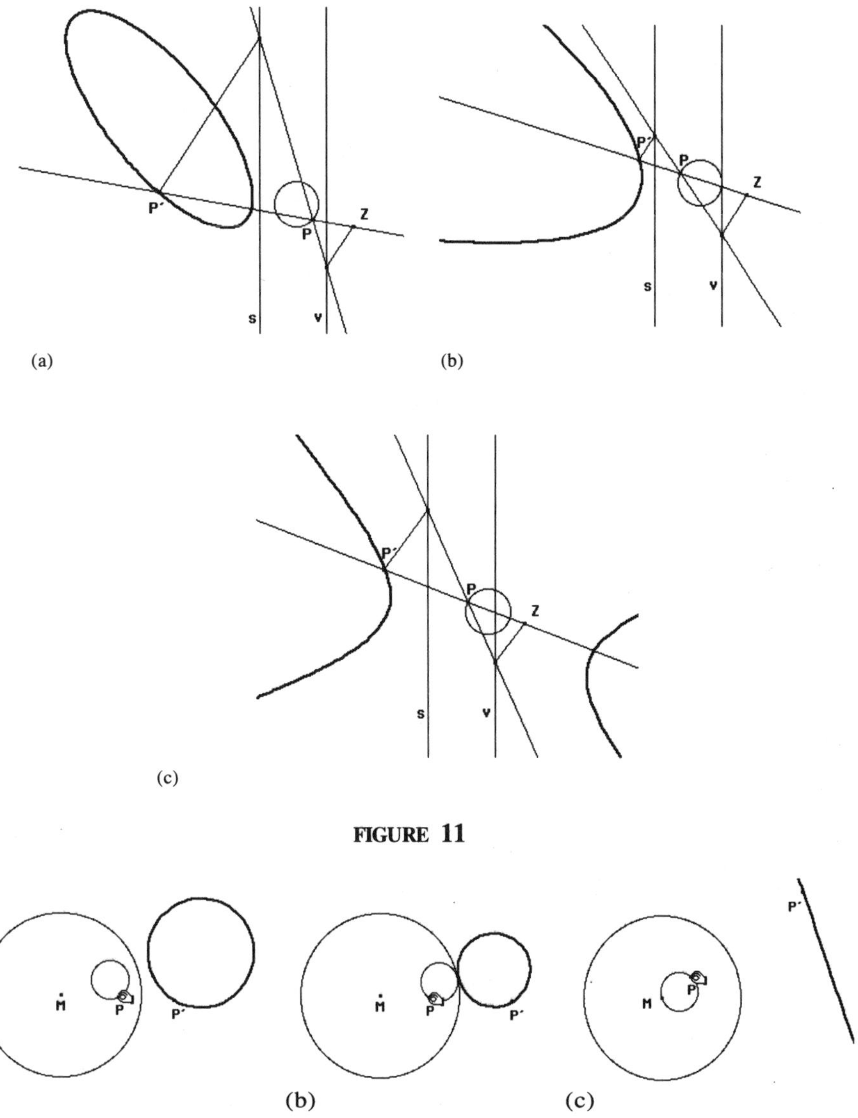

FIGURE 11

FIGURE 12

Example 10. Figure 12 shows images of circles under an inversion. (We limit ourselves here to circles which lie within the inversion circle.) If the circle is tangent to the inversion circle, then so is the image circle, at the same point (Figure 12(b)). If the circle passes through the center of the inversion circle, its image is a straight line, i.e., a circle of infinite radius (Figure 12(c)). With this technique, a circular movement can be converted into a straight-line movement: this produces the construction of Peaucellier's Cell. In Figure 13(a) dotted lines indicate construction lines that are normally hidden, as in Figure 13(b). We can test our screen Peaucellier Cell, and see that we do obtain a straight line as the locus of P' as P moves round the circle (Figure 13(c)).

Generation of Conic Sections

In Examples 8 and 9, conic sections were produced as images of circles under certain transformations. Here are some additional common constructions of conic sections.

Example 11. The perpendicular axis construction of an ellipse: through the basic point B on a circle, construct a perpendicular to the diameter of the circle. Let A be the midpoint of the segment of the perpendicular that joins B to the diameter (Figure 14(a)). As B is dragged around the circle, A traces out an ellipse (Figure 14(b)).

Example 12. The vertex-circle construction of an ellipse: here two concentric circles are given, with basic point B on

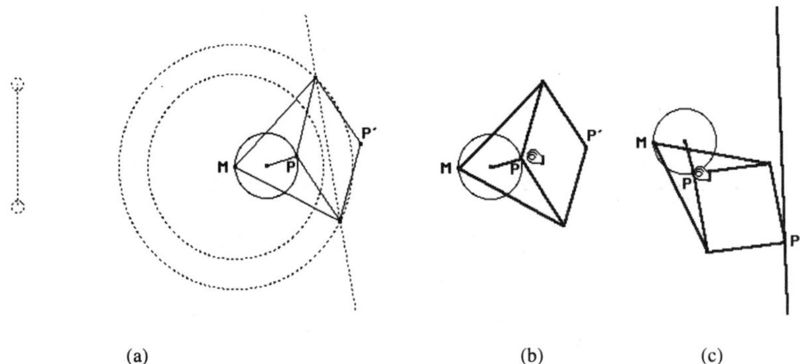

(a) (b) (c)

FIGURE 13

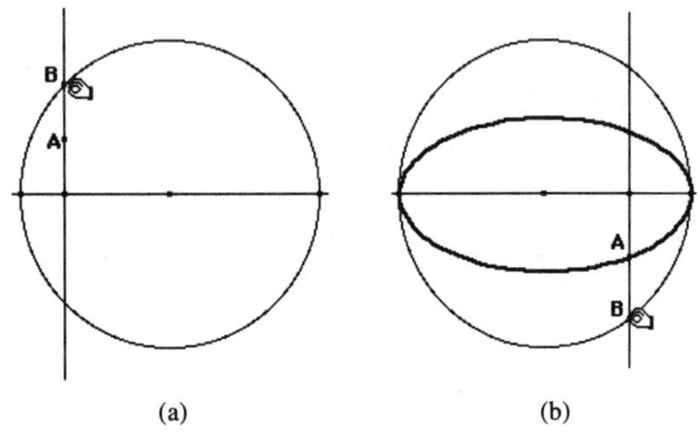

(a) (b)

FIGURE 14

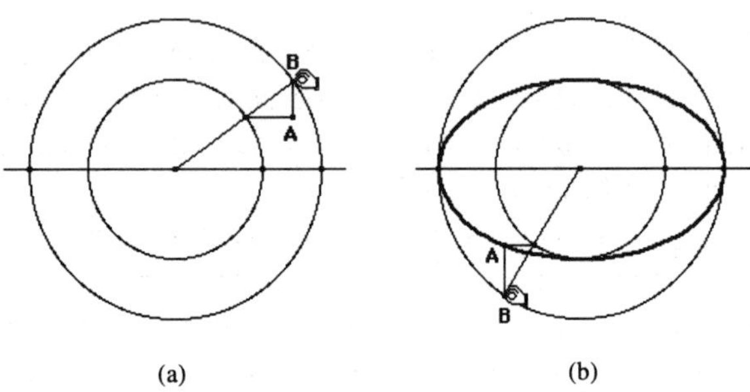

(a) (b)

FIGURE 15

the outer circle. The radius from the center to B is drawn, and a right triangle is constructed as shown in Figure 15(a). The locus of A (at the right angle of the triangle) as B travels around the circle, is an ellipse (Figure 15(b)).

Example 13. The tangent construction of conic sections: in each of these constructions, the foci of the conic are F_1 and F_2 (or F).

For the ellipse, the guide-circle is centered about the focus F_2 and has the length of the major axis of the ellipse as its diameter. Basic point B is on the guide-circle, and the perpendicular bisector of BF_1 is tangent to the ellipse at A, the point of intersection of the perpendicular bisector and BF_2 (Figure 16(a)). As B travels around the circle, A traces out the ellipse (Figure 16(b)). The tangent construction for the hyperbola is analogous (Figure 17).

In the tangent construction of the parabola the guide-circle is a "circle" with "infinitely large" radius, i.e., a straight line. The perpendicular bisector of FB is the tangent through A to the parabola, and the distance from each point on the parabola to the guide-line l equals its distance to the focus F (Figure 18).

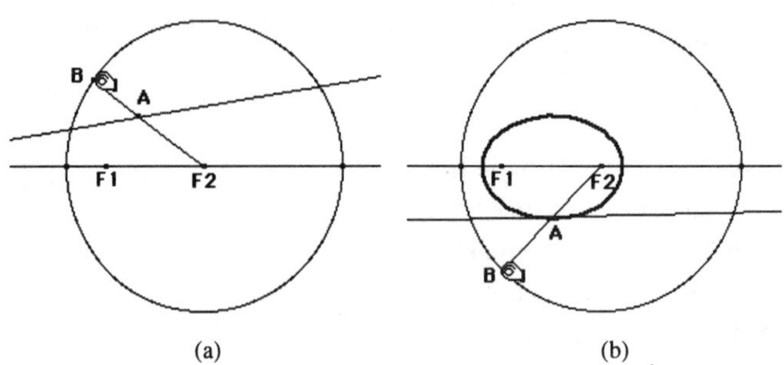

(a) (b)

FIGURE 16

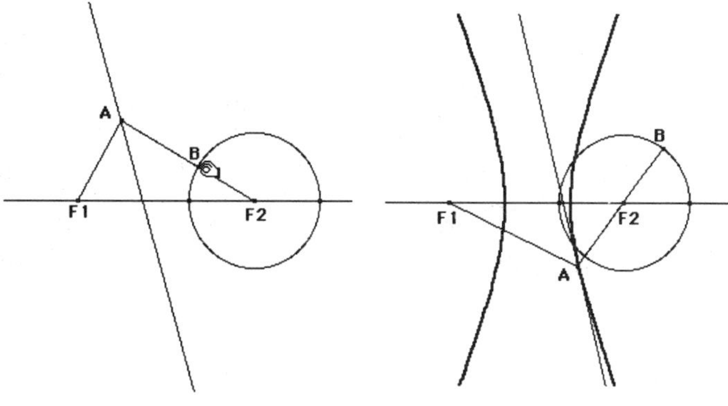

FIGURE 17

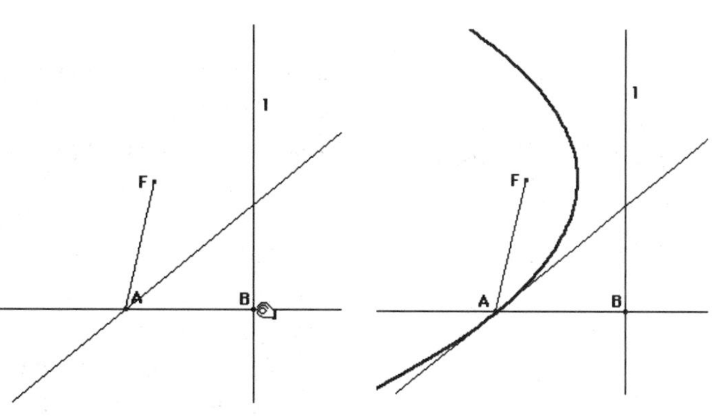

FIGURE 18

Generation of Algebraic Curves as Loci

Cabri-géomètre is particularly suitable for the cissoidal generation of algebraic curves.

General definition of a cissoid. Two curves k_1 and k_2 and a point P (the center of rotation) are given. A straight line l through P intersects k_1 at B and k_2 at G. As the line l rotates about P, the intersection points B and G vary. The length of the variable line segment BG is copied on l on both sides of P to produce PA_1 and PA_2 (see Figure 19). The locus described by A_1 (or A_2) as B moves along k_1 is a curve called a cissoid. In our next example, we consider an historically important case of the cissoid.

Example 14. We take a circle as curve k_1 and a straight line as curve k_2. Point P lies on k_1, and the perpendicular from P to k_2 passes through M, the center of the circle (Figure 20(a)). PA is constructed here in the same direction as BG. The position of the circle k_1 relative to the line k_2 affects the shape of the curve generated by A. Figure 20 illustrates several different cissoids generated in this manner. We begin with the circle to the right of the line and then move the circle to the left. Figure 20(c) (in which the line is tangent to the circle) shows the cissoid of Diocles, which was used in antiquity for doubling the cube—it is the "oldest" algebraic curve other than those of order 2.

Figure 21 shows some other special cases. In Figure 21(a), the line k_2 passes through the center of circle k_1; the curve generated by A is Newton's strophoid. In Figure 21(b), the line k_2 passes through the midpoint of the radius of circle k_1; the curve generated by A is Maclaurin's trisectrix, which was used for angle trisection.

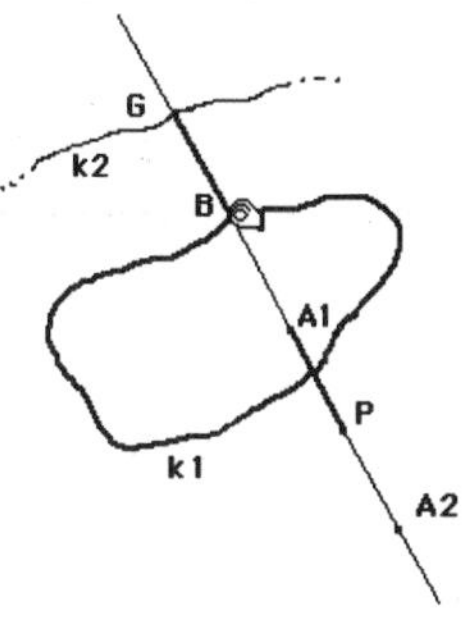

FIGURE 19

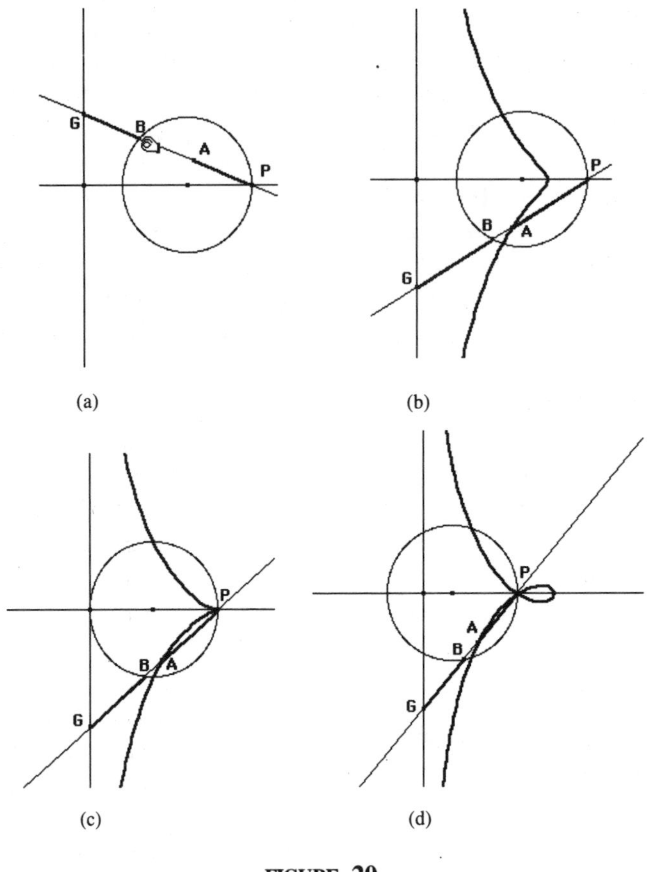

FIGURE 20

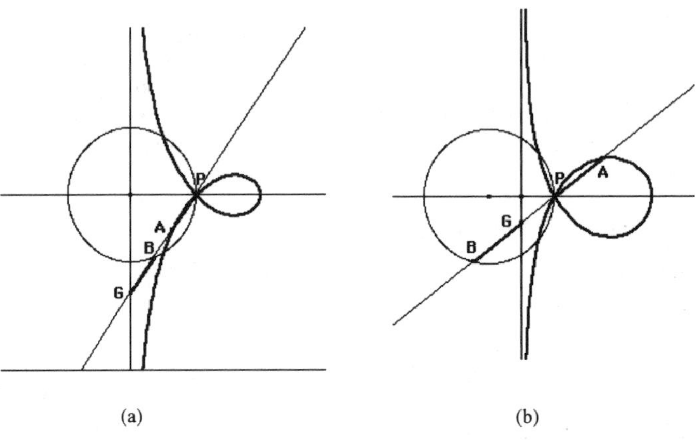

FIGURE 21

Other variations in the position P, the center of rotation, can produce an algebraic curve that has two branches, which has k_2 as asymptote. If we take both k_1 and k_2 as circles, then other curves (such as Pascal's limaçon) can be produced.

Concluding Remarks

In this presentation we have restricted ourselves to loci that are traced by a single point dependent on a another point which is moved. It is possible, however, to produce loci of several dependent points *at the same time*; thus, loci of loci can be produced. In addition, the basic (independent) point can draw its own trail as it is moved; thus, freehand lines and their images can be generated.

In cases where curves cannot be produced by direct compass and straightedge construction, as with algebraic curves for instance, we have shown how such curves can be produced as loci. This same technique can be employed in exploratory situations in which a dependent point in some construction is chosen and its locus is generated by moving some other basic point. One then asks the question: *Is an interpretable locus produced?* In some examples, the answer will be *Yes*!, and in others, the answer is ambiguous, or a definite *No*! In all the examples we have given, the locus was always a definite curve, clearly related to the guide-path of a moving independent point. When the guide-path is not specified, but an independent point is freely dragged, then an uninterpretable locus is likely to occur. It is a good exercise to explore what happens to the locus of a dependent point when different guide-paths are specified for the independent point that is dragged.

At present, with *Cabri-géomètre I* (and other dynamic geometry software) it is not possible to generate a referable locus with the software. The locus is only a state of the screen, and the system has no knowledge of its nature—it is merely composed of individual pixel points with no internally recorded structure. It would be desirable to process a locus in one of the following ways:

— Numeric solution: Curve fitting to (some of) the locus points. (By linear interpolation *Cabri-géomètre II* allows now the automatic drawing of referable loci.)
— Algebraic solution: Derivation of an exact analytic geometric representation of the locus, using an expert system.

We also note that not all loci can be produced with *Cabri-géomètre* and other dynamic geometry software. The reasons for the restriction on possible loci lie in the types of transformations of geometric configurations that are possible in drag-mode. There is a real danger in such restrictions: we can cover only those constructions which are feasible with the given tool—an instance of "tool-conditioned selectionism."

The interactive generation of loci should not entirely replace pencil-and-paper (or hands-on) constructions of loci by students. With the interactive generation of a locus using software, only one locus point actually needs to be constructed, and the rest of the locus-points are generated as the independent point is moved. In the traditional ruler-and-compasses method, the same construction must be repeated point by point in order to have sufficient points to connect by freehand interpolation to produce a continuous curve. Computer-supported generation of loci certainly has advantages: it is time-saving, and it has modification possibilities. It makes exploring considerably easier. But the effects of practice by hand and the internalization of the method of construction of loci are partly lost. To draw loci using *Cabri-géomètre* and to do so using conventional tools on paper involve quite different experiences and skills. It is wrong to discredit the traditional approach—both it and computer-based methods are valuable. The choice of tool depends on the objectives.

Bibliography

Schumann, H., and Green, D. (1994). *Discovering Geometry with a Computer–using Cabri-géomètre*, Bromley, Kent, England: Chartwell-Bratt.

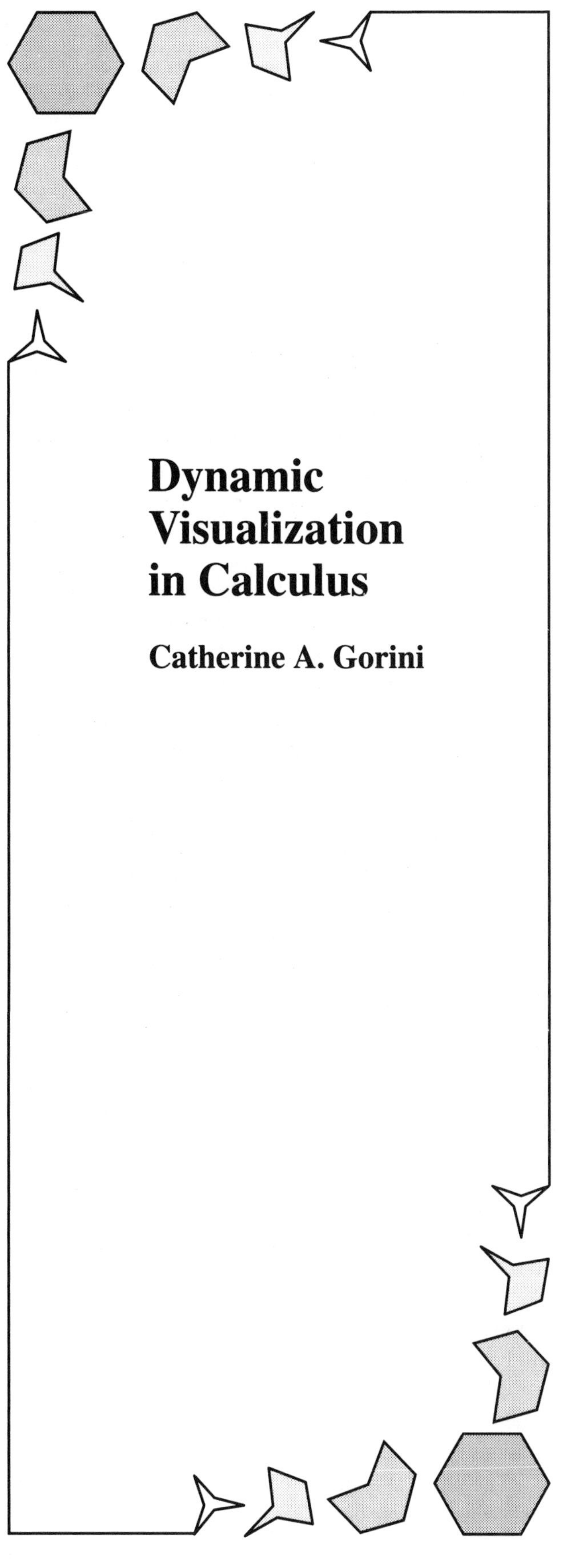

Dynamic Visualization in Calculus

Catherine A. Gorini

As the study and analysis of continuous change, calculus is one of the most important and widely applied branches of mathematics. But, because of the ephemeral nature of change, calculus can also be very challenging for students, who must learn to feel comfortable with using calculus to model the kinds of changes that come up, particularly in science and engineering. For this, students need to develop an intuitive understanding of the nature of change. An obvious pedagogic challenge is to *help* students understand change. To that end, this paper will deal with one approach to this objective and will describe how *Geometer's Sketchpad* can be used to help students with maximization/minimization and related rates word problems.

Solving word problems in calculus depends on taking a verbal description of a physical situation, translating it into a geometric model, deriving algebraic equations relating the quantities of interest, and then applying the appropriate techniques of calculus to get an answer. In the process of setting up the geometric model, students should be thinking about the problem in an intuitive or conceptual way and should be developing some estimate of the correct answer to the problem. The value of dynamic geometry for students in calculus is in giving them the tools to create a good geometric model for the problem that can lead them to an intuitive understanding of the problem and good estimates of the solution.

Maximization and Minimization Problems

Maximization and minimization problems include some of the most significant applications in calculus. For example, determining the path a beam of light will take in certain situations is a minimization problem because of Fermat's principle which says that light takes the path requiring the least amount of time.

In a maximization and minimization problem in first-year calculus, there is some independent quantity that can take on a continuum of values as well as some other quantity, given by an objective function, that we are particularly interested in and that depends on this independent quantity. The goal of the problem is to find the value of the independent quantity that maximizes or minimizes the dependent quantity or objective function. The technique of calculus is, of course, to find the local maximum or minimum of an appropriate function using derivatives, and students are generally very good at finding the points where the derivative of a function is zero. When the independent quantity is restricted to a certain closed interval, as it is in many applied problems, the answer may be any one of the points located by setting the derivative equal to zero, or it may be at an extreme value of the independent quantity.

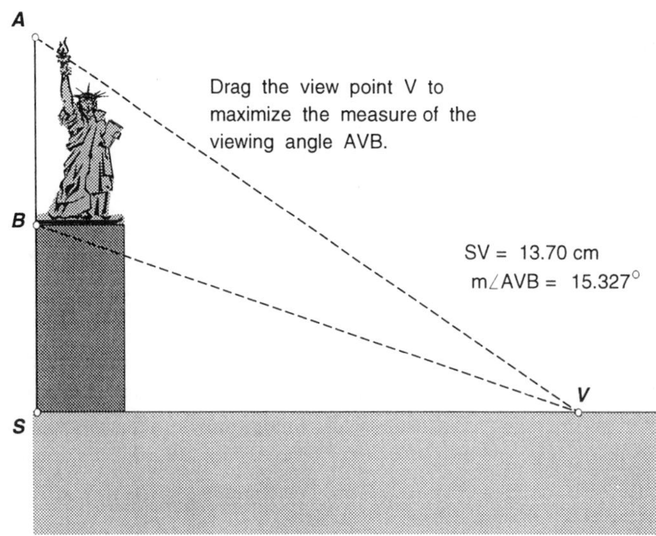

FIGURE 1

For most calculus students, the challenge in maximization and minimization problems is to translate the wording of a problem into a function. The easiest way to do this is to begin with a geometric model or diagram of the situation described in the problem; an appropriate objective function can then be derived from relationships inherent in the diagram. It is important that students also understand why a maximum or minimum value of the dependent variable can be located at an extreme value of the dependent variable and why they must check all possible values when several potential answers can be located.

Let us consider a maximization problem from the Calculus Consortium based at Harvard [1] to see the value of this approach. We are looking at the Statue of Liberty, which is 46 meters high and stands on a pedestal and base which together are 47 meters high. We want to get the best view, which will be when our viewing angle is maximum. How far away from the base should we be?

In order to understand the viewing angle and its relationship to the dimensions of the statue, we must draw a picture. This picture can, of course, represent only one of a continuum of possible locations for the viewer. To understand how the viewing angle changes as the distance from the statue changes, we can use dynamic geometry. The sketch shown in Figure 1, which includes clip art of the Statue of Liberty, shows very clearly how the viewing angle changes with distance. As we drag the position of the viewer along the surface of the water, we are able to conceptualize all possible positions and viewing angles. It is easy to see that very close to the statue and far away from the statue, the viewing angle gets smaller. The best viewing angle must be at some medium distance. To find the exact value of this distance, we must find an equation relating the viewing angle to the distance from the statue and then use the techniques of calculus. Students may also wish to graph the equation they get using a graphing program to see that the maximum viewing angle does indeed occur when the derivative of the viewing angle function is zero. Alternatively, students can use the graphing feature of *Sketchpad* to get a graph of the viewing angle as a function of distance directly from their sketch (see Figure 2). This graph depends on the dimensions used in the sketch and will give an intuitive feeling for the solution rather than an exact answer.

Using these three different approaches—sketches, graphs, and derivatives—can really help students grasp what is going on in the problem and how all the parts of the problem—viewer, statue, distance, and viewing angle—are connected to each other.

The derivation of Snell's law from Fermat's principle of least time is a good example of the power of calculus as applied to physics. It is a common experience to see light bend as it goes from one medium (air, water, glass, and so on) to another. One of the earliest laws about the behavior of light, Fermat's principle of least time, says that light will take the path that minimizes the total time traveled. Thus, in a medium where the speed of light is constant, its path will be a straight line. When light goes from one medium to another, it will bend to minimize time. Determining the angle of bending based on the velocities of light in the two mediums is a straightforward minimization problem.

The physical situation of this problem can be nicely modeled with *Geometer's Sketchpad.* A ray of light must go

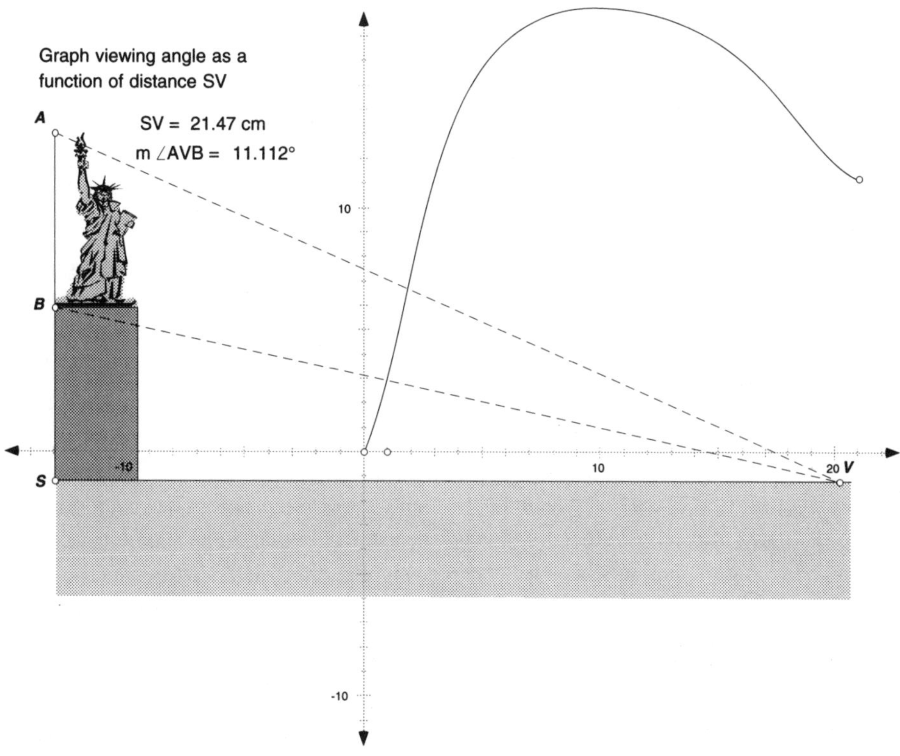
Graph viewing angle as a function of distance SV
A
SV = 21.47 cm
m ∠AVB = 11.112°
10
B
-10
S
10
20 V
-10

FIGURE 2

from a point A in air to a point B in water (see Figure 3). The light will travel in a straight line in the air or water, but will bend when going from the air into the water. Our sketch of this will use 3 cm/sec for the speed of light in air and 2 cm/sec for the speed of light in water in order to simplify the calculations and focus on understanding the principles involved. (The speed of light in air is about 3×10^8 m/sec, and in water it is about 2.2×10^8 m/sec, so the ratio is about the same and large numbers are avoided.) The calculator in *Sketchpad* computes the total time the light takes in going from fixed point A to fixed point B through variable point V on the surface of the water, as well as the sines of the angle of incidence and the angle of refraction. Moving point V to get a minimum time convinces us that there is a point

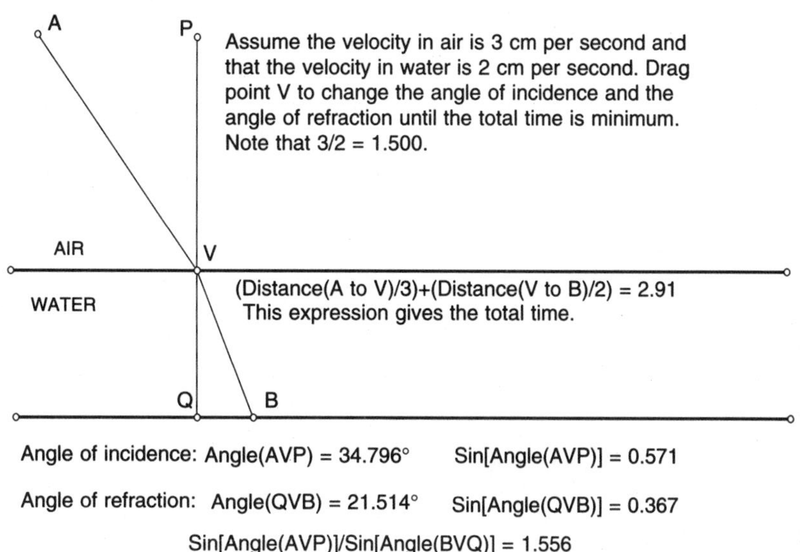
A
P
Assume the velocity in air is 3 cm per second and that the velocity in water is 2 cm per second. Drag point V to change the angle of incidence and the angle of refraction until the total time is minimum. Note that 3/2 = 1.500.
AIR
V
WATER
(Distance(A to V)/3)+(Distance(V to B)/2) = 2.91
This expression gives the total time.
Q
B
Angle of incidence: Angle(AVP) = 34.796°
Sin[Angle(AVP)] = 0.571
Angle of refraction: Angle(QVB) = 21.514°
Sin[Angle(QVB)] = 0.367
Sin[Angle(AVP)]/Sin[Angle(BVQ)] = 1.556

FIGURE 3

where the total time is a minimum. By comparing the ratio of the sines and the ratio of the velocities of light in the two media, students can conjecture Snell's law, which says that the ratio of the sines of the angles of the paths with respect to a vertical line will be equal to the ratio of the velocities of light in the different mediums. They can then verify Snell's law using the techniques of calculus.

Related-Rates Problems

Related-rates problems deal with quantities that are related to each other in some physical way and that are both changing with respect to an independent variable, usually time. Typically, the rate of change of one of the quantities is given, and the rate of change of another quantity is asked for. In general, these problems seem to be more difficult for students than maximization/minimization problems, perhaps because related rates generally are more complicated and students must conceptualize a changing physical system. For these problems, the animation feature of *Sketchpad* is ideal.

To introduce related-rates problems in the classroom, one may wish to use a standard related-rates problem such as a man pulling up a weight using a pulley. The height of the pulley, the length of the rope, and the speed of the movement of the man are all given, and we want to find the speed of the weight when it is a certain height above the ground. An animated sketch (see Figure 4) shows students the situation described in the problem in a very realistic way. They can watch it over and over as often as they like, and they can stop the man anywhere along his path. By carefully observing the sketch, they can see that, because the length of the rope stays constant, the rate of change of the height of the weight and rate of change of the length of the diagonal piece of rope are equal. By comparing the distance travelled by the man and the distance travelled by the weight, it is clear that the man must be moving more quickly than the weight.

This problem can also be discussed in terms of the value of a machine, in this case, the pulley. The usefulness of a pulley in this example is two-fold. It allows the force the man is exerting to be applied in a different direction on the weight, and it allows a heavy weight to be raised with less force than if the direction of the force were vertical. To compensate for the lesser amount of force the man is using, he must pull through a longer distance in the same amount of time. This agrees with our earlier observation that the man moves more quickly than the weight. Once this intuitive analysis of the problem has been done, it is a simple matter to use the Pythagorean theorem to relate the height of the weight and the distance travelled by the man. This equation can be differentiated with respect to time and solved for the speed of the moving weight.

Another example of a related-rates problem is pulling a boat into a dock (see Figure 5; this animated sketch is included with version 3 of *Geometer's Sketchpad*). We want to know the velocity at which the boat is moving if we know the constant rate at which the rope is being pulled in. As the sketch is animated, the boat is pulled into the dock, and it is clear that the rope must have a greater velocity than the boat. Furthermore, as we see all possible configurations of the problem, it is obvious that the length of the rope that is out, the distance of the boat from the dock, and the height of the dock are again all related by the Pythagorean theorem.

In other related-rates problems, analysis of an animated sketch helps in understanding how the rates of change are

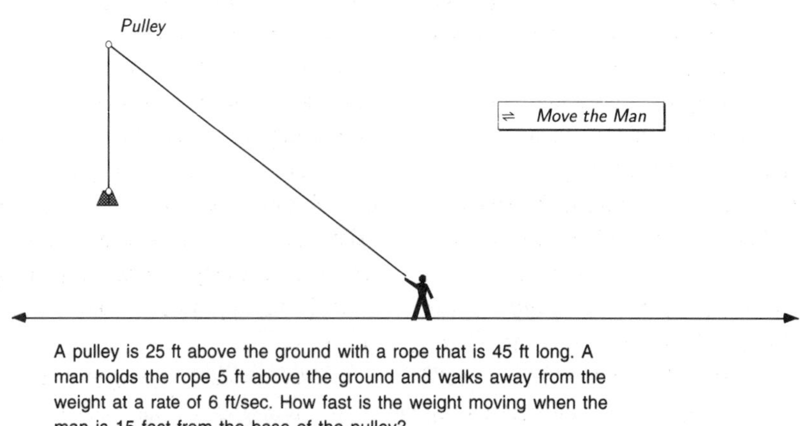

Pulley
Move the Man
A pulley is 25 ft above the ground with a rope that is 45 ft long. A man holds the rope 5 ft above the ground and walks away from the weight at a rate of 6 ft/sec. How fast is the weight moving when the man is 15 feet from the base of the pulley?

FIGURE 4

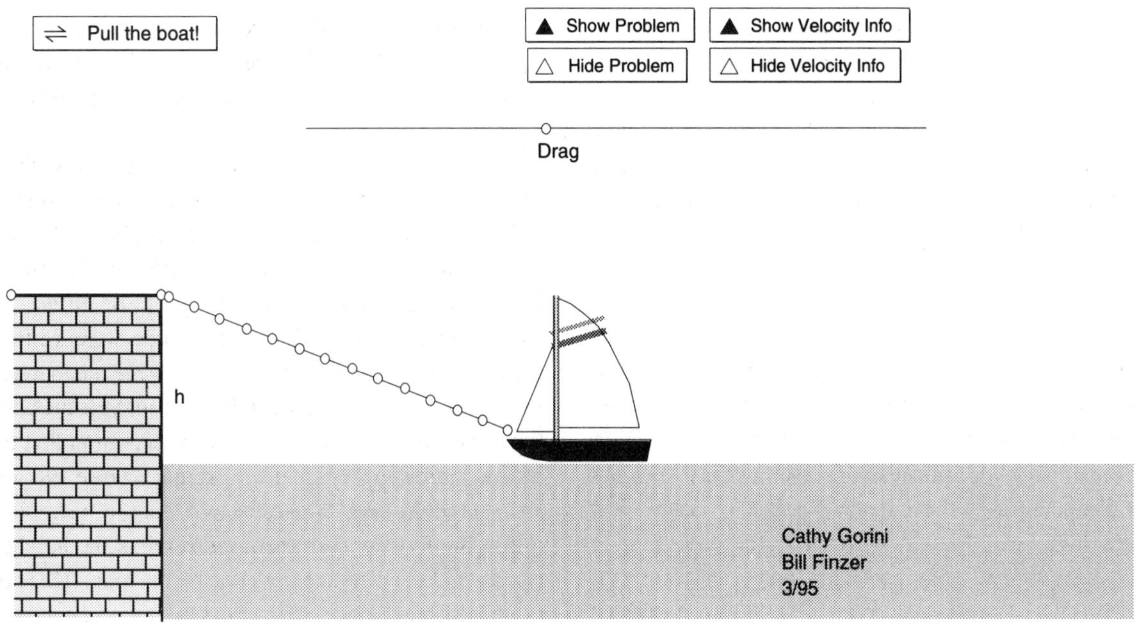

FIGURE 5

related to each other and in estimating the answer. After that, it is simply a matter of expressing that relationship of the quantities in the problem as an equation, taking its derivative with respect to time, inserting the known quantities, and solving for the problem's unknown.

Teaching Ideas

In introducing a new type of problem, I demonstrate a few problems with sketches I have already done, analyze them as thoroughly as possible with the students, get an estimate for the answer, and then solve the problems using calculus. After that, students solve some problems for which sketches are already prepared, using the diagram to produce an algebraic equation themselves. Finally, students can solve problems on their own, using the computer as they wish.

I find that this format allows me to ask additional questions along with standard textbook problems, such as: "Is the boat's speed greater than the car's speed? Why or why not?" I then require students to hand in a diagram with their work if they want to receive full credit. I sometimes assign problems without numbers and require students to describe what is happening purely in terms of "faster than" and "slower than." I often ask "What if . . .?" questions, such as "Could the man's speed ever be the same as the weight's speed?" "What if Alaina can speed up going through the park and attain a speed faster than going on the sidewalk?" "What does it mean if the independent variable takes values outside of this region?" In addition, I provide students the opportunity to create their own "What if?" questions.

Some Helpful Technical Points

Students can learn to use *Sketchpad* very quickly and enjoy making and demonstrating their own sketches. I have a separate file with clip art for students to use in making their sketches more elaborate. These pictures include people, a boat, a car, and, of course, a lighthouse.

There are many features of *Sketchpad* which are particularly useful for sketching calculus problems. The built-in calculator can be used to compute areas, volumes, times, or any other quantities that depend on the independent variable. The animation feature is essential for related-rates problems. A separate line or circle with a point on it is a good way to get the kind of motion you want. In *Geometer's Sketchpad,* an animated point moves the same number of pixels per frame, so animation on a fast computer is realistic. The graphing capabilities of version 3 allow students to create a graph relating any two of the changing quantities in their sketch. This can be a very valuable way for students to learn to connect changing physical situations to graphs in the cartesian plane.

I have found that to get a good, reliable sketch doesn't really take a lot of time, but I have had to start over two or three times before I get the correct relationships between constructed objects so that movement of a particular object gives the desired effect on other objects. Color can be an effective way to distinguish regions of the sketch. To color a half-plane, just make a large rectangle that extends beyond the viewing window and color it; from the window, it will look as if the half-plane is colored.

Not all maximization/minimization or related-rates problems are suitable for sketching with *Sketchpad*. For example, *Sketchpad* will not be able to illustrate the special features of a problem about the volume or surface area of a sphere. *Sketchpad* can be used in such problems as a "dynamic calculator," which can be interesting for some students, but the geometry of the problem will not be illustrated. The problems that work the best are those with a two-dimensional layout and quantities or rates that are easily seen or computed from the diagram.

Conclusion

One of the most important ideas for students to learn in solving applied problems is that they must *draw sketches, pictures, or diagrams* and *construct models,* so that they can understand what is going on in the problem. Before they solve the problem, they should be able to learn from their diagrams and models what kind of answer they will get—large, intermediate, or small. Dynamic geometry is an extremely valuable tool because it makes the process easy, systematic, and lots of fun.

Reference

1. Hughes-Hallett, Deborah, and Andrew M. Gleason et al., *Calculus*, New York, Wiley & Sons, 1994.

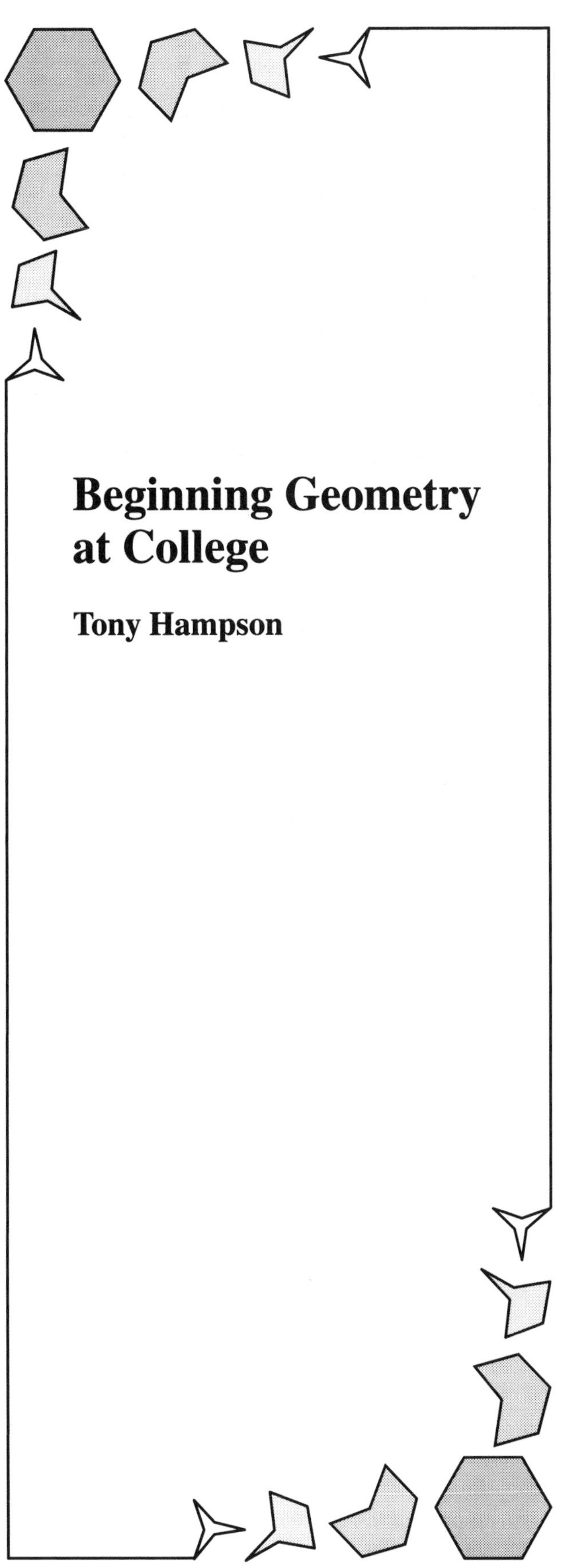

Beginning Geometry at College

Tony Hampson

At the 1994 Undergraduate Mathematics Teaching Conference, held at Nottingham University, England, one of the themes was geometry. Participants were invited to share their thoughts on the place of geometry in the undergraduate curriculum, and a working group was set up to report on this issue. In his brief to this working group, Keith Austin of the University of Sheffield summarised the situation in the United Kingdom as follows:

> The place of classical Euclidean geometry in the school curriculum came under severe threat in the 1960s and the subject had all but disappeared by 1980. Since then there have been calls for the reintroduction of this and other geometries as a significant component of the first year undergraduate course. Among the reasons for this have been aesthetic considerations, that geometry is in itself good mathematics, and the fact that geometrical language and a geometrical way of thinking pervade so much of modern mathematics. There is also an increasing need for undergraduates to cultivate an appreciation of the interaction between intuition and rigour. Geometry is seen to be an especially fertile area in which they can do this. [1]

Implicit in this statement is a recognition that many British students have limited experience of geometry before coming to College and may benefit from an opportunity to begin the study of geometry as part of their undergraduate programme. This was certainly the view expressed by the working group in its report.

At Cheltenham and Gloucester College our experience certainly supports this national picture, and most of the students embarking on our degree programmes in Mathematics could be described as beginners in geometry. Their school experience consists predominantly of algebraic formulae and techniques, and they have little sense of any mathematics beyond this.

In order to begin to remedy this situation we devised a course entitled Geometry Workshop. This was offered to first year students within the Cheltenham and Gloucester College Undergraduate Modular Scheme from 1990 to 1995. Since then mathematics has been deleted from our Cheltenham programme, but the materials produced and approach described have been incorporated in a module entitled Approaches to Geometry which is offered within our overseas programme.

The Geometry Workshop module ran for twelve teaching weeks with two hours of class contact each week. It was assessed by two coursework assignments containing a mixture of routine exercises and investigative work.

The students taking the course had a variety of backgrounds, maths being a major or minor subject in modular B.Sc. and B.Ed. Primary and Secondary Teacher Education programmes. Their knowledge of geometry was generally very limited, often consisting of isolated facts about triangles, parallelograms, or circles. Some may have worked with coordinates, vectors, or plane transformations but in such a way as to bring out the algebra rather than develop any geometrical insight. Few had any experience of geometry as an area of exploration and investigation.

Aims of the Geometry Workshop Course

In the light of the above, the aims of the course were:

To consolidate understanding of basic plane geometry through practical work and experiment.

To improve geometrical awareness, knowledge, and confidence.

To develop an appreciation of and an ability to engage in the process of investigation in a geometrical context.

When I was introduced to *The Geometer's Sketchpad* in 1992, it struck me immediately that this could be an ideal medium through which to teach Geometry Workshop. However, this introduced a new problem in that many students were as unfamiliar with computers as they were with geometry so that the first two weeks of the course had to be given over to using the computer and playing with *Sketchpad*. In this time the students were expected to work through the four Introductory Tours in the *Sketchpad* Guide.

That left only ten weeks for geometry and investigation. This was hardly enough time for serious in-depth study, but it gave students an opportunity to engage in a variety of activities and through them to begin to appreciate the place of geometry in Mathematics.

Structure and Content

In order to impose some structure on the course, I decided to trace the historical development of the subject and relate this to *Sketchpad* via the sequential introduction of the commands Construct, Trace Locus, and Transform. We looked first at some concerns and contributions of the Greek geometers such as the construction of various figures, incommensurable lengths, the solution of "equations", conic sections, axioms, and proof. Next, we discussed some developments during the Renaissance period including techniques for perspective drawing. This led on to projection and projective invariants which we then used to prove some basic results in Projective Geometry. Finally we explored the invariance properties of some standard plane transformations and investigated symmetry patterns and composition products.

I had some difficulty in finding suitable published material for this course. While all of the topics are quite standard and are covered in the literature, my approach was not. I did not want my students to be following a book in which all the results were cut and dried and presented in a neat logical framework. I wanted them to explore for themselves, make conjectures, and seek explanations. In the end I prepared a booklet for the course containing background notes and examples to support the work done in class.

Sketchpad (Version 2.0 for Macintosh) was used throughout the programme. I shall now go on to describe some of the activities used at various stages of the course.

Properties of figures

In the course booklet, properties of figures were presented as concise propositions which students were invited to investigate using *Sketchpad*. For example, the properties of isosceles triangles were framed in two propositions (a) and (b) along with instructions for investigation as follows.

Proposition (a). For the triangle *ABC* shown in the diagram, the following statements are equivalent:

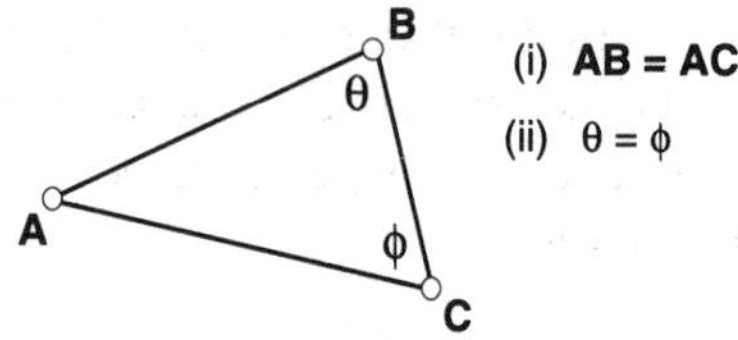

1 Devise a way of constructing the given figure so that $AB = AC$ no matter how the figure is changed by dragging.
Verify that the angles at *B* and *C* are equal and remain equal as the figure is dragged.

2 Now construct the figure using only the condition that $\theta = \phi$. Make sure that these angles remain equal as the figure is changed by dragging.
Verify that the lengths AB and AC are equal and remain equal as the figure is dragged.

Proposition (b). In the triangle DEF shown in the diagram, the point *M* lies on the segment DF. If any two of the following statements are true, so are the other two.

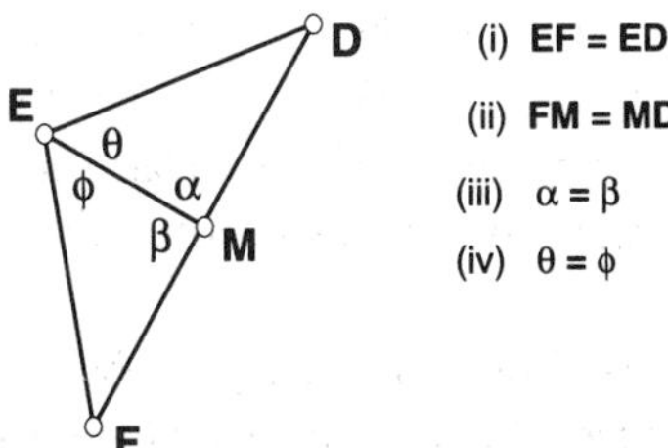

3 Choose any two of the properties (i) to (iv) and construct the figure in such a way that these two properties hold. Verify that the other two properties also hold and continue to hold as the figure is changed by dragging points or lines.

In exercises such as these, students began to use *Sketchpad* creatively to attain a limited geometrical objective. In devising their constructions, they had to make decisions based on their geometrical knowledge and intuition. They were encouraged to share ideas and compare approaches.

Sketchpad demonstrations are very convincing, and most students firmly believed that the results they demonstrated were true. This led naturally to a consideration of proof. They often found it very difficult to switch off the computer and look instead for explanations and relationships expressed in logical argument. Euclid's axiomatic approach was introduced and some examples of proofs worked through.

Many of Euclid's propositions involve the presentation of constructions followed by proofs. This treatment models quite closely the position that the students found themselves in. The aim here, however, was not to teach Euclid in any systematic way but to encourage an appreciation of the role of proof in mathematics.

Geometric algebra

Incommensurability was discussed and methods devised for constructing sides of squares with integer areas. Students then attempted some exercises in geometric arithmetic involving irrational numbers. They were encouraged to seek "elegant" construction methods, which they usually interpreted to mean "use the least number of steps possible". In some exercises strict Euclidean construction methods were required. This is equivalent to restricting the use of the *Sketchpad* Construct menu to **linear objects through a pair of points, circle by centre and point** and **point at intersection.**

Example 1. On a line l fix two points A and B. Take the distance AB as your unit of length.

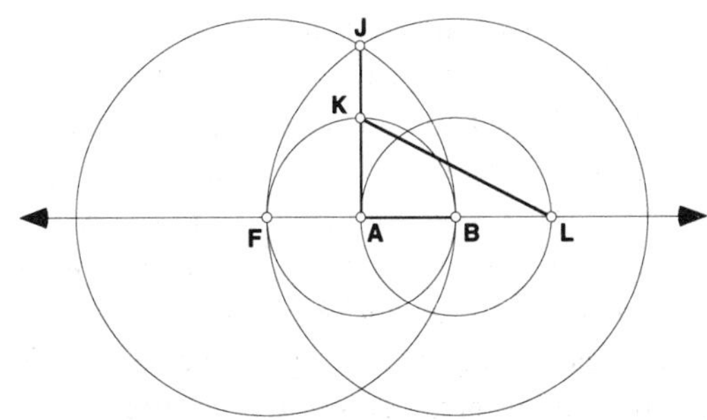

FIGURE 1

(a) Construct line segments with lengths $\sqrt{2}+\sqrt{3}$ and $1+\sqrt{5}$.
(b) By modifying or developing your construction in (a) (if necessary), find points X and Y on l such that $|AX| = \sqrt{2}+\sqrt{3}$ and $|AY| = 1+\sqrt{5}$ and hence determine which of these magnitudes is the greater.

A wide variety of approaches is possible to such questions, and students were expected to compare their methods with those of others and explain to each other why various approaches worked. A common approach began with the *Sketchpad* construction shown in Figure 1.

In this sketch the line segments AB, BK, AJ, and KL have lengths 1, $\sqrt{2}$, $\sqrt{3}$, and $\sqrt{5}$ respectively. A solution is then easily obtained if one is allowed to use the *Sketchpad* command **circle by centre and radius**. If, however, strict Euclidean methods are required, the construction must be modified or developed in order to bring the various lengths together. One way of doing this is illustrated in Figures 2, 3, and 4.

In Figure 2, two further circles have been drawn, one with centre A and radius $\sqrt{3}$, and the other with centre B and radius $\sqrt{2}$. These circles intersect at the point W. It is then a simple matter to construct a segment AX on line l with length $\sqrt{2}+\sqrt{3}$ as in Figure 3.

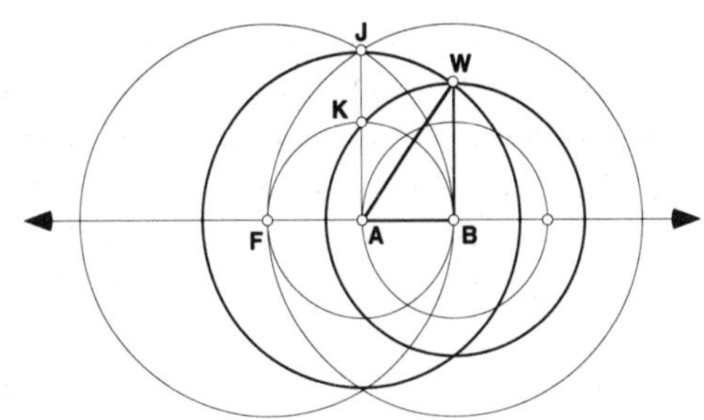

FIGURE 2

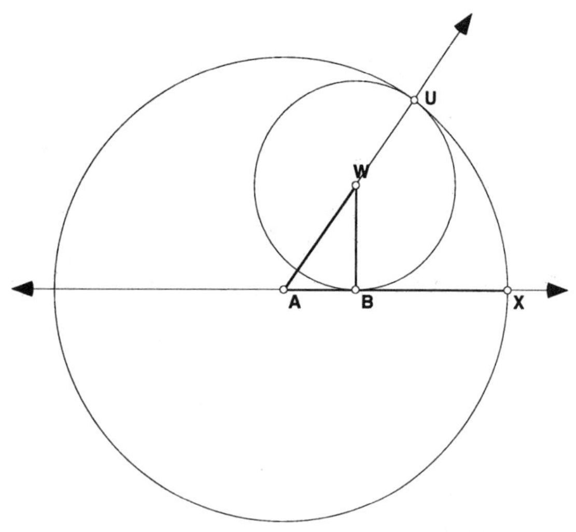

FIGURE 3

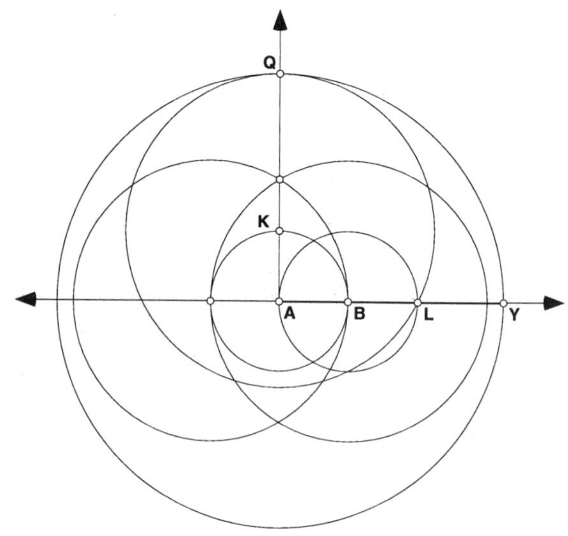

FIGURE 4

For $1+\sqrt{5}$, a segment AQ can be constructed as indicated in Figure 4, and rotated onto line l to produce the required segment AY.

Part (b) can now be answered by performing these two constructions on the same sketch.

We next looked at the geometrical interpretation of quadratic equations and considered methods for solving them as proposed by Euclid and also an approach attributed to Descartes. This is illustrated in the next example.

Example 2. Use Descartes's method to find the roots of the equation $x^2+5x-3=0$.

We need first a line segment with unit length; this is OM in Figure 5. From this we construct a perpendicular segment PM with length $\sqrt{3}$, where 3 is the constant term in the equation.

The construction in this case is particularly straightforward since $\sqrt{3}$ is the altitude in an equilateral triangle of side length 2. The construction lines are shown as dashed lines in Figure 5.

The next part of the construction is to locate a point N on OM extended such that MN has length $\frac{5}{2}$, 5 being the x coefficient in the equation. This can be done in a variety of ways. In Figure 6 a circle has been drawn with radius $\frac{3}{2}$ and centre D.

Finally a circle is drawn with centre N to pass through M, and the points X and Y in which the line through P and N intersects this circle are located as in Figure 7.

Then PX represents the positive root of the equation. We can show this by applying the Pythagorean Theorem to triangle PMN using the fact that in the construction, $PM=\sqrt{3}$, $MN=\frac{5}{2}$, and $XN=\frac{5}{2}$.

$$(PX+XN)^2 = PM^2+MN^2$$
$$PX^2+2PX\cdot XN+XN^2 = PM^2+MN^2$$
$$PX^2+2PX\cdot\left(\frac{5}{2}\right)+\left(\frac{5}{2}\right)^2 = (\sqrt{3})^2+\left(\frac{5}{2}\right)^2$$
$$PX^2+5PX-3=0$$

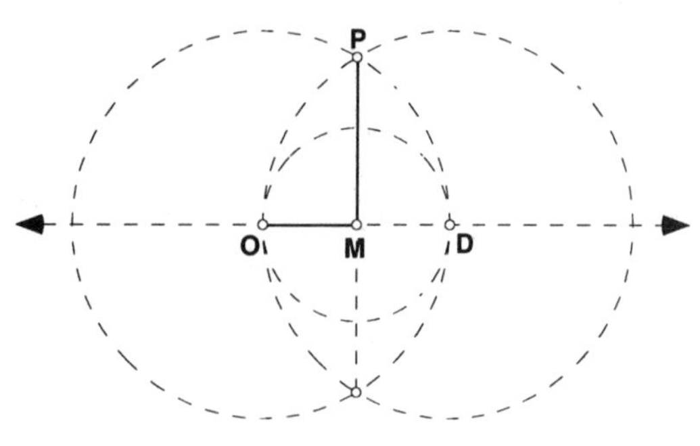

FIGURE 5

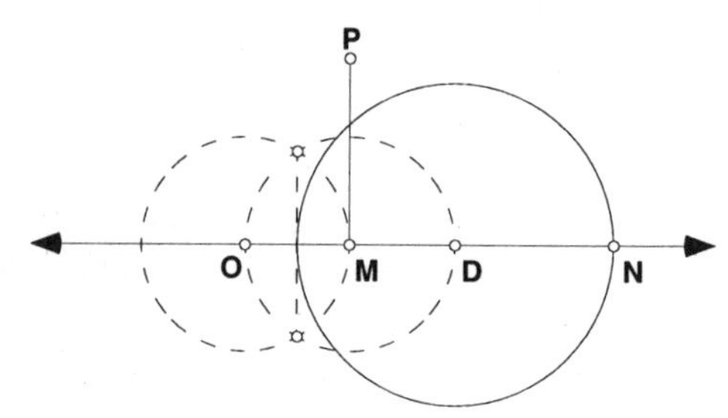

FIGURE 6

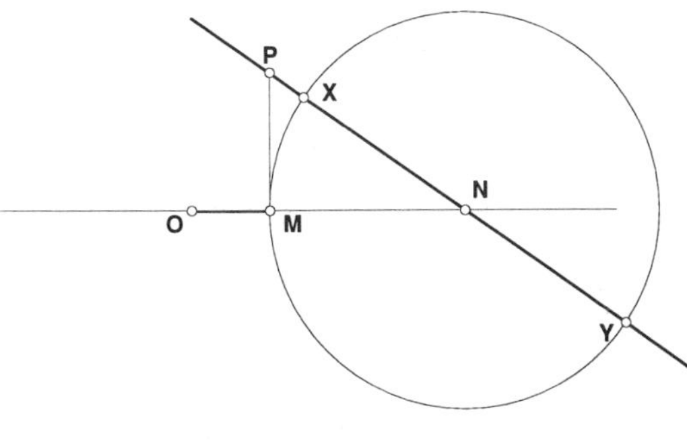

FIGURE 7

Similarly it can be shown algebraically that PY represents the negative root (though of course negative roots were not recognized by geometers even as late as Descartes).

One advantage of performing these constructions with *Sketchpad* is that very accurate measurements can be made of the various lengths and substituted into the equation. This arithmetical verification helps to make the geometry more convincing.

An extension to this exercise is to devise a modification of the above approach so as to solve the equation $x^2 - 5x + 3 = 0$.

Other such construction methods can be found in texts on the history of mathematics such as those by Eves and Burton. Kline's *Mathematics: A Cultural Approach* is also a useful source.

Conics

The conics were largely unfamiliar curves for my students. A small number had met them in school courses in analytic geometry or calculus. Even so, they generally had no prior knowledge of any geometrical properties of these curves. With *Sketchpad*, conics could be created as loci using the **trace locus** feature.

Our exploration of properties of conics was limited by the inability of Version 2 of *Sketchpad* to produce a permanent image of a locus which could be manipulated on the screen or printed. Still, the production of conics of various shapes and sizes and in various locations was attempted, and students experienced great satisfaction from seeing the curve being traced out as they moved the mouse.

We defined the conics using the constant ratio of distances from focus and directrix. As an illustration let us consider the ellipse. Figure 8 shows an ellipse with focus F and directrix l. It is created as the locus of the point P as the point X is dragged on the ray r. The key to the construction is the triangle XYZ at the bottom of the figure. In

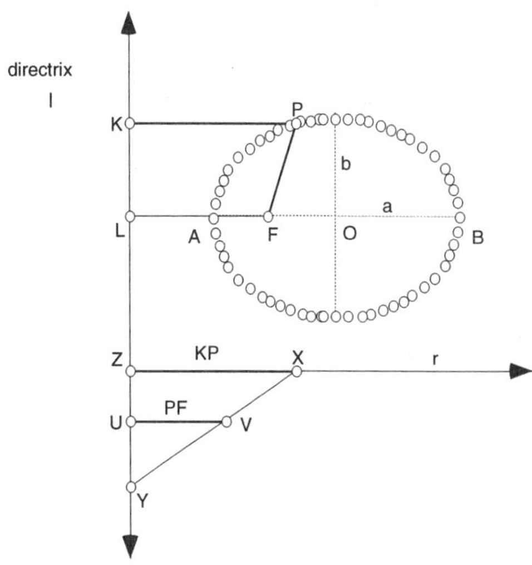

FIGURE 8

this triangle the segment UV is parallel to r so that the ratio UV/ZX is constant as X moves along r. P is constructed as the intersection of the line through X parallel to l and a circle centre F with radius UV. Consequently, $PF = UV$ and $PK = ZX$ so that, when X is dragged, P moves in such a way that the ratio PF/PK remains constant.

As a classroom activity, students were asked to experiment with this sketch to find the effect of changing the distance of the focus from the directrix and/or the position of the point U on YZ. The ensuing discussion led to a realisation that the *shape* of the ellipse is governed by the ratio $\frac{PF}{PK}(= k)$ and its *size* by the distance $FL(= f)$. This motivated the following sequence of pencil-and-paper tasks.

Example 3.

(i) Show that the lengths of the segments AF, AL, BL, and AB can be expressed in terms of f and k as follows:

$$AF = kf/(1+k),\ AL = f/(1+k),$$

$$BL = f/(1-k),\ AB = 2kf(1-k^2).$$

(ii) The midpoint, O, of AB is called the centre of the ellipse. Reflection in a line through O perpendicular to LF interchanges A and B and creates images F' and l' for F and l. Write down expressions for AO, FO, and LO.

(iii) Show that F' and l' are an alternative pair of focus and directrix for the ellipse. (Take any point P on the ellipse and compare the distance of P from l' with its distance from F'.)

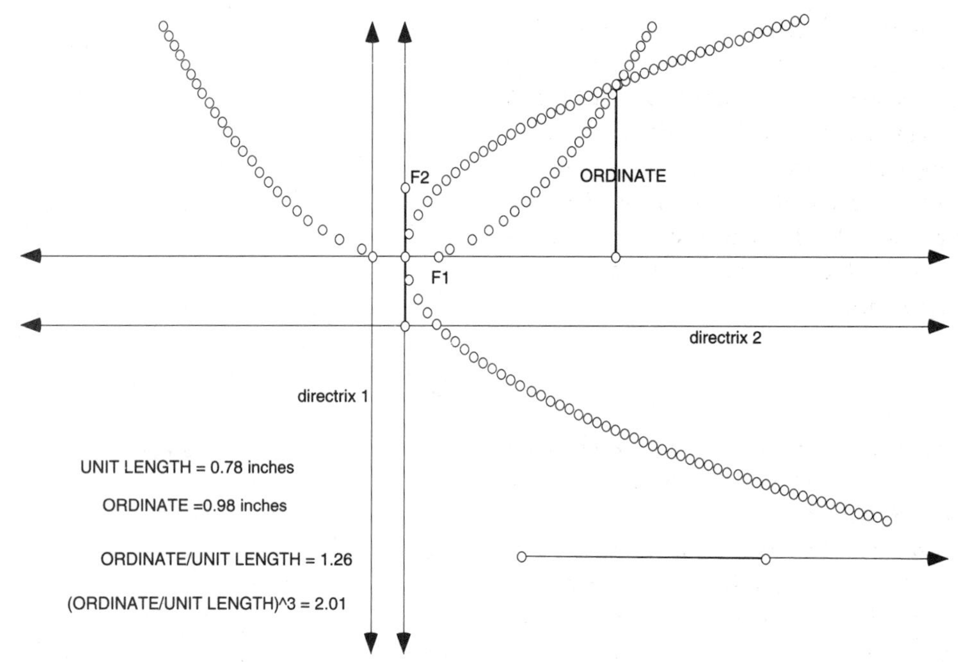

FIGURE 9

(iv) Show that $FF' = 2k^2 f(1-k^2) = kAB$.
(v) Show that $PF + PF' = AB$.

The prime objective of this exercise was to establish result (v). This could then be verified by *Sketchpad* demonstration and used in the development of a method for constructing an ellipse when F, F', and the length of AB are given.

A similar approach was adopted for the parabola and hyperbola, making appropriate changes to the construction in Figure 8. For the parabola UV and ZX must be equal, while for a hyperbola $UV > ZX$. To get the second branch of the hyperbola, another line at a distance ZX from the directrix has to be drawn, this time to the left of the directrix.

A link with the previous work on geometric algebra was made by considering construction methods for finding cube roots using conics. We had already seen that the Greeks had devised geometrical methods for constructing square roots and solving quadratic equations. They were unable, however, to devise a Euclidean method for constructing cube roots. In puzzling over this problem, they discovered that cube roots could be found if curves other than lines and circles were permitted. This was demonstrated using parabolas.

In Figure 9 a length representing the cube root of 2 is constructed by finding the point of intersection of two parabolas. The directrices of the parabolas are perpendicular, and their focal lengths are 1 and 2 units respectively. We can think of them as corresponding to the equations $y^2 = x$ and $x^2 = 2y$. Where they cross, $y^4 = x^2 = 2y$ so $y^3 = 2$. The construction gives the cube root of 2 as 1.26 to 2 decimal places.

On this sketch, all the construction lines and circles have been hidden. The segment with a point in the lower right corner controls the simultaneous tracing of the two parabolas.

Another motivation for the study of the conics can be found in their optical properties and particularly the significance of the foci. These can be exploited in the construction of tangents to the conics using **angle bisector** and **perpendicular line** from the construct menu. Figure 10 shows a sketch in which a tangent to an ellipse has been drawn by this method.

Much of the work we did on conics would have been easier if Version 3 of *Sketchpad* had been available with its

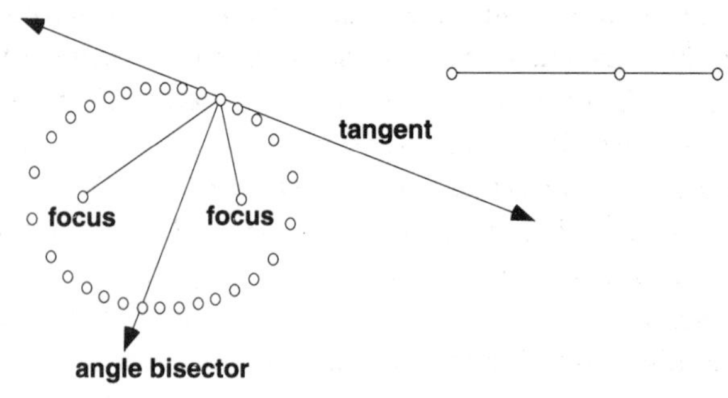

FIGURE 10

ability to retain and print images of loci and to permit work with them. The coordinate facility of Version 3 also allows links between the geometry and algebra of conic sections to be explored. Many properties of conics can be established by *Sketchpad* investigation or demonstration and then proved algebraically. It was beyond the scope of the Geometry Workshop course to consider algebraic models for geometry. This was taken up in a subsequent course in the second year of the degree programme.

Example 4.

(i) Let Q lie on the tangent at P to a parabola with focus S so that the angle SQP is 90°. What happens to Q as P moves on the parabola?

(ii) Let the tangent and normal at P meet the axis of the parabola at T and G respectively. Show that P, T, and G are equidistant from the focus of the parabola.

(iii) The two tangents from the point V to a parabola are at right-angles to each other. Describe the locus of V.

For Geometry Workshop students these were essentially exercises in the use of *Sketchpad.* The construction of the figures described was sufficient challenge. It was enough at this stage that they were able to arrive at conjectures through experiment. Those with a background in coordinate geometry were invited to attempt proofs; others were assured that we would review these results the following year.

Perspective

Sketchpad can also be used to illustrate basic drawing techniques. Examples of applications to art can be found among the Sample Activities provided with the software. We began with the drawing of a tiled floor.

Figure 11 shows the first steps in creating the tiling pattern. Points A and B are selected on a horizon line and rays k, m, p, and q drawn using the tool box on *Sketchpad.* A quadrilateral is produced and one of its diagonals produced to cut the horizon at M. Further lines are introduced in the order indicated, either as sides of tiles or diagonals. Eventually an effective tiling pattern is obtained and may be enhanced using the **polygon interior** feature as in Figure 12.

It is possible to drag any of the initial objects, A, B, p, q, k, or m and observe changes in the pattern. The impressive thing, of course, is that the drawing remains realistic in that the perspective effect is maintained. Another surprise comes when the other set of diagonals is drawn—they all intersect at a fourth point on the horizon.

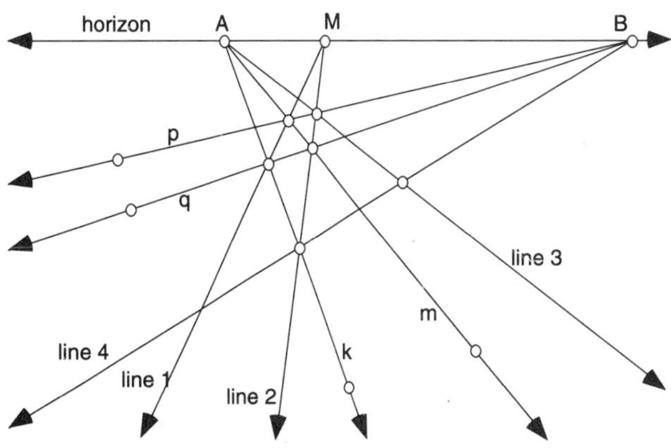

FIGURE 11

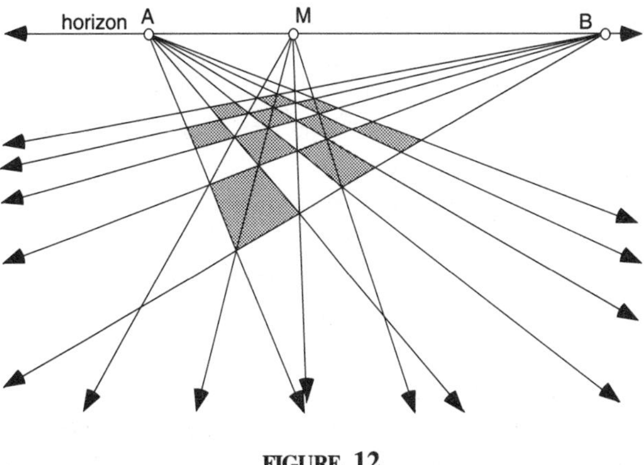

FIGURE 12

Following discussion of the tiling pattern, the basic principles of perspective drawing were explained and students were invited to experiment with a drawing of their own. They were later asked to attempt the perspective drawing of a rectangular building with a pitched roof.

The other use for *Sketchpad* in this part of the course was in demonstrating results such as Pappus's Theorem and Desargues's Theorem.

Pappus's Theorem. *Let l and m be two lines. Let A, B, and C be any three points on l, and a, b, and c be any three points on m. Let Ab and aB meet at X, Ac and aC meet at Y, and Bc and bC meet at Z. Then X, Y, and Z all lie on a straight line.*

This can be demonstrated very effectively as in Figure 13.

In the *Sketchpad* sketch any of the points on l and m can be moved and the effect on X, Y, and Z observed. Similarly the lines l and m can be moved.

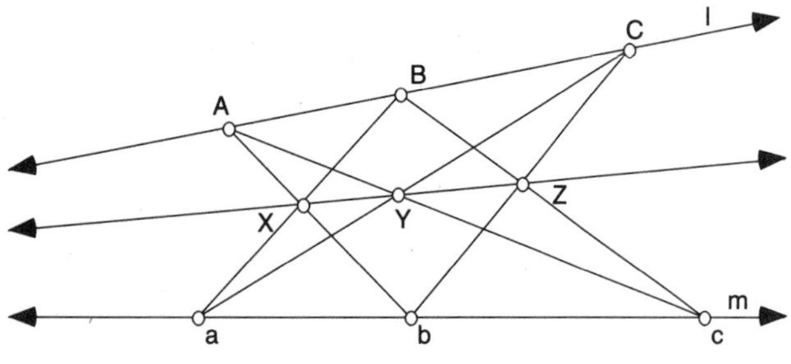

FIGURE 13

Desargues's Theorem. *Let l, m, and n be three lines which are all concurrent at a point O. Let A and a be two points on l, B and b be two points on m, and C and c be two points on n. Let the lines AB and ab meet at X, AC and ac meet at Y, and BC and bc meet at Z. Then the points X, Y, and Z all lie on a straight line.*

This is illustrated in Figure 14, in which the triangles ABC and abc have been shaded.

Proofs of these theorems were obtained by projecting the figure so as to map an appropriate line "to infinity." This produces a new figure in which there are sets of parallel lines and similar triangles. The configuration in Figure 14, for example, can be transformed into that of Figure 15 by mapping the line abX to infinity.

Desargues's Theorem now corresponds to a claim that AB is parallel to ZY. This can be proved by considering the pairs of similar triangles $\triangle AOC \approx \triangle YcC$ and $\triangle BOC \approx \triangle ZcC$ and showing that it must follow that $\triangle ABC \approx \triangle YZC$.

We deduced that an equivalent result must hold for the original figure since incidence properties are invariant under projection.

This was the first time the students had seen the invariance properties of a transformation used in this way, and it led naturally to the discussion of invariance which figured in the final section of the course.

Transformations of the Plane

Students had used items from the construct menu at earlier stages of the course. In this final section of the course they were encouraged to investigate aspects of the algebra of transformations and explore the pattern-making possibilities of *Sketchpad*.

We concentrated for the most part on isometries and began with a class discussion of quite simple problems such as describing the composite of two half-turns or reflections. The basic pattern of the work followed the sequence: explore, conjecture, experiment, explain. Students were encouraged to look for geometrical arguments, based on properties of figures and the invariance characteristics of the transformations to support their conclusions.

Example 5. A, B, C, and D are four points in the plane. What single transformation has the same effect as the composite of consecutive half-turns in A, B, C, and D?

Conjugacy provided another rich area for investigation.

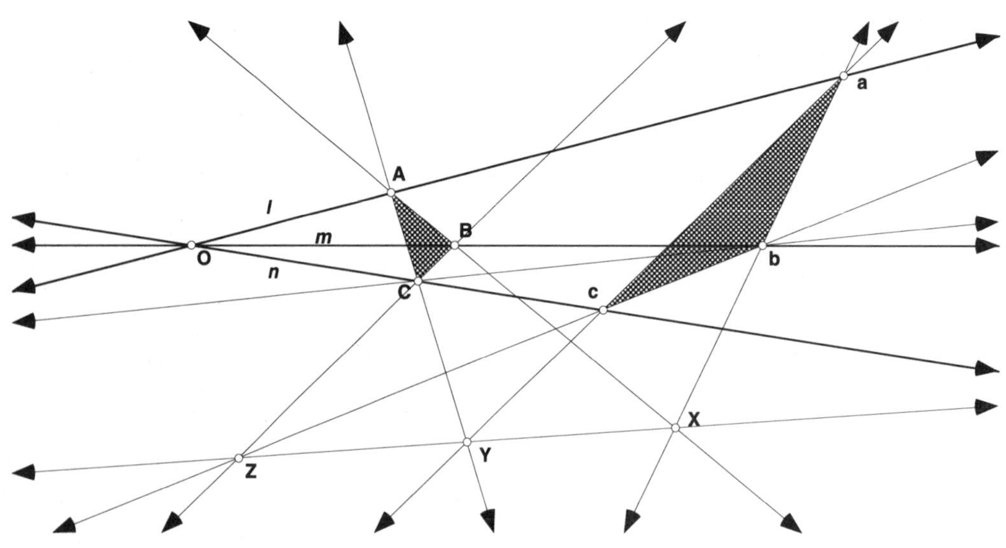

FIGURE 14

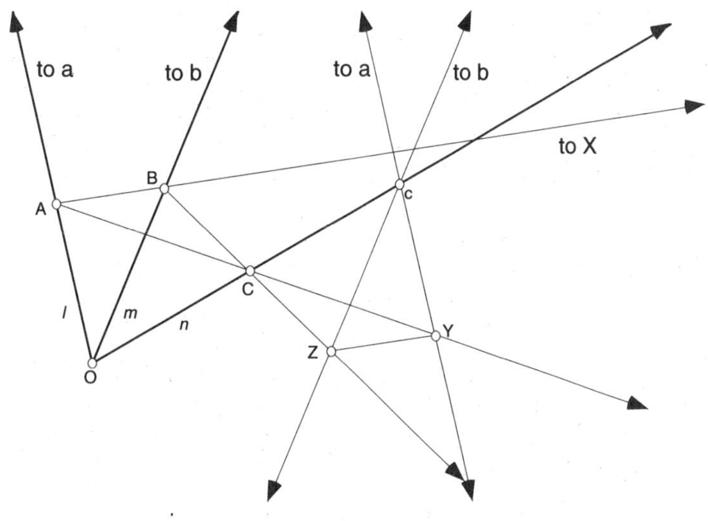

FIGURE 15

Example 6. Let A be a fixed point in the plane, let R be the rotation with centre A and angle θ, and let T be a fixed translation of the plane. Describe the single transformation having the same effect as the composite $R^{-1}TR$.

The final area for investigation was symmetry. Strip and plane patterns were constructed using generating transformations. The use of *Sketchpad* made the generation of patterns from a motif very simple. Students experimented with combinations of transformations to produce patterns from basic motifs and were encouraged to undertake some analysis of their patterns along the lines indicated by the next example.

Example 7. In Figure 16, 90° rotations about the point E combined with reflection in the mirror line q produce the entire pattern from the single motif F.

Denote the 90° rotation about E by **R** and the reflection in q by **Q**.

(i) Express in terms of **R** and **Q** the transformations that take (a) F to X, (b) F to Y, (c) X to Y.

(ii) Express in terms of **R** and **Q** the following symmetries of the pattern

(a) reflection in the line s,

(b) 180° rotation about the point A.

Evaluation

The course outlined above ran for three years and was taken by around 140 students. Only two students failed to complete the course successfully.

Even so the student reaction to the course was mixed, with some students finding it very demanding and quite stressful. From each run of the course, student evaluations were obtained, and these were used to guide the gradual revision of both delivery and assessment.

The difficulties some students faced in mastering the computer meant that they took feelings of insecurity into the study of the geometry. More individual support in the early weeks helped solve this problem. While shared use of computers was beneficial in later parts of the course, in the first few weeks this proved to be an obstacle to learning. Eventually I was able to base the lectures in a computer laboratory so that each student had his/her own machine. No longer was it possible for weak or timid students to spend most of the time watching a colleague learn how to use *Sketchpad*. Once they found that they were not alone in needing help, they made more use of the support available, and progress improved.

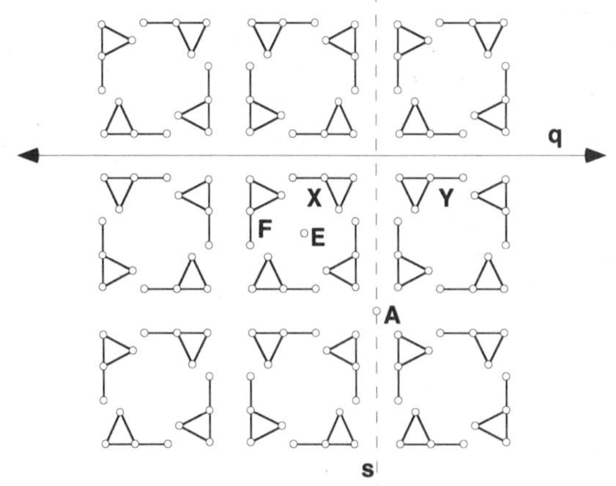

FIGURE 16

Given the time available and the diversity of the student intake, it proved unrealistic to expect them all to cover the wide range of topics described above. Indeed, to achieve the aims of the course it was not necessary to cover a vast amount of geometry as long as student involvement led to some feel for the subject and a development of investigative skills. In fact, the mathematically more able could follow the entire programme, and they enjoyed the opportunity to pose and investigate their own questions. However, many of the less confident students missed the assurance of a more didactically presented course with specific techniques to be rehearsed and a closed style of assessment. They found the requirement to explore a problem area and explain what they observed quite foreign. The most frequent plea was for more computer time and support outside normal classes, and, if the course had continued, I would have argued for greater class contact time.

All this had to be reflected in the assessment of the course. The first assignment covered the Greek Geometry and was undertaken by everyone. In the second assignment, which covered the other topics, students were allowed some choice in the topic they investigated. The assignments were not completely open-ended, however. Some routine questions were included so as to allow weaker students in particular to obtain some credit for work done on specified tasks.

As I mentioned at the beginning of this paper, the College has now deleted the Mathematics Field from its Cheltenham programme. However, mathematics is still taught within our overseas programme, and the lessons learned from teaching the Geometry Workshop course have influenced the design of courses currently being offered in Hong Kong and Malaysia. We have accepted the need for geometry in undergraduate mathematics and believe that, through the use of dynamic geometry software, geometry can be made relevant for this generation of students.

There remains the need to persuade mathematics educators in other Colleges and Universities of the potential for curriculum development of dynamic geometry software. Geometry has been perceived for years as cut, dried, and dead. I believe that, given an appropriate treatment, geometry can resume its position as the foremost area in which to observe, experience, and understand the processes of mathematics in action.

References

1. Austin, K. et al. (1994). Geometry in the Undergraduate Syllabus. *Proceedings of the 1994 UMTC.* Shell Centre for Mathematical Education, University of Nottingham, pp. 67–75.
2. Burton, D.M. (1995). *The History of Mathematics. An Introduction*, 3rd ed., Dubuque, W.C. Brown.
3. Eves, H. (1992). *An Introduction to the History of Mathematics*, 6th ed., Philadelphia, Saunders.
4. Hampson, A. (1993). *Geometry Workshop: Background Notes and Examples.* Cheltenham, Cheltenham and Gloucester College of Higher Education.
5. Kline, M. (1962). *Mathematics: A Cultural Approach.* Reading Mass., Addison-Wesley.

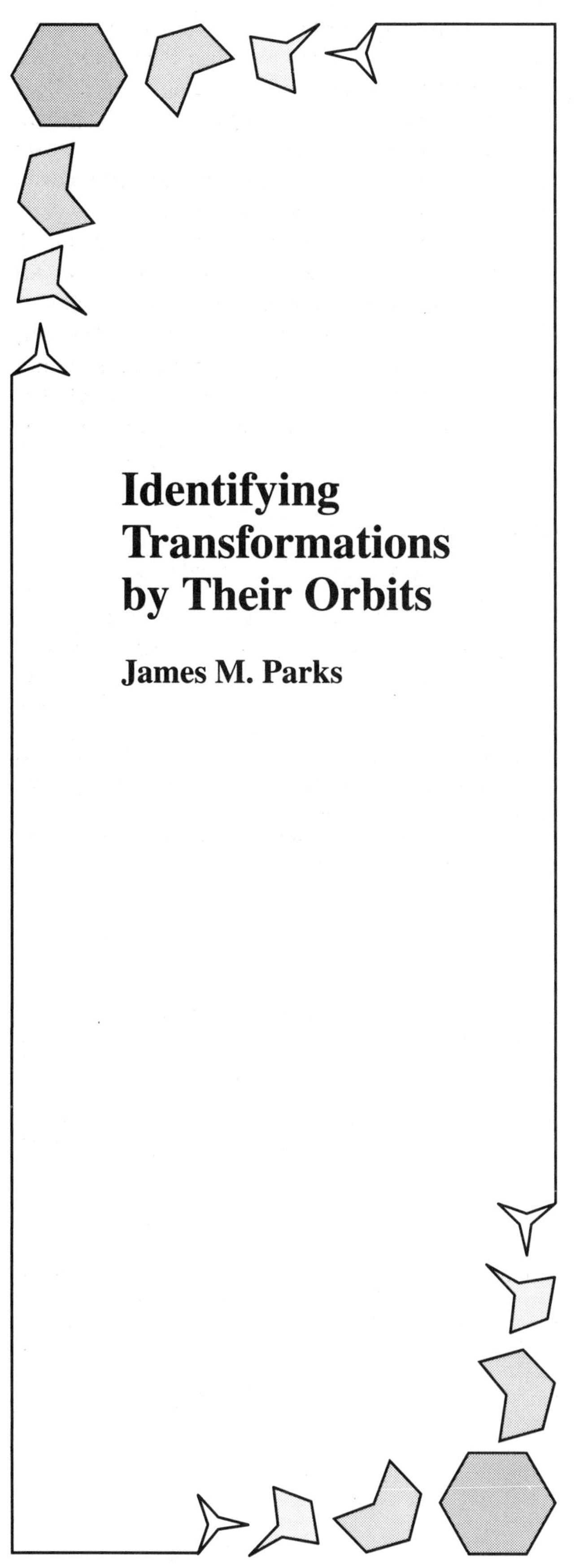

Identifying Transformations by Their Orbits

James M. Parks

We present an alternate classification scheme for isometries (distance-preserving transformations) of the plane that uses a computer and dynamic geometry software (like *Geometer's Sketchpad* [6]). We have found this technique to be very useful when discussing isometries and applications of isometries in the classroom. In addition, the students find this approach much easier to use and understand than the "traditional" method of classifying isometries.

Usual Method of Classifying Isometries of the Plane

The Classification Theorem for isometries of the plane states that there are exactly four types of isometries: translations, rotations, reflections, and glide-reflections. (The identity transformation can be considered to be a rotation of 360°.)

The usual proof given of this Classification Theorem considers compositions (products) of reflections (see [1] or [5]). The key to this approach is to prove that any isometry of the plane is the composition of at most three reflections, and then to show that these compositions exhaust the list above of the four known types of isometries of the plane.

While this proof is very efficient and satisfying mathematically, it can be very difficult to use in applications. If the student is confronted with applications that require the identification of isometries which arise in constructions or even just wishes to apply isometries to find other results, then the "composition of reflections" arguments can be a confusing experience.

For example, suppose T is a composition of a reflection and a rotation (see example B below). Then T must be an isometry by the Classification Theorem, since it is a distance-preserving transformation, but which one is it? It is not at all obvious how to change T to a composition of reflections in order to identify it as a single isometry.

An alternate classification scheme for isometries of the plane and a computer with dynamic geometry software like *Geometer's Sketchpad* gives us the tools to handle problems of this type easily.

Alternate Method of Classifying Isometries

Another method of classifying isometries of the plane which seems to be more "natural" and which is often easier to use, particularly for students, depends on the orientation properties and the fixed points of the isometries.

A transformation T of the plane is said to be *orientation-preserving* if T takes any triangle $\triangle ABC$ with vertices oriented in a counterclockwise manner to a triangle

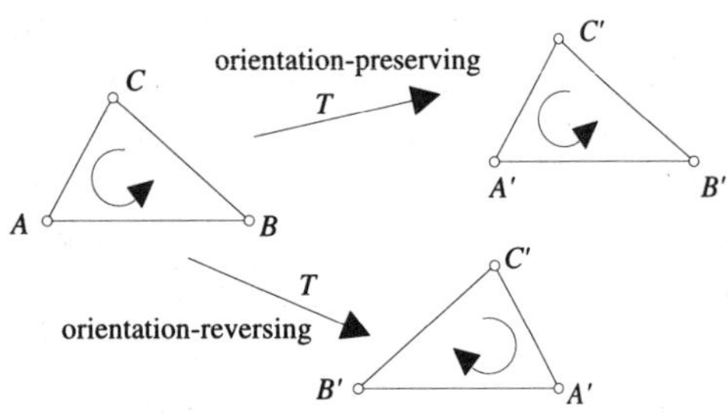

FIGURE 1

$\Delta A'B'C'$, with vertices also oriented in a counterclockwise manner (see Figure 1). If the image triangle $\Delta A'B'C'$ has vertices oriented in a clockwise manner, then we say that T is *orientation-reversing.* Also, if a transformation T of the plane satisfies $T(P) = P$ for some point P in the plane, then we call P a *fixed point* of T.

The orientation properties and fixed points of the four isometries of the plane uniquely characterize each isometry and thus may be used for classification (see also [3], [4] & [5]). This information is summarized in the chart below.

ORIENTATION	FIXED POINTS	ISOMETRY
Preserving	no (none)	Translation
	yes (one)	Rotation
	yes (all)	Identity
Reversing	yes (a line)	Reflection
	no (none)	Glide-reflection

It is possible to prove the Classification Theorem for isometries of the plane based solely on their orientation and fixed-point properties, as listed in the chart above. However we will not give this result here. (This is a good exercise in transformation geometry for students.)

Examples

In the examples below it is assumed that the student is aware of the following facts about isometries:

a. A composition of two isometries is an isometry.
b. A composition of two orientation-preserving or orientation-reversing isometries is an orientation-preserving isometry.
c. The composition of an orientation-preserving isometry with an orientation-reversing isometry is orientation-reversing.

In this setting, the method used to classify a given isometry T is accomplished in two easy steps:

1. Check to see whether T preserves or reverses orientation.
2. Check T for fixed points.

The key to this method is in the second step, which is carried out with the aid of a computer and a dynamic geometry software program like *Sketchpad.*

Given a transformation T and a point X, if we apply T to X and then apply T to that image $T(X)$, and so on, a sequence of iterates of X is produced, $\{X, T(X), T(T(X)) = T^2(X), T^3(X), \ldots\}$. This sequence is called the *orbit of X* generated by T (it is sometimes called the "orbit of X under the group generated by T"). When T is understood, we will denote this sequence by $\{X, X_1, X_2, X_3, \ldots\}$.

Given a point X and an isometry T, the *Sketchpad* program will quickly produce many terms in the orbit of X and display them visually to give the viewer an idea of what the orbit of the point X looks like. It is then possible to deduce whether or not the transformation T has a fixed point. In addition, *Sketchpad* allows us to check quickly a large number of examples by dragging the objects involved in the example, which in turn helps to determine the invariant properties of the isometry.

A. Determine the isometry T, if T is a rotation of 45° about point P followed by a rotation of 30° about point Q, $P \neq Q$.

Solution. Since T is the composition of two rotations, it must be an orientation-preserving isometry, so T is either a translation or a rotation (see the chart at left). We next check T for fixed points. To get an idea of the action of T on points in the plane, we choose a point X and use *Sketchpad* to produce several iterates of X under the action of T. Connecting these successive points in the orbit of X under T with line segments, we can see clearly the effect of T on X (see Figure 2).

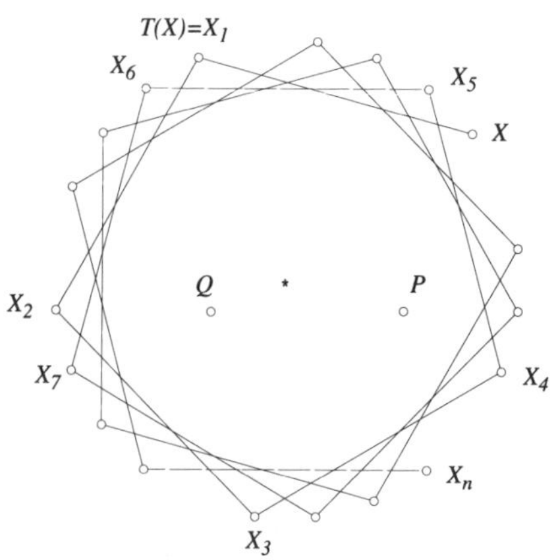

FIGURE 2

We observe that the points in the orbit appear to have a circular pattern, and, by dragging the point X to a new position, we can instantly check to see whether the choice of X has any effect on this result (it doesn't). We can also move the points P and Q to see whether their positions affect the result (although positions of points in the orbit change, a circular pattern is maintained). By moving X towards the region of the star $*$ between P and Q in Figure 2, it appears that this point is a fixed point of T^n (the software allows us to locate the approximate location of the fixed point!). We conjecture that T has a fixed point near the star $*$, and the following fixed point theorem [2] confirms our conjecture.

Fixed Point Theorem: Let $f\colon S \to S$ be a function on a set $S \neq \phi$ to itself. If $f^n = f \circ f \circ \cdots \circ f$ (n times), and f^n has a single fixed point P for some $n > 1$, then f also has the single fixed point P.

Proof. Suppose f^n has a single fixed point P. Then $f(f^n(P)) = f(P)$. But $f(f^n(P)) = f^n(f(P))$, therefore $f(P) = P$.

The technique we have illustrated to identify the composition of two particular rotations can also be used to help students discover the answer to more general questions: When is the composition of two rotations a rotation? When is it a translation?

B. Suppose T is the composition of a 45° rotation about point P followed by a translation with vector AB, for $A = (0, 0)$ and $B = (1, 1)$. Identify the isometry T.

Solution. Since T is an orientation-preserving isometry, T must be either a translation or a rotation. Let X be a point in the plane and iterate the action of T on X to see that the orbit of X begins to form a circular pattern. In Figure 3,

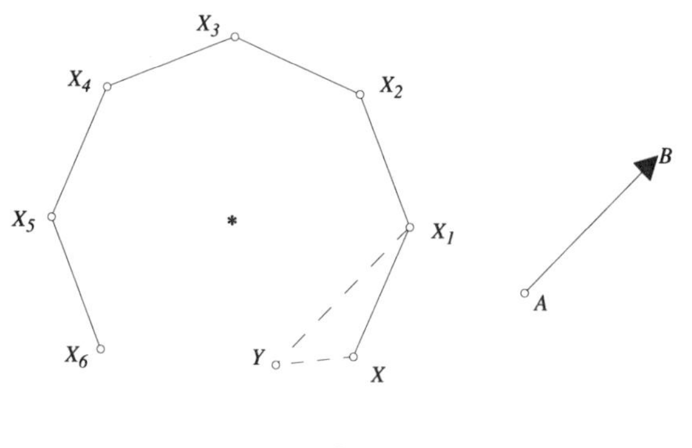

FIGURE 3

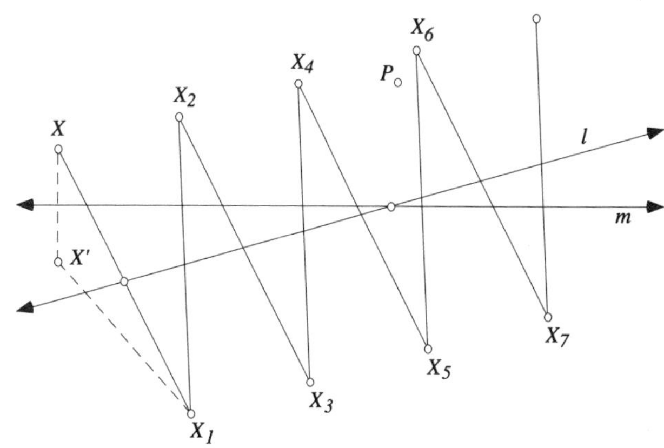

FIGURE 4

line segments connect the first 6 points in the orbit of X. (The dashed lines indicate the intermediate step in the first transformation: from X to Y by rotating about P and then to $T(X) = X_1$ by translating Y.) Clearly T must have a single fixed point near *, thus T must be a rotation.

C. Let T be the composition of a reflection about line m followed by a rotation of 30° about a point P, where P is not on m. Determine the isometry T.

Solution. Clearly T is an orientation-reversing isometry, so it must be either a reflection or a glide-reflection. We need only check T for fixed points, so we choose a point X in the plane and iterate the action of T on X to obtain points in the orbit of X. In Figure 4, X is reflected about line m to X', and then X' is rotated about P to X_1. A line segment connects X and $T(X) = X_1$; the iteration process repeats to get the orbit of X: $\{X, X_1, X_2, X_3, \ldots\}$. These points are connected in succession by line segments. Clearly T has no fixed points, so it must be a glide-reflection about some line l. By dragging the point X, the choice of position of X may be shown to have no effect on this observed pattern (recall X is not on m), and in fact we can approximate the location of the glide-line l.

As a companion example we can consider the question: "What if P is on the line m?" With *Sketchpad*, the point P can be dragged to the line m, and it is then easy to see what type of isometry T must be. (We leave this to the reader.)

Additional Transformations

The above techniques may also be used for problems of compositions that involve dilations. Dilations are orientation-preserving transformations that have a single fixed point (the center of dilation). If the dilation factor is not very close to 1,

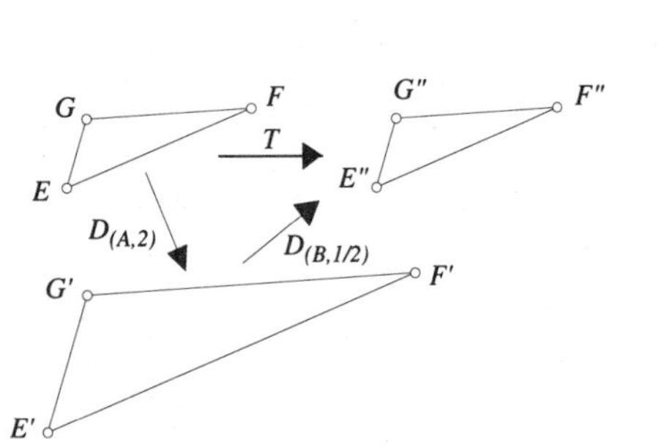

FIGURE 5

it is easy to recognize such transformations by their images. The following example gives a simple illustration of this.

D. Suppose $D_{(A,2)}$ and $D_{(B,1/2)}$ are dilations about points A and B with dilation factors of 2 and 1/2, respectively. Show that the composition of $D_{(A,2)}$ with $D_{(B,1/2)}$ is an isometry.

Solution. Let $T = D_{(B,1/2)} \circ D_{(A,2)}$. Since each dilation is orientation-preserving, T is also orientation-preserving. Choose a triangle $\triangle EFG$ and apply T to it to see that the image of $\triangle EFG$ under T appears to be congruent to $\triangle EFG$ and that there are no fixed points (see Figure 5). Since dragging $\triangle EFG$ to change its location and shape does not affect these properties, it appears that T is a translation. To verify this with *Sketchpad*, fill the interior of $\triangle EFG$ and apply to it the translation defined by the vector GG''. The translated interior of $\triangle EFG$ lands exactly inside $\triangle E''F''G''$, as predicted. This conclusion suggests a more general result about the composition of two dilations with dilation (scale) factors whose product is 1, and students can be expected to conjecture the result.

Conclusions

In Example A we discovered that the method of displaying several points in the orbit of a point under a composite of rotations could be used to find the location of the fixed point of the resulting rotation. We may then ask the students: "Why is the fixed point at this particular spot?" The software allows the students to measure the distances and angles relative to the given points P and Q and the given angles of rotation, and to experiment with other choices of points and angles. This leads them, in a natural way, to discover what the relations are between the resulting isometry and the given isometries. In a similar way the other examples lead them to discover "new" mathematics on their own. Through the use of dynamic geometry software, they are encouraged to think mathematically and to discover patterns through examples. This leads them to make conjectures about the results, and then they may proceed on to discover the mathematical reasons behind these results.

References

1. Greenberg, M.J. *Euclidean and Non-Euclidean Geometries*, 2nd ed., W.H. Freeman & Co., 1980.
2. Kolodner, I. "Fixed Points," *Amer. Math. Monthly*, 71(64), 906.
3. Parks, J.M. "Transformations, Fixed Points, and Orientations," *New York State Teachers' Journal*, vol. 45(95), 117–121.
4. Rees, E. *Notes on Geometry*, Springer-Verlag, 1983.
5. Wallace, E.C. and West, S.F. *Roads to Geometry*, Prentice Hall, 1992.
6. *Geometer's Sketchpad* v3.0, Key Curriculum Press, 1995.

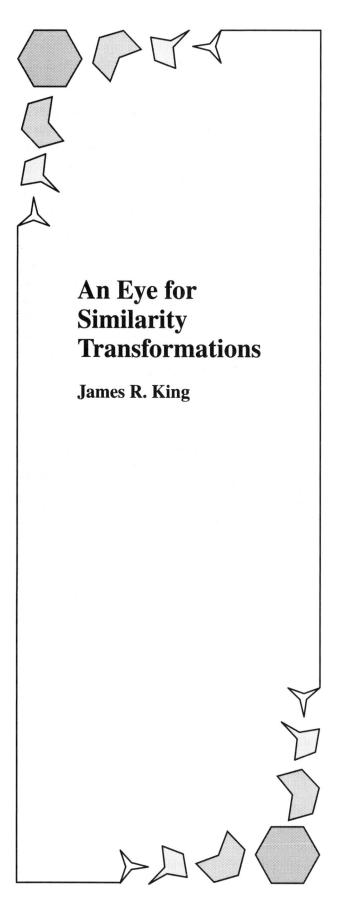

An Eye for Similarity Transformations

James R. King

Similarity transformations are familiar in our daily lives: copy machines scale drawings up or down, and television screens show swirling logos to excite us. These transformations can also be a valuable part of a geometer's tool kit for solving problems. Dynamic geometry software makes it possible to work with geometrical transformations in a concrete, interactive manner that makes them accessible as never before. In this paper we focus our attention on how similarity transformations of the plane can be studied with dynamic geometry software and how such similarities can be used to solve geometry problems.

An eye for similarities may give insights into problems that first appear to have no connection with transformations. For example, given a point A and a circle c, consider the locus of midpoints M of segments AB, where B is a point that travels on the circle c. When this locus is traced using dynamic geometry software, one observes that the locus appears to be a circle; the radius of the circle stays constant as A is moved from place to place.

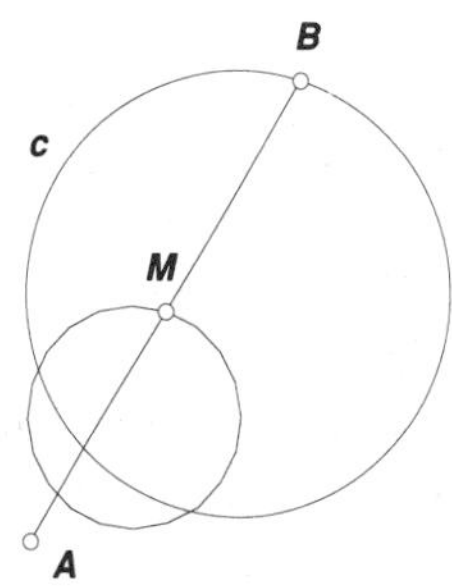

FIGURE 1

To many, this may be surprising and even a bit mysterious, and, if one tries to prove it using similar triangles, it may seem even more awkward and mysterious, with special cases required for A inside and outside the circle c. But, if one is familiar with dilations, Figure 1 is seen as a picture of a dilation of the circle c; the center of the dilation is A, and its ratio is $\frac{1}{2}$. The midpoint M is the dilation of point B, and the locus of the points M is merely the image of c under the dilation.

In the next section we give some definitions and begin to look at more geometry problems, with an eye for similarities.

1. Similarities and Dilations: Definitions and Properties

Throughout this paper, we use the notation PQ in two distinct ways: PQ can mean the segment with endpoints P and Q, or PQ can mean the directed distance from point P to point Q. The context in which the notation is used should make its meaning clear. The length of the segment PQ will be denoted as $|PQ|$. We also often will use the shortened phrase "the line PQ" to mean the line through the points P and Q. All of the properties of similarity transformations that we note in this section can be found in the references [3], [16].

Definition. A transformation F of the Euclidean plane is a ***similarity transformation with ratio of similitude k*** if, for all points P and Q, $|F(P)F(Q)| = k \cdot |PQ|$. In other words, F scales lengths by a constant factor $k > 0$.

In this paper, we will use the word **similarity** as a short form for "similarity transformation." It follows from the definition that a similarity F maps a circle to a circle, a segment to a segment, and a line to a line. Since similarities necessarily preserve ratios of distances, it is also true that the image $F(M)$ of the midpoint M of a segment AB is the midpoint of the segment $F(A)F(B)$. It is clear that the composition of similarities with ratios k and l is a similarity with ratio kl; and, if F has ratio k, then F^{-1} is a similarity with ratio k^{-1}. An ***isometry*** is a similarity with ratio of similitude $k = 1$.

The basic similarity is the dilation.

Definition. Given a point O and a real number $k \neq 0$, the ***dilation*** with center O and ratio k is the transformation that carries a point P to the point P' on the line through O and P for which $OP' = k \cdot OP$. (A dilation is also called a *dilatation*, a *central similarity*, or a *homothety*.)

The dilation ratio $k = OP'/OP$ can be either positive or negative. It is positive if P and P' are on the same side of O, and it is negative if P and P' are on opposite sides of O. From the definition, it follows that the inverse of the dilation with center O and ratio k is the dilation with center O and ratio k^{-1}.

The fundamental figure for illustrating a dilation with center A is a triangle ABC, along with the image points B', C', of B and C, respectively. For historical reasons, we will call this figure the *Thales triangle figure*. Two such figures are shown in Figure 2; in Figure 2(A) the ratio k is between 0 and 1, and in Figure 2(B) the ratio k is negative.

In Figures 2(A) and 2(B), triangle ABC is similar to triangle $AB'C'$ by side-angle-side, since angle BAC equals angle $B'AC'$ and $AB'/AB = AC'/AC = k$. Since $|B'C'| = |k||BC|$ for all points B and C, it follows that the dilation is a similarity. (This fact was not part of the definition of dilation but is a consequence of the geometry of the Thales triangle figure.) In the similar triangles, the corresponding angles ABC and $AB'C'$ are equal, so lines BC and $B'C'$

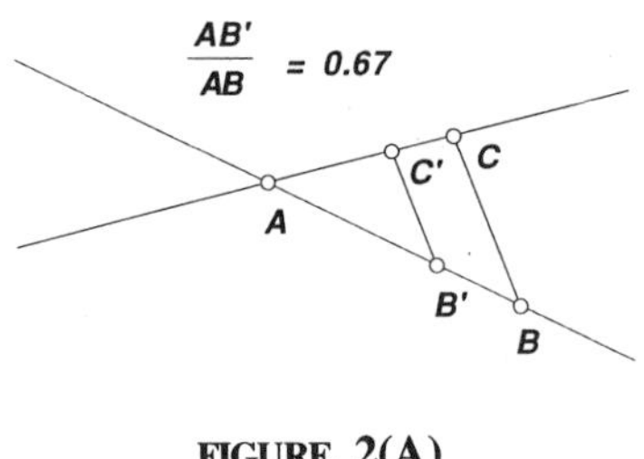

FIGURE 2(A)

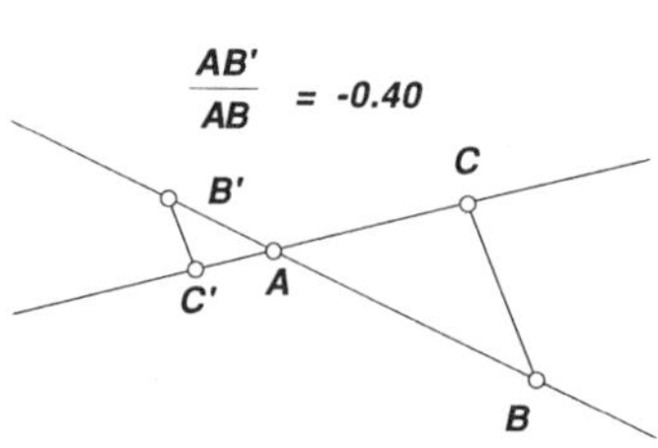

FIGURE 2(B)

are parallel; the converse also holds. This is a fundamental property of dilations: a segment $B'C'$ is the image of a segment BC under a dilation if and only if $B'C'$ is parallel to BC (except for the special case when BB' is parallel to CC'; then the mapping is a translation rather than a dilation).

Our arguments on the Thales triangle figure (Figure 2) can be reversed to construct a dilation with straightedge and compass. Begin with points A, B, C and point B' on line AB. Construct point C' as the intersection of line AC with the line through B' parallel to line BC. If we define k as AB'/AB, then $AC'/AC = k$ also, and C' is the dilation of C with center A and ratio k. (Note: This construction breaks down when C is on the line AB. In this case C' can be constructed with the aid of an auxiliary point D.) If this construction is carried out with dynamic geometry software, point B' can be dragged along the line AB, so that different positions of B' such as those in Figures 2(A) and 2(B) can be obtained with a single construction. This underscores the fact that dilations with either positive or negative ratios are special instances of a single general concept.

This construction can be extended to dilate any finite set of points. For example, to construct the image of a polygon $ABCD$ by the dilation with center O that takes A to A', construct B' by the method above. Then take B' as a new starting point and construct C' as the intersection of the line OC with the line through B' parallel to BC. Continue the process for the additional points. A' can then be dragged to make the ratio k positive (Figure 3(A)) or negative (Figure 3(B)).

Instead of using the straightedge and compass construction, there is an easier way to carry out dilations in both *The Geometer's Sketchpad* and *Cabri Geometry II:* Dilation is a

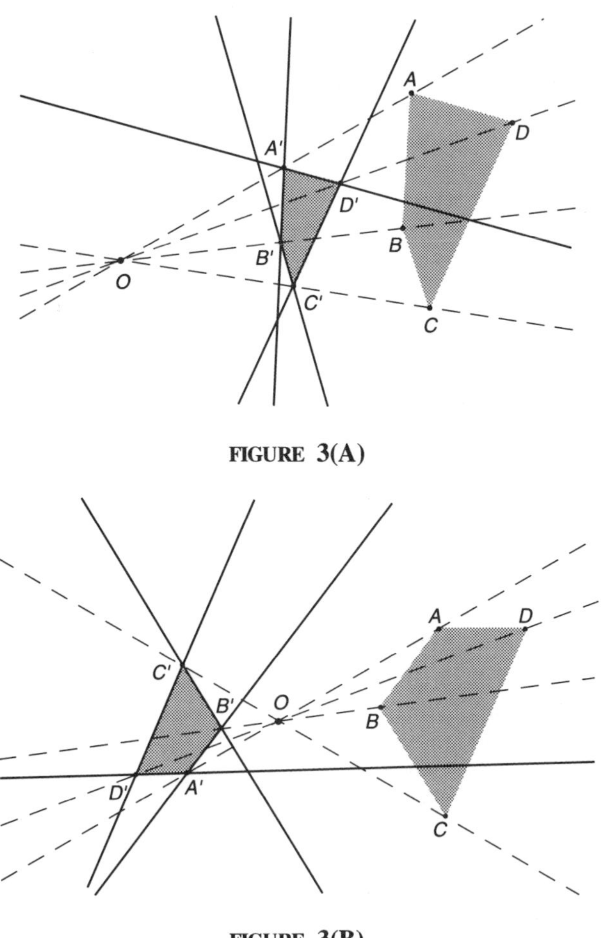

FIGURE 3(A)

FIGURE 3(B)

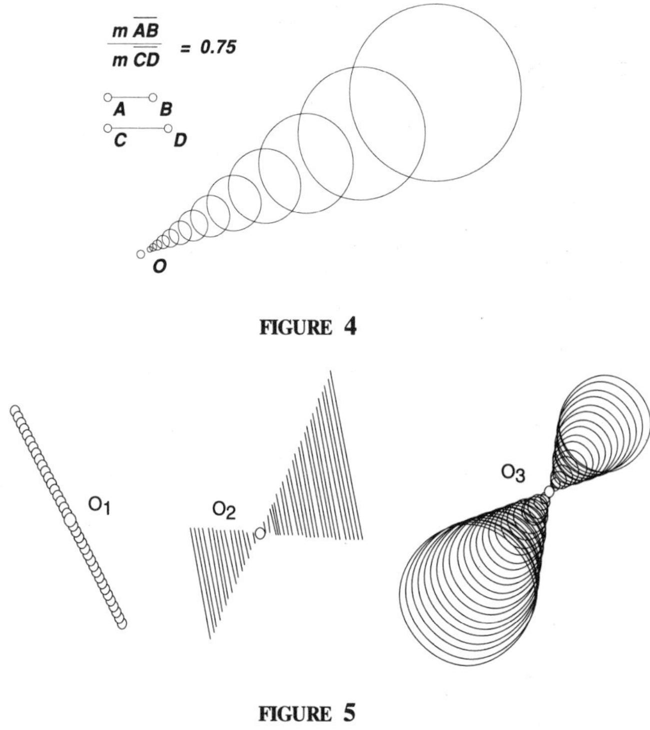

FIGURE 4

FIGURE 5

built-in transformation. In fact, dilation can be carried out in a variety of ways; this flexibility allows one to choose a method that fits with the desired geometrical and pedagogical goals of a construction:

- Free-hand dilation of an object by dragging it;
- Dilation of an object with respect to a center point C and a specified fixed ratio;
- Dilation of an object with respect to a center point C and a ratio specified as some measurement in a figure, for example, the ratio of two segments.

In Figure 4, a circle has been dilated repeatedly by a dilation whose ratio is AB/CD. If an endpoint of one of the segments AB or CD is dragged, the circles all enlarge or shrink accordingly.

2. Problem-Solving Using Images of Dilations

With dynamic geometry software one can trace an object S as it is dilated repeatedly with a fixed center O and variable ratio k. Figure 5 shows the traces of a point, a segment, and a circle under repeated dilation about a fixed center.

The set of dilations F_k with a fixed center O and a variable ratio k forms a group G_O, with group composition $F_k F_l = F_{kl}$. Traces such as in Figure 5 represent the orbit of a geometric object S under the action of such a group. This orbit is just the set of images $F_k(S)$, for all $k \neq 0$. Such an orbit trace can be very helpful in solving geometric problems.

The simplest orbit under the group G_O is the orbit of a point A, which is the line OA (with the point O removed). This simple observation can be surprisingly powerful. Here are two examples.

Example 1. Midpoints in the Thales triangle. For any dilation with center O, the image M' of the midpoint M of a line segment AB must be the midpoint of the image segment $A'B'$. Also, the orbit of M is the line OM. Thus, as A' is dragged, M' moves along line OM (Figure 6). From this we conclude that, if $A'B'$ is the image of AB under a dilation, then the center of dilation is on the line through the midpoints of AB and $A'B'$.

As a corollary, this means that in Figure 7, if AB and DE are parallel, C is the intersection of lines AD and BE and F is the intersection of lines BD and AE, then the line CF passes through the midpoints of AB and DE. (Thus CF is a median of triangles CAB and CDE.) We see this by observing that C is the center of a dilation that takes AB to

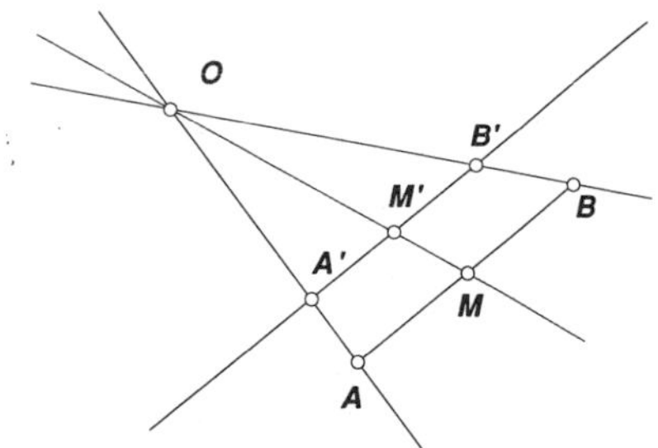

FIGURE 6

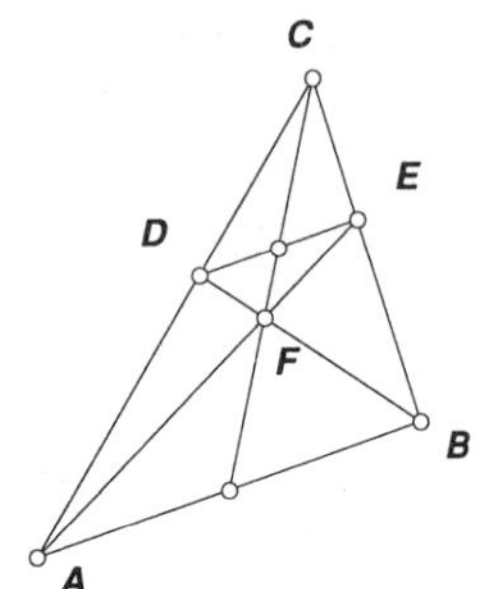

FIGURE 7

DE and that F is the center of a dilation (with negative ratio) that takes AB to ED. This also proves that the diagonals of a trapezoid such as $ABED$ intersect at a point on the line through the two midpoints of the parallel sides.

Example 2. Circle tangent to two given intersecting lines, through a given point. The orbit of the circle in Figure 5 is the trace of the set of circles tangent to two given lines that intersect at the center of dilation. We use this fact to solve the following problem: *Given two lines through O and a point P distinct from O, construct a circle through P tangent to both lines.*

We first find a partial solution by constructing a circle c tangent to the two lines, without regard for P. To do this, take any point T on OB, and construct a circle c through T whose center is the intersection of the angle bisector of AOB with the perpendicular to OB at T (see Figure 8).

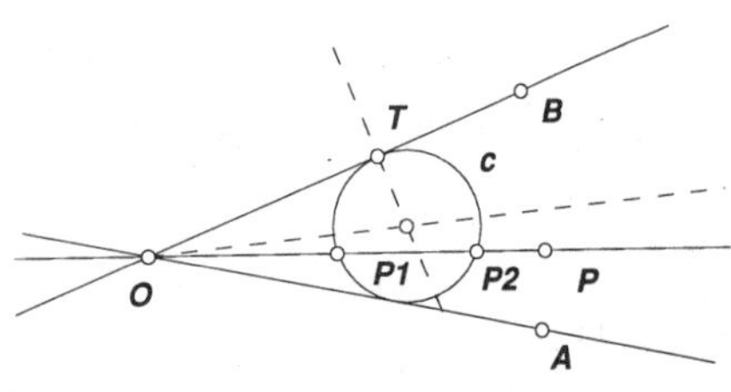

FIGURE 8

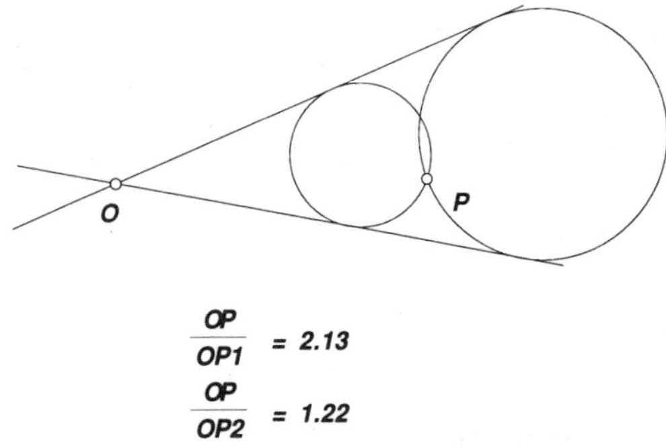

$$\frac{OP}{OP1} = 2.13$$

$$\frac{OP}{OP2} = 1.22$$

FIGURE 9

By dragging T, c can be moved so that it will pass through P. This experiment will reveal that two such circles pass through P. Now trace the locus of c as T is dragged; this is the same trace as in Figure 5. This suggests that the two required circles can be constructed by dilating c once we figure out the appropriate ratios that will make the image of c pass through P.

For a dilation F with center O, if the circle $F(c)$ passes through P, then $F^{-1}(P)$ is on c. We know that $F^{-1}(P)$ must be on the line OP, so points of the form $F^{-1}(P)$ are the intersections of the line OP with c. In general there are two such points, P_1 and P_2, and each produces a circle passing through P by dilating c by ratio OP/OP_1 or by ratio OP/OP_2. See Figure 9.

3. Problem Solving Using Centers of Similitude

In the previous section, problems were solved by constructing the images of an object by a dilation. Other problems can be solved by finding a center of dilation. Given two similar figures S_1 and S_2, a point which is the center of a dilation that takes S_1 to S_2 is called a ***center of similitude*** of the two figures.

For example, if AB and CD are any two parallel segments in the plane, the center E of the dilation that carries A to C and B to D is the intersection of the lines AC and BD. A center of similitude I for the segments AB and DC is found in the same manner. Thus, if orientation is ignored, two parallel noncongruent segments have two centers of similitude, one with a positive ratio k and the other with the negative ratio $-k$. See Figure 10.

If point D is dragged so that the segments AB and CD are approximately congruent, one of the centers, E or I, will remain between the segments and the other will move off the screen as the figure $ABDC$ approximates a parallelogram. This represents the exceptional case in which there is one

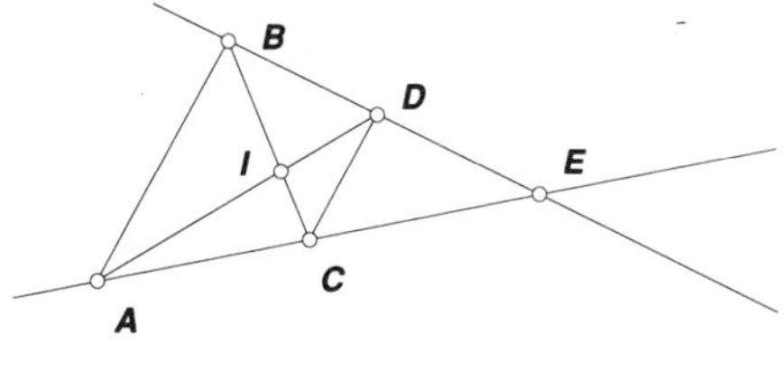

FIGURE 10

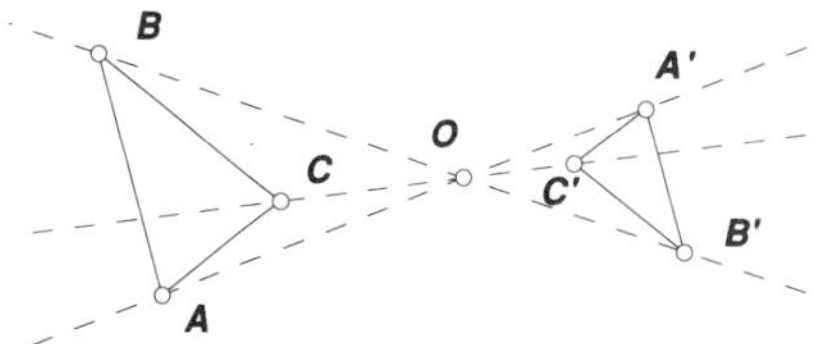

FIGURE 12

dilation with ratio -1 (a half turn), that carries AB to DC, and there is also a translation that carries AB to CD, but not a second dilation. We can imagine the second center of similitude as a point at infinity, in the direction of the dilation. (It is interesting to explore to what extent theorems about dilations remain true if one includes translations as a kind of generalized dilation with center at infinity.)

We give two elementary applications that use a center of similitude. In each case, it is instructive for a student to manipulate the figure to get insight into general relationships and the way they look in special cases. It is even more instructive if the student is challenged to create the figure in the first place. In both examples, the proof remains valid if the center of similitude is a point at infinity.

Example 3. We will prove the following: *Given two parallel lines, with points A, B, C on one line and A', B', C' on the other line, if $AC/AB = A'C'/A'B'$, then the lines AA', BB', and CC' are concurrent.*

The proof is simply to let O be the center of similitude of AB and $A'B'$. The dilation which carries AB to $A'B'$ must take C to a point D on line $A'B'$ with $AC/AB = A'D/A'B'$ (since dilations preserve signed ratios). Thus $D = C'$, since D is determined by this ratio. But D is on line OC, so C' must be on this line as well. See Figure 11.

Example 4. We will prove the following: *Given two triangles ABC and $A'B'C'$ with corresponding sides parallel, the lines AA', BB', and CC' are concurrent.*

To prove this, let O be the center of similitude of AB and $A'B'$. If F is the dilation which takes AB to $A'B'$, it carries line AC to the line through A' which is parallel to AC, so this must be line $A'C'$. Likewise the image of line BC is line $B'C'$. Since C is the intersection of AC and BC, point $F(C)$ must be the intersection of $A'C'$ and $B'C'$, namely C'. Thus O, C and C' are collinear. See Figure 12.

Example 5. Inscribing a given triangle inside a triangle. An interesting application of the construction in Example 4 solves the following problem: *Given two triangles ABC and DEF, construct a triangle $D'E'F'$ inscribed inside triangle ABC (extended) so that the sides of $D'E'F'$ are parallel to the sides of DEF.*

We first construct triangle $A'B'C'$ with sides parallel to ABC with DEF inscribed in $A'B'C'$. To do this, construct the line through D parallel to BC, the line through E parallel to AC, and the line through F parallel to AB. These three lines intersect to form the desired triangle $A'B'C'$. Next, find the dilation T from triangle $A'B'C'$ to triangle ABC as in Example 4. Then the image triangle $T(DEF)$ is inscribed in triangle ABC (Figure 13).

To find other triangles similar to DEF inscribed in triangle ABC (with sides extended), just use the rotation tool to rotate triangle DEF. As triangle DEF moves to a new position, the constructed triangle $A'B'C'$ and the center of dilation will move and change. The vertices of the image $T(DEF)$ will move along the sides of the stationary triangle ABC, so that the sides of the image remain parallel to those of the rotating DEF. If at the same time we also rotate a selected point in DEF (for example, the centroid G), the

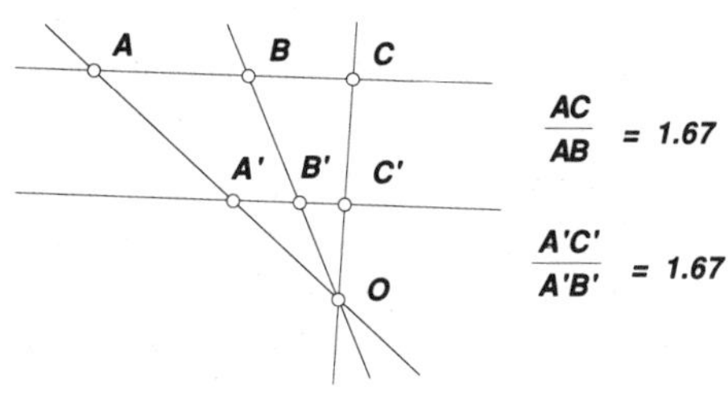

FIGURE 11

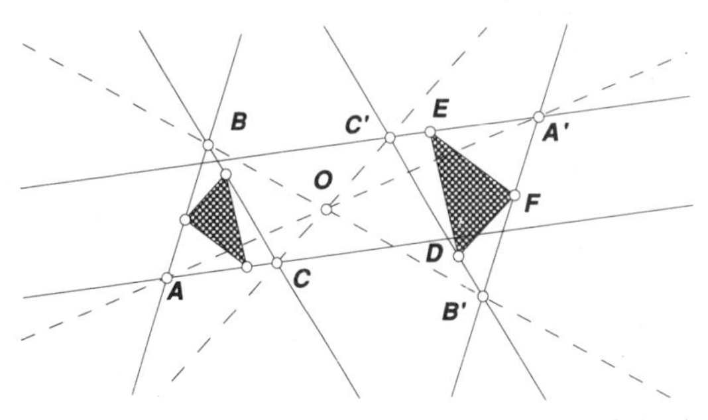

FIGURE 13

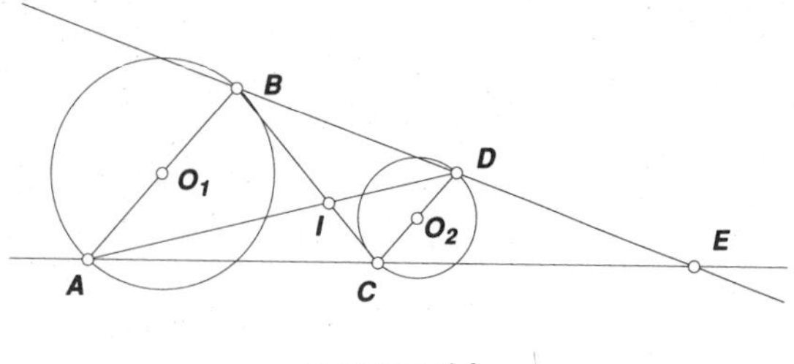

FIGURE 14

trace of the image point $T(G)$ will be a line despite the apparent "rotation" of $T(DEF)$. [16, pp. 71–73]

4. Applications that Use Centers of Similitude of Two Circles

Given two circles, one can construct two centers of similitude by the same method that worked for two segments. Just construct two parallel diameters and find the two centers of similitude, E and I, of the diameters (Figure 14).

The centers E and I will always exist unless the circles are congruent or concentric, in which case there is only one center. Centers can be inside, outside, or on the circles, but, if a center is exterior to one circle, it must be exterior to the other. If the circles are tangent, the point of tangency is a center of similitude. Figure 15 shows a few of the possibilities.

In order to work with these centers, it is useful to hide the construction lines and then create a script or macro that will construct the two centers automatically for two given circles.

Here are some examples of problems or situations where these centers appear.

Example 6. Common tangents to two circles. We solve the construction problem: *Given two circles, construct the lines tangent to both circles.*

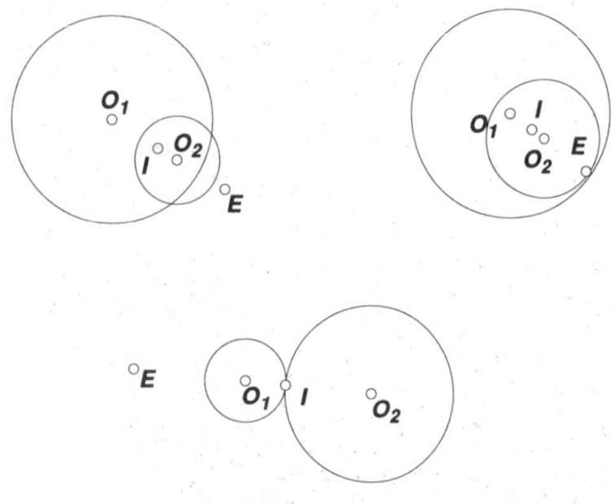

FIGURE 15

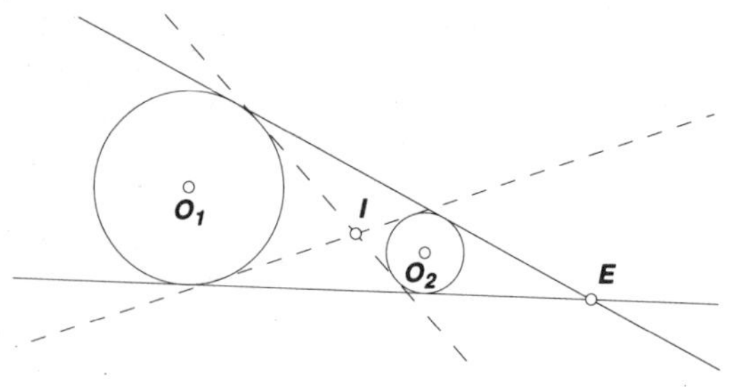

FIGURE 16

A line that is tangent to both circles must pass through a center of similitude that is exterior to (or on) both circles. So, to construct such tangents (if they exist), start with a center of similitude P which is exterior to one circle and construct two tangent lines to that circle through P. These lines will automatically be tangent to the other circle as well. (To construct the tangents to a circle with center A through a point P, construct the points of tangency as the intersection of the given circle with the circle with diameter AP [1, pp. 183–84], [8, pp. 50–52].) Figure 16 shows the case in which four tangents can be constructed.

Example 7. Points on the Euler Line. Given a triangle ABC, construct the circumcircle, which passes through A, B, and C, and has as its center O the intersection of the perpendicular bisectors of the sides of the triangle. For this same triangle ABC, construct the medial triangle $A'B'C'$, whose vertices are the midpoints of the sides of ABC. Finally, construct the circumcircle of triangle $A'B'C'$, and call its center O'. See Figure 17.

If the centers of similitude E and I of these two circles are constructed, where are these points located in triangle ABC?

The center I with negative ratio is not only a center of similitude of the two circles; it is the center of similitude

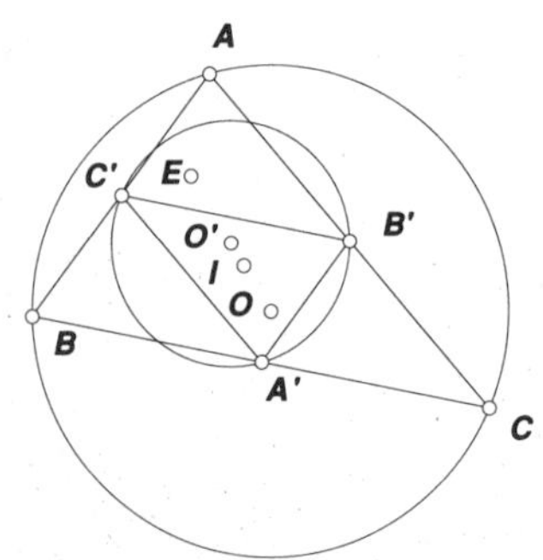

FIGURE 17

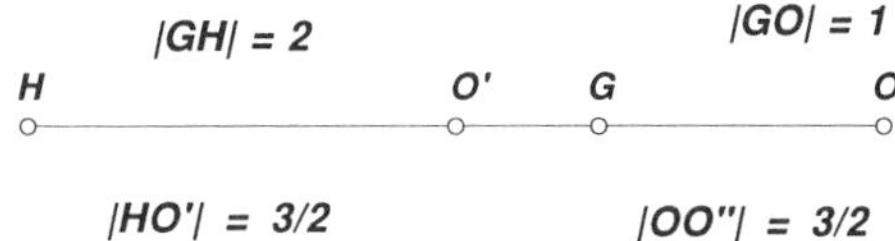

FIGURE 18. If $|GO| = 1$, then $|GH| = 2$, $|HO'| = 3/2$, and $|O'O| = 3/2$.

of the two triangles as well. This is an example of the center of similitude of two triangles with corresponding sides parallel, illustrated in Figure 12. The center is the point of concurrence of the lines AA', BB', CC'. But these lines are the medians of triangle ABC; so I is the *centroid* G of ABC. The ratio of similitude for this dilation is $-\frac{1}{2}$, since the ratio of the lengths of the triangle sides is $\frac{1}{2}$ and the directions are reversed.

Each perpendicular bisector of triangle ABC is an altitude of triangle $A'B'C'$. Thus the circumcenter O of triangle ABC is the orthocenter of triangle $A'B'C'$ (the orthocenter of a triangle is the point of concurrence of the altitudes). But $I = G$ is a center of a dilation with ratio -2 that carries triangle $A'B'C'$ to ABC; therefore, the point O is mapped by this dilation to H, the orthocenter of ABC. The ratio $GH/GO = -2$. Also, $GO'/GO = -\frac{1}{2}$, so the four points H, O', G, O are arranged on a line as in Figure 18.

This means that $HO'/HO = +\frac{1}{2}$, so H is the center of a dilation that carries O to O' with ratio $\frac{1}{2}$. Since the ratio of the radii of the circles is also $\frac{1}{2}$, this dilation must carry the large circle to the small one. This means that H is the other center of similitude, E.

The line containing the four points H, O', G, and O is called the *Euler line* of the triangle. The small circle in Figure 17 is called the Nine Point Circle of triangle ABC. There is a beautiful and rich relationship among dilations and special points and lines of the triangle ABC. For more about the nine-point circle, see [3, pp. 20–21]; for special points of the triangle, see [7, pp. 6–13], [6]; and, for more about dilations and triangles, see [4, pp. 165–69].

5. Compositions of Dilations

When two dilations are composed, the new transformation is necessarily a similarity. The following properties of such compositions (found in [16, pp. 28–29]) can be used to prove geometric theorems.

- The composition of a dilation with ratio k and a dilation with ratio l is a dilation with ratio kl (with the exception of the case $kl = 1$, when the composition is a translation).

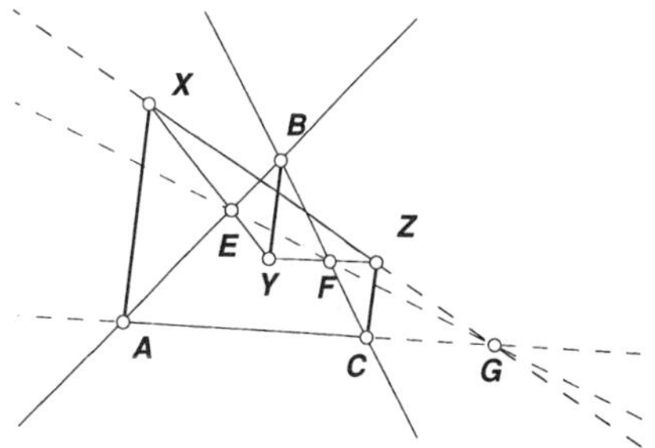

FIGURE 19

- If D_E is a dilation with center E and if D_F is a dilation with center F, then the center G of dilation $D_G = D_F D_E$ is on the line EF. (If $E = F$, then $G = E$ also.)

How can these composition relationships be visualized with a dynamic figure? Begin with dilation centers E and F. There are several possible ways to introduce the ratios of dilation into the figure. Here is one way.

Begin with a point A distinct from E and F; then construct a point B on line AE. Let D_E have ratio $k = EB/EA$ so that $B = D_E(A)$. Likewise, construct C as a point on line BF and define D_F to have ratio $l = FC/FB$ so that $C = D_F(B)$.

Take another point X in the plane and construct $Y = D_E(X)$ and $Z = D_F(Y)$ using the ratios k and l defined by A, B, and C. Observe that segment AX is parallel to segment BY in a Thales figure; also segment BY is parallel to CZ, so AX is parallel to CZ, and the center of similitude G can be constructed as the intersection of line AC with line XZ. Define the dilation D_G to have the ratio kl, so D_G takes AX to CZ. See Figure 19.

Now drag X around the plane. G will not move, and the line EF will always pass through G as expected. Measure GC/GA and compare it with kl as any of the points move.

The similarity $S = D_G^{-1} D_F D_E$ is an isometry which takes any line m to a line parallel to m (so S is a translation), and S fixes points A and X (so S must be the identity). Thus the composition $D_F D_E = D_G$ is a dilation. Also, for any point P on line EF, the image point $Q = D_F D_E(P)$ is also on EF. By definition, G is on line PQ, which is the same as line EF. (In the special case when $E = F$, then $E = G$ also.) For example, in Figure 19, let P be the intersection of line AX and line EF. Then Q is the intersection of line CZ and line EF by the Thales triangles in the figure. The location of G can be computed most easily by setting $P = E$.

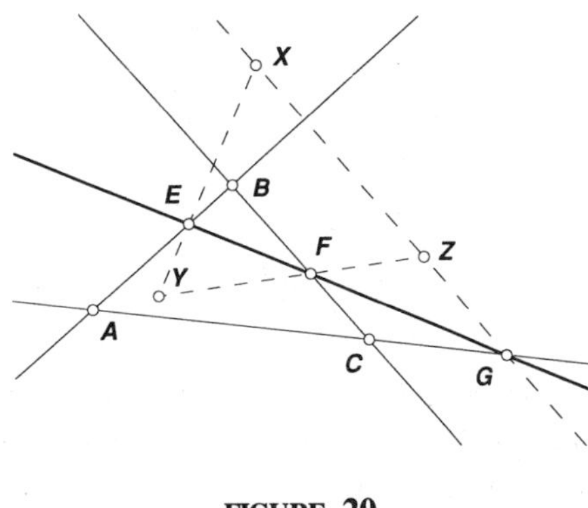

FIGURE 20

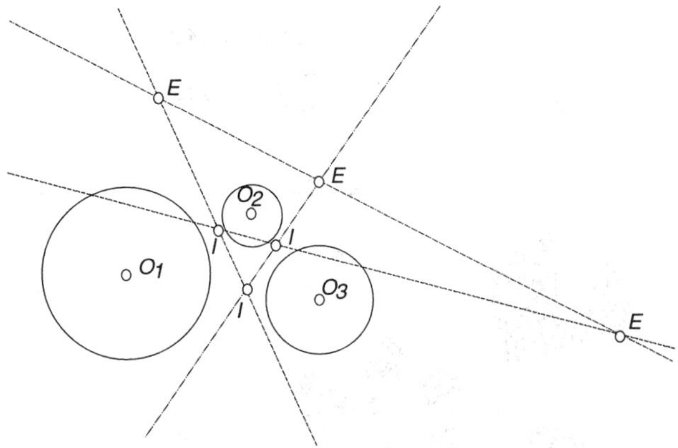

FIGURE 21

Example 8. Menelaus's Theorem. The previous composition relation can be interpreted as a theorem about points on the sides of a triangle: *Given triangle ABC, let E, F, G divide the sides AB, BC, and CA (extended), respectively. The points E, F, G are collinear if and only if*

$$(EB/EA)(FC/FB)(GA/GC) = 1.$$

See [3, pp. 66–67].

The points A, B, and C in Figure 20 are defined as in Figure 19. The dilations with centers E, F, and G are defined as before, so G is the intersection of line EF with line AC.

The product of the first two dilation ratios equals the third:

$$(EB/EA)(FC/FB) = (GC/GA),$$

or

$$(EB/EA)(FC/FB)(GA/GC) = 1.$$

Example 9. Three-Circle Theorem. We will prove the following: *Given three circles $c1$, $c2$, $c3$, each pair of circles has two centers of similitude. The six centers of similitude are arranged on four lines, three points to a line.*

If F is a dilation that takes $c2$ to $c2$ and G is a dilation that takes $c2$ to $c3$, then GF takes $c1$ to $c3$, so the center of similitude of GF must lie on the line through the centers of F and G. (See Figure 21). The theorem follows when this reasoning is applied to each case in the figure. [16, p. 29]

Since the product of two negative ratios is a positive ratio, there is at least one center E with positive ratio on each of the four lines. If two circles are tangent, the point of tangency is a center. This leads to some interesting special cases, which can be explored qualitatively by dragging the circles in the figure so that they appear tangent. For more theorems about three circles, see [16, pp. 31–32].

6. Beyond Dilations

What other similarity transformations are there besides dilations? One way to obtain a new similarity is to compose two similarities that are already known. The two kinds of similarities already mentioned in this paper are dilations and isometries. The isometries are all classified; any isometry is either a translation, a rotation, a line reflection, or a glide reflection [15, pp. 60–67], [11]. Thus we can investigate what similarities result from composing a dilation with each type of isometry.

In fact, this method produces all the similarities of the plane, for any similarity S is the product of a dilation D and an isometry F. To see this, let k be the ratio of similitude of S. If D is any dilation with ratio k, then $D^{-1}S$ has ratio 1; so $D^{-1}S$ is an isometry F, and $S = DF$. Thus we can use the classification of isometries to classify similarities. [2, pp. 67–76], [4, pp. 158–61]], [16, pp. 53–56].

A basic type of similarity transformation is the spiral similarity.

Definition. A ***spiral similarity*** with center O, ratio k, and angle θ is the composition of a rotation by angle θ about center O and a dilation with ratio k about the same center O. (A spiral similarity is also called a *rotary dilation.*)

If the center O is given, along with two points A and A' distinct from O, there is a unique spiral similarity F with center O for which $F(A) = A'$. The ratio of F is $k = |OA'|/|OA|$, and the angle of rotation equals the angle from AO to OA'. See Figure 22. When the angle of rotation is a multiple of 180°, the spiral similarity is a dilation; and, when $k = 1$, it is a rotation. It is clear from the definition that two spiral similarities F and G with the same center commute.

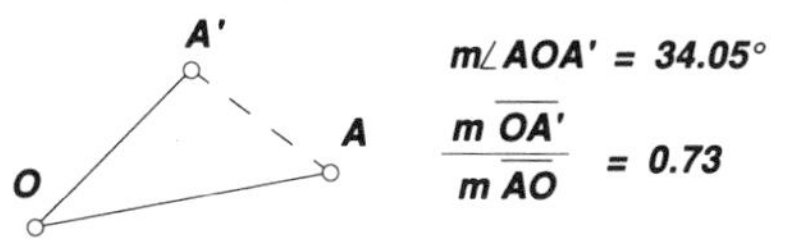

FIGURE 22

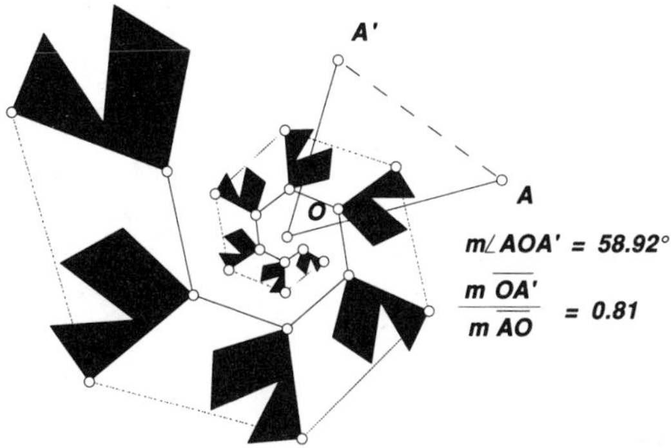

FIGURE 23

With dynamic geometry software, a spiral similarity can be defined as a composition of built-in transformations. Once the transformation is defined, it can be applied to any shape. The image shapes will be arranged around logarithmic spirals that connect corresponding points. In Figure 23, a spiral similarity has been applied iteratively to a polygon, and two spirals connecting corresponding points have been drawn. One can manipulate the shape of the spiral by dragging A or A'. Exploring a spiral in this way provides insight into the relationship between the shape of the spiral and the angle and ratio of the spiral similarity.

What other types of similarities can arise as compositions? Students can conjecture and experiment with these transformations using dynamic geometry software. They can define a similarity by composing built-in transformations, a dilation D with an isometry F. One tool that can be used is to trace the orbit of a point P under repeated iterations of the similarity $S = DF$, a technique described for isometries by Parks [10]. In the context of dynamic geometry, this means that a sequence of points $S^n(P)$ can be computed and connected by segments. Then P can be dragged around, and one can observe the nature of the orbit of P. Some key cases to consider are:

- Composition of a dilation and a translation;
- Composition of a dilation and a rotation;
- Composition of two spiral similarities with different centers.

The surprise will be that the latter two cases are both spiral similarities. The orbit of P, as it moves around, can

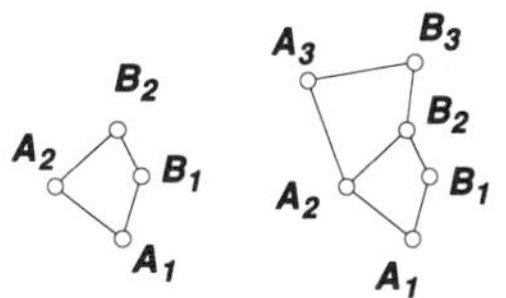

FIGURE 24

be observed experimentally to be a logarithmic spiral. The center of the spiral can be approximated by looking for a position of P which is fixed by the transformation. The center can also be constructed, as we will see later. The nature of these similarities is no longer a surprise after one proves that any orientation-preserving similarity which is not an isometry must be a spiral similarity [16, pp. 53–54].

Orientation-reversing similarities that are compositions of dilations with line reflections are also interesting to study, but, for the rest of this paper, we will limit ourselves to the orientation-preserving transformations: dilations, translations, rotations, and their compositions.

Example 10. Spiral Growth. Similarities appear in modeling certain kinds of growth in nature. An organism may grow an object such as a shell or a branch by replicating the same shape over and over. In some cases, such as the chambers of a honeycomb, the shape is the same size each time. In other cases, such as a snail shell, as the organism grows bigger, each new object is proportionately bigger as well [12, pp. 748–79].

For example, a shell such as the shell of a chambered nautilus can be modeled by starting with a single chamber and then building a new chamber, similar to the first, which is attached to a side of the original chamber. For example, in Figure 24, quadrilateral $A_2A_3B_3B_2$ is similar to $A_1A_2B_2B_1$.

If this process is repeated, by attaching a new quadrilateral to edge A_3B_3 of $A_2A_3B_3B_2$ and then continuing to add chambers in the same way, a shell-like spiral of quadrilaterals is produced, as in Figure 25. The edges of this shell lie on logarithmic spirals.

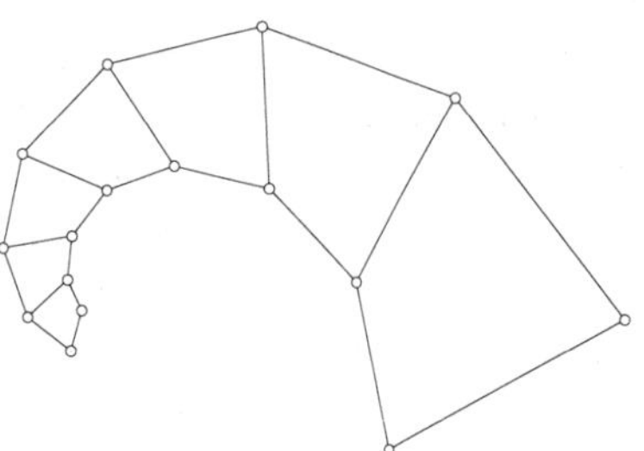

FIGURE 25

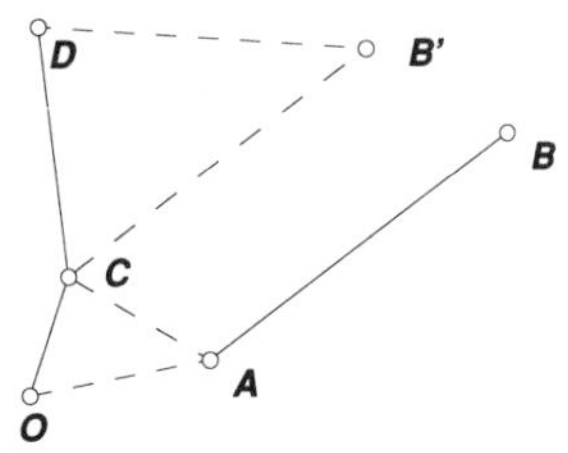

FIGURE 26

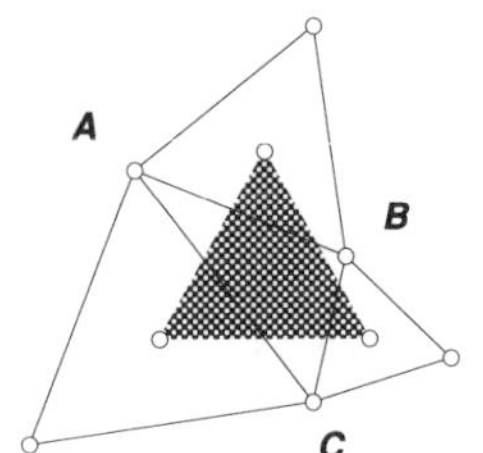

FIGURE 27

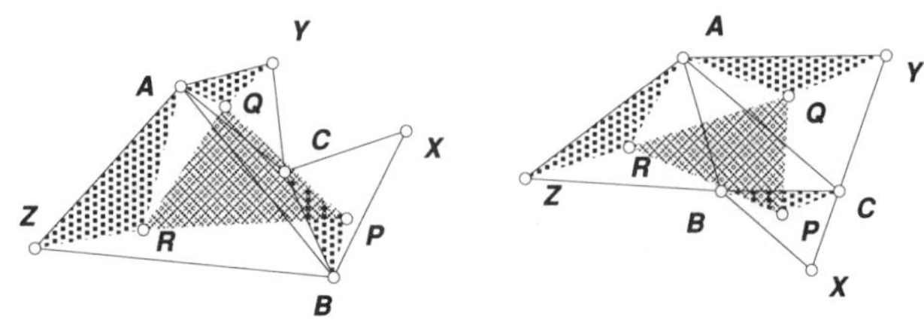

FIGURE 28

To build this model with dynamic geometry software, the original chamber $A_1A_2B_2B_1$ is constructed and then the second and additional chambers can be constructed using a similarity. In this example, an orientation-preserving similarity that takes segment A_1B_1 to segment A_2B_2 carries the first quadrilateral to the second one; the same transformation also carries the second quadrilateral to the third, etc.

Here is a way to construct this similarity and also to see why logarithmic spirals appear. It reflects the following key fact about similarities [16, pp. 43–44, 53–54]:

- Given a segment AB and a segment CD, there is a unique orientation-preserving similarity transformation $T_{AB,CD}$ that carries AB to CD. This similarity is either a spiral similarity or a translation.

One method for constructing $T_{AB,CD}$ is to compose two transformations. Let H be the translation that takes A to C, and let G be the spiral similarity with center C that carries $B' = H(B)$ to D. Define $T_{AB,CD}$ to be GH.

In the case when $B' = D$, then G is the identity and $GH = H$ is a translation. When $B' \neq D$, one can construct the center O of the spiral similarity GH using a different similarity: let O be $T_{B'D,AC}(C)$. Then GH is the spiral similarity with center O that takes A to C.

When AB is not parallel to CD, this can be seen in Figure 26; in this case triangle $B'CD$ is carried by $T_{B'D,AC}$ to the similar triangle AOC, and the angle of rotation $B'CD$ of GH equals angle AOC. Also, the ratio $|AB|/|CD| = |OC|/|OA|$. When AB is parallel to CD, then GH is a dilation (a special case of a spiral similarity), and the definition of O is the same, since $B'CD$ is similar to AOC even if the points $B'CD$ are collinear.

Example 11. A generalization of Napoleon's Theorem. Napoleon's Theorem concerns an arbitrary triangle with equilateral triangles constructed on its sides, as in Figure 27. It states that, no matter what the shape of triangle ABC, the centers of the equilateral triangles also form an equilateral triangle. This is a fascinating figure to explore with dynamic geometry. Proofs of Napoleon's Theorem can be given using isometries [15, p. 93] and tessellations [14, pp. 49–51].

We will use properties of similarities to prove the following theorem of de Villiers.

Theorem [13]. Given a triangle ABC and a point X, consider the triangle XBC on one side of ABC. Construct points Y and Z so that the triangles CAY and BZA are similar to XBC. Then for any point P in triangle XBC, the corresponding points in the two similar triangles form a triangle PRQ which is similar to BZA.

(By a point corresponding to P, it is meant that a similarity transformation carrying triangle XBC to the other triangle will carry P to the corresponding point. Thus, special points of each triangle, such as the centroid, the orthocenter, etc., will be corresponding.)

In Figure 28, after the point P was chosen, the triangle *PBC* was lightly shaded, as well as the corresponding triangles in the other two triangles, so as to see the similarity relationships. The triangle *PRQ* is shaded a darker shade.

These triangles *CAY* and *BZA* can be constructed dynamically using the segment-to-segment transformation $T_{AB,CD}$ described above. Let γ be the transformation $T_{BX,AC}$, and let β be the transformation $T_{CX,AB}$. Then $\gamma(C) = Y$ and $\beta(B) = Z$. Also, $\gamma(P) = Q$ and $\beta(P) = R$.

Now we need to find why triangle PRQ is similar to the others. To do this, we need to find a relationship connecting the spiral similarities β and γ.

Note that we can arrive at A by two paths, since it is a vertex of two attached triangles: $A = \beta(C)$ and $A = \gamma(B)$. But $C = \gamma(X)$, and $B = \beta(X)$, so $A = \beta\gamma(X) = \gamma\beta(X)$.

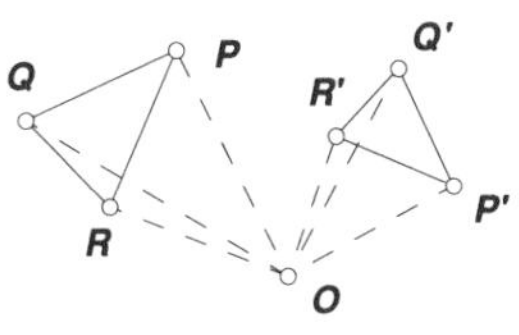

FIGURE 29

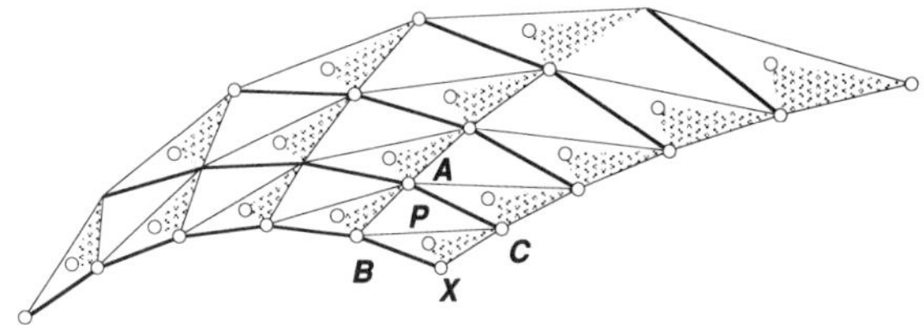

FIGURE 30

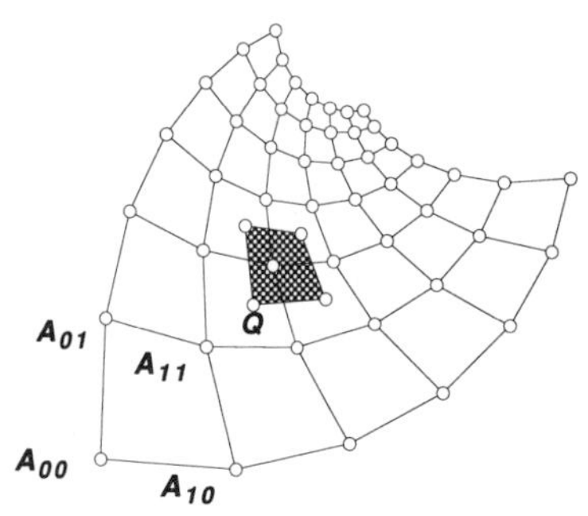

FIGURE 31

A Lemma (stated and proved below) implies that β and γ commute for all points and have the same center. Let's see how this fact proves the Theorem.

The triangle PRQ is the same as triangle $P\beta(P)\gamma(P)$. To prove the theorem we must prove that, for any point P' distinct from O, the triangle $P'R'Q' = P'\beta(P')\gamma(P')$ is similar to $P\beta(P)\gamma(P)$. If δ is the spiral similarity with center O that takes P to P', then triangle $P'\beta(P')\gamma(P') = \delta(P)\beta(\delta(P))\gamma(\delta(P))$. But, since all these spiral similarities commute, this is the same as the image under δ of the triangle $P\beta(P)\gamma(P)$. Thus, the triangles are similar. See Figure 29.

Lemma. *Let β and γ be spiral similarities. If for some point X, $\beta\gamma(X) = \gamma\beta(X)$, then the two similarities commute for all X (and thus have the same center).*

Proof. Let $U = (\gamma\beta)^{-1}(\beta\gamma)$. U is an isometry, since the product of the similarity ratios is 1. The angles of rotation of these spiral similarities add up to a multiple of 360 degrees, so this means that U must be a translation. But we know that for some X, $U(X) = X$. Since the only translation with a fixed point is the identity, U is the identity. Now we show that any commuting spiral similarities γ and β have the same center. Assume neither transformation is the identity. Let M be the center of γ. Then $\gamma\beta(M) = \beta\gamma(M) = \beta(M)$, so $\beta(M)$ is a fixed point of γ, so it is equal to the center M of γ. But this says $M = \beta(M)$, which can happen only if M is also the center of β. QED

Now that we have seen that the figure in the de Villiers's generalized Napoleon's Theorem can be constructed from spiral similarities, let's see what happens when these transformations are iterated.

Beginning with triangle XBC, we apply combinations of β and γ. Since these transformations commute, the order does not matter, only the number of applications of each transformation. The result is akin to Pascal's triangle. Each β moves a step in the B direction, and each γ moves in the C direction. The whole figure is made up of pieces of two families of logarithmic spirals, all with the same center. The β-spirals are indicated by the heavy lines.

In Figure 30, de Villiers's generalized Napoleon's Theorem figure is contained in a spiral tessellation based on the quadrilateral $ABXC$. The quadrilaterals of the tessellation are built by the Chambered Nautilus construction of Example 10, carried out for both pairs of opposite sides rather than just one. This figure can be found in [14, pp. 52–53].

For any point Q, the quadrilateral $Q\beta(Q)\gamma\beta(Q)\gamma(Q)$ is similar to the original quadrilateral and each quadrilateral tile. An example is the gray quadrilateral in Figure 31, which lies in a tessellation figure based on quadrilateral $A_{00}A_{10}A_{11}A_{01}$. The shape of the spirals depends on the angles and side ratios of the quadrilateral. In Figures 31 and 32, we see how the spiral pattern can change dramatically when the fundamental quadrilateral changes.

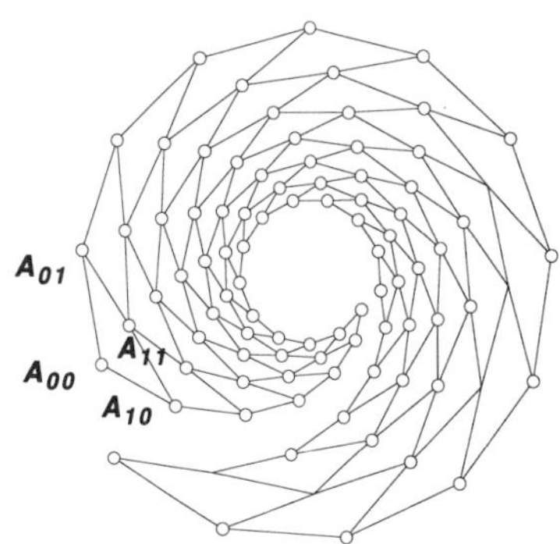

FIGURE 32

Conclusion. We have seen that similarities appear in geometry in many guises. There are many interesting directions to follow from here. More applications of similarities to geometry can be found in the book of Yaglom [16], for example. Some approaches to the foundations of geometry are based entirely on transformations [4]. There is an important link between complex number mappings, spiral similarities, spirals, and complex exponentials [5]. Also, complicated patterns of logarithmic spirals appear in many fractals generated by the recursive composition of similarities [9]. The study of these topics can be enriched by exploring dynamic geometry figures.

References

1. Birkhoff, G.D. and Beatley, R. (1959). *Basic Geometry*, Chelsea, New York.
2. Coxeter, H.S.M. (1961). *Introduction to Geometry*, Wiley, New York.
3. Coxeter, H.S.M. and Greitzer, S.L. (1967). *Geometry Revisited*, New American Library #19, Mathematical Association of America.
4. Guggenheimer, H. (1967). *Plane Geometry and Its Groups*, Holden-Day, San Francisco.
5. Hahn, Liang-shin (1994). *Complex Numbers and Geometry*, Mathematical Association of America, Washington, D.C.
6. Hofstadter, Douglas R. (1996). "Discovery and Dissection of a Geometric Gem," *Geometry Turned On*, Mathematical Association of America, Washington, D.C., pp. 3–14.
7. Honsberger, R. (1995). *Episodes in Nineteenth and Twentieth Century Euclidean Geometry*, New Mathematical Library #37, Mathematical Association of America, Washington, D.C.
8. King, J. (1995). *Geometry Through the Circle with The Geometer's Sketchpad*, Key Curriculum Press, Berkeley.
9. Lauwerier, H. (1991). *Fractals: Endlessly Repeated Geometrical Figures*, Princeton.
10. Parks, James M. (1996). "Identifying Transformations by their Orbits," *Geometry Turned On*, Mathematical Association of America, Washington, D.C., pp. 105–108.
11. Schattschneider, D. (1996). "Visualization of Group Theory Concepts with Dynamic Geometry Software," *Geometry Turned On*, Mathematical Association of America, Washington, D.C., pp. 121–127.
12. Thompson, D'Arcy W. (1942). *On Growth and Form*, Cambridge.
13. de Villiers, M. (1996). "The Role of Proof in Investigative Computer-based Geometry: some Personal Reflections," *Geometry Turned On*, Mathematical Association of America, Washington, D.C., p. 15.
14. Wells, D. (1988). *Hidden Connections, Double Meanings*, Cambridge University Press, Cambridge.
15. Yaglom, I.M. (1962). *Geometric Transformations*, New Mathematical Library #8, Mathematical Association of America, Washington, D.C.
16. Yaglom, I.M. (1968). *Geometric Transformations II*, New Mathematical Library #21, Mathematical Association of America, Washington, D.C.

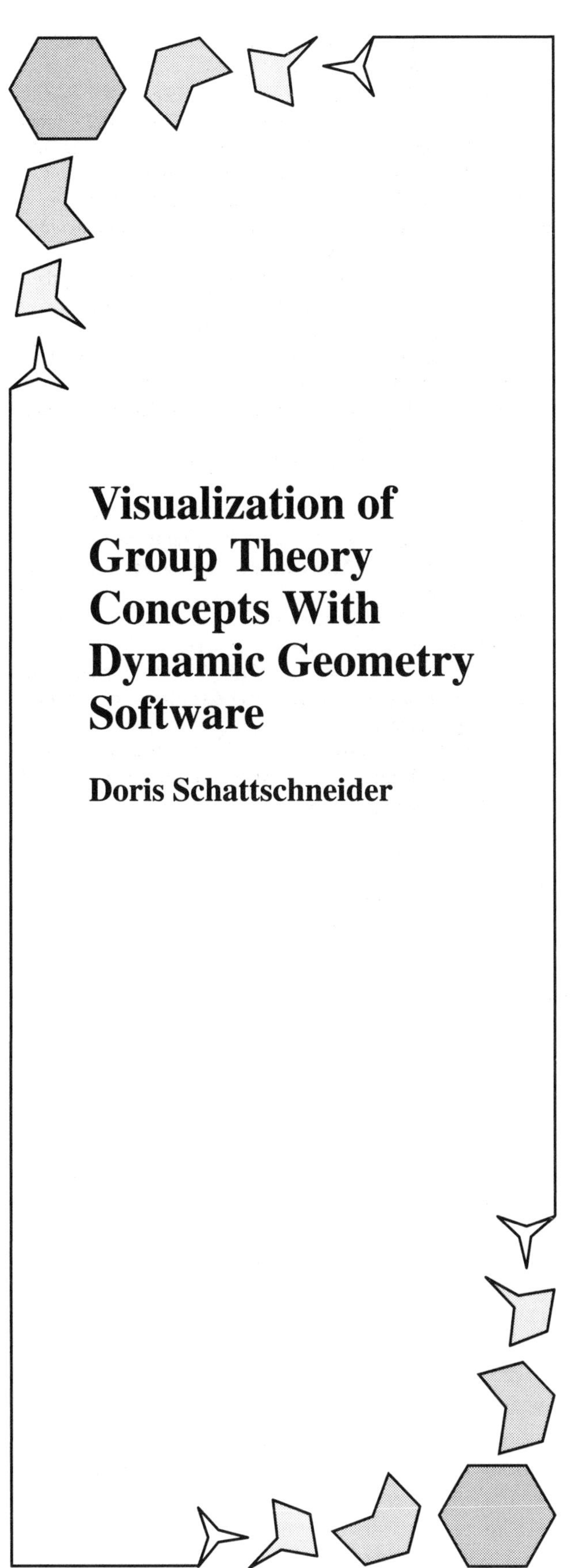

Visualization of Group Theory Concepts With Dynamic Geometry Software

Doris Schattschneider

In most colleges and universities, geometry in the Euclidean plane is not a significant part of the curriculum—a bit is done in the calculus sequence, a bit is done in linear algebra, and perhaps one course for prospective school teachers of mathematics is primarily devoted to this topic. Yet the modern algebra course (a staple for most math majors) with its study of groups provides a wonderful opportunity to use dynamic geometry software to visualize many abstract concepts.

Historically, groups were studied not as abstract algebraic systems, but as sets of actions—what we today call groups of transformations. Groups of plane isometries are elementary examples of such groups, and they are complex enough to embody almost all of the basic concepts and properties encountered in a first course in group theory. Some texts (for example, [3]) begin by investigating simple transformation groups such as the group of symmetries of an equilateral triangle or square. One advantage of this approach is the resulting de-emphasis of number systems as examples of groups, allowing students to see that not every group has all the nice properties that number systems do. I also like this approach since the actions of plane isometries can be visualized, their properties can be witnessed, conjectures can be made, and theorems can be verified with a dynamic geometry program like *The Geometer's Sketchpad.* With this tool, seeing can most often lead to understanding and believing.

A fundamental theorem about plane isometries states that the action of an isometry on a single triangle completely determines the action of that isometry on all points in the plane. Certainly this is not true of most functions and is a surprise in itself.[1] The theorem provides justification for our visual approach: by looking at how an isometry (or composition of isometries) acts on one of the simplest of plane figures, we know its full properties. So the first theorem that is proved is the "Three Points Theorem."

Three Points Theorem. *The images of three noncollinear points completely determine a plane isometry. That is, if* A, B, C *are noncollinear points in the plane and the images* $T(A) = A'$, $T(B) = B'$, $T(C) = C'$ *under an isometry* T *are known, then the image* $T(X)$ *of any other point* X *in the plane is determined by the condition that all distances are preserved.*

The proof of the theorem can be carried out by hand with a compass (which I have students do). It can also be carried out with *Sketchpad* (which I have students do after the "by hand" effort), and in this case the point X can be dragged at will, and the point $T(X) = X'$ will be seen to be uniquely determined as it moves in tandem with X. Here

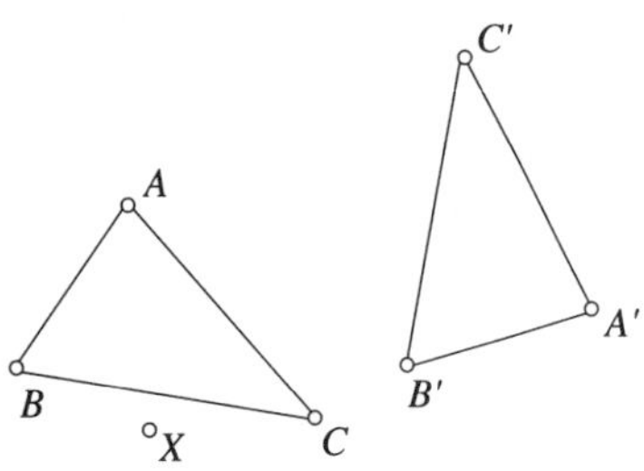

FIGURE 1

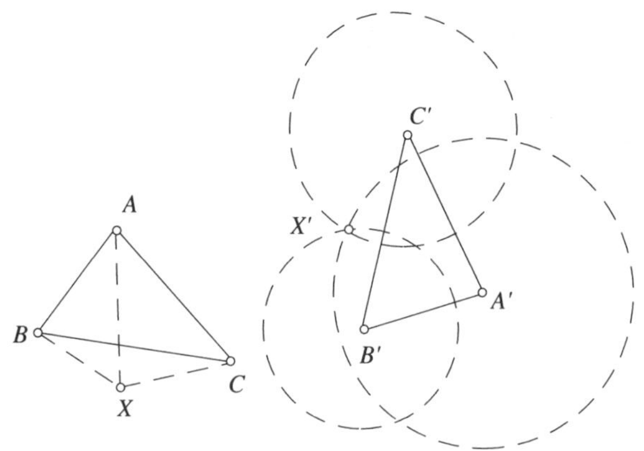

FIGURE 2

are the student instructions for the proof, augmented by two figures. (Figures are not provided to the students; they are to construct them.)

Proof. Place 3 noncollinear points A, B, C. Choose any point A' in the plane to be $T(A)$; choose $B' = T(B)$ on the circle with center A' and radius AB. (Since T preserves distances, we must have $|AB| = |A'B'|$.) Finally, choose $C' = T(C)$ to be one of the two intersection points of the circle centered at A' with radius $|AC|$ and the circle centered at B' with radius $|BC|$. (Why must C' be chosen this way?) Now place a point X (somewhere near triangle ABC to make the picture easier to see). [See Figure 1.]

Show there is only one point X' such that the distances $X'A', X'B', X'C'$ are equal to the corresponding distances XA, XB, XC, so this point X' must be chosen as $T(X)$. The point X' is constructed as the unique intersection point of three circles: the circle centered at A' with radius AX, the circle centered at B' with radius BX, and the circle centered at C' with radius CX. [See Figure 2.] When you have done the construction with *Sketchpad*, drag the position of X to see what happens. Why is X' unique? (How many points of intersection are possible for three circles?) What happens to the proof if the points A, B, C are collinear?

Once this theorem has been proved, students can investigate properties of isometries by simply investigating their actions on triangles. It is best to use scalene triangles since it is easier to identify corresponding points with their images, and there is no possibility of having more than one isometry send a figure to its image. We all know the danger of inference from a single picture, even if students easily make such inferences. But with *Sketchpad*, the shape of a triangle on which an isometry acts is easily changed by dragging: as it moves through a seemingly infinite number of shapes, its changing image can be observed. The justification for making solid conjectures is that you see not one, nor even a small number of instances of the observed behavior, but what seems to be a continuum of examples. With this technique, the evidence for conjectures is not only compelling visually, but also mathematically. If certain observed properties are "always" true for a particular isometry—if it holds for "all" shapes of the triangle as it is dragged—it is likely always true.

Isometries themselves in *Sketchpad* can also be dynamic. For example, the vector that defines a translation can be dragged to produce new vectors with different slopes and/or lengths, and, as this is done, the image of a triangle on which the translation acts is simultaneously changed to a new image under the newly defined translations. Similarly, reflection mirrors can be dragged to new positions in the plane, centers of rotation can be dragged to new positions, and angles of rotation can be dragged to change their measure and/or direction. Using these dynamic features, students can infer whether classes of isometries have particular properties—e.g., is composition of two translations always commutative? By manipulating the defining elements of an isometry, they can also discover particular cases in which properties hold that are not true for all cases, for example, they can determine precisely when two reflections commute.

The concepts we illustrate for groups of isometries are fundamental in any introductory course studying groups. To begin the study of a group, we must be told how the elements combine and when two elements are considered to be equal. For isometries, elements combine using composition of functions. The composition of two isometries is a two-step process. It can be witnessed as the first isometry acts on a figure to produce an image and then the second isometry acts on the image to produce a second image. This defines a new isometry whose action takes the original figure to the final image obtained by the two-step process. To illustrate this convincingly with *Sketchpad*, the first image can be colored with one color or shade, and the second image colored with a different color or shade. It is also possible to hide the first image and retain only the final image. These visual aids help students to follow the action of the composition. *Sketchpad* also allows a user-defined composition of isometries to be added to its built-in menu of transformations so it can be invoked for further use.

The definition of equality of two isometries is often not treated explicitly in texts, and this can cause confusion for students. In groups of numbers with the usual rules of arithmetic, students readily accept that the numbers $2+3$ and $4+1$ are equal or that $\frac{2}{3}$ and $\frac{8}{12}$ are equal. But without previous knowledge of arithmetic, these equalities are not obvious! Since students usually are not familiar with isometries, the definition of equality must be clearly given. Two isometries are equal if and only if they are equal as functions; that is, for every point in the plane, the image of the point under the first isometry is the same as the image of the point under the second isometry. The fact that any isometry must be a translation, rotation, reflection, or glide-reflection is useful to know, but not necessary in this context. In a geometry course, we prove (with constructions using *Sketchpad*) the fundamental theorem that every isometry is the composition of at most three reflections. The fact that there are only four types of isometries then follows as a corollary in which one examines all possible compositions of three or fewer reflections. In the group theory course, this fact can be assumed when needed.

Just as with numbers, two isometries constructed via completely different compositions may be equal. For example, a rotation of 360° about any center is equal to the composition of a reflection with itself in any mirror—both are equal to the identity isometry that sends each point to itself. This example is easy to imagine, but with more complex examples, it may not be so easy to decide. For such examples, thought experiments alone are not enough; checking them out with *Sketchpad* is extremely helpful in seeing and understanding. The Three Points Theorem implies that, to see whether two transformations are equal, students need only to check that the image of a triangle under the first transformation coincides with the image of that triangle under the second transformation.

In what follows, I give several examples of how group theory concepts (for groups of isometries) can be discovered or verified with the use of *Sketchpad*. For most, I briefly describe exercises that can be performed by students, and in a few cases I provide the actual exercise. Several of the basic concepts of transformations, such as composition and equality (discussed above), commutativity, and inverses, are ones that are encountered in elementary courses before the formal study of group theory. These exercises can be helpful in those first encounters.

Inverses. Students draw a scalene triangle and act on that triangle by an isometry (a dynamic translation, or rotation, or reflection, or glide-reflection). They are asked to fill in the interior of the image triangle and find another isometry that takes this back to the original triangle. It is a bit of a game—what action will take the filled-in interior of the image and plop it inside the outline of the original triangle? They quickly discover that every isometry has an inverse, and it is the same type of isometry; in fact, they can give a complete description of the inverse in relation to the original isometry.

Order of an Element. Students are asked to act on a scalene triangle by an isometry, fill the interior of the image triangle, then repeat the action of the same isometry on the interior of the image triangle and iterate this process several times, each time acting by the same isometry on the most recent image. After how many iterations does the image coincide with the original triangle? For a reflection, "one" is the answer—a reflection is its own inverse; as a group element, it has order 2. For a dynamic rotation in which the angle can be adjusted through a seeming continuum of values, the answer will vary from "one" to "never"—further investigation and measurement of the angle (or specification of a numerical value for angle measure) will reveal and confirm that the order of the element is directly related to the measure of the angle. This exercise can only be completed accurately by going beyond the visual and noting that, if the rotation has order n, then $nx = 2\pi k$, where x is the angle's measure and k is an integer. For a translation or a glide-reflection, the answer will be "never"—these are group elements of infinite order. In every case where the order is finite (and in theory, when the order is infinite), the number of distinct images of the triangle generated by endless repeated application of the isometry is exactly the order of the isometry. Other concepts that are implicit (and can be made explicit) in this exercise are generator of a cyclic group and closure.

Commutativity. Several investigations can show that composition of two isometries is sometimes commutative and sometimes not. For each investigation, students are directed to sketch a scalene triangle, act on it by a specified isometry R, then act on the image of the triangle by a second specified isometry S and hide the intermediate image. Then they are directed to fill the interior of the original triangle and transform that interior by S and transform its image by R and hide the intermediate image. The two isometries commute if and only if the two final images agree. To commute, the transformed interior must land exactly in the transformed outline of the triangle. Students can be asked to drag a vertex of the triangle to change its shape and see whether what they observe still holds; they also can be asked to drag the defining elements of the isometries (translation vectors, rotation centers and angles, reflection and glide mirrors) to see whether their observation holds in "all" cases. With this technique, students will find that translations always commute, as do rotations about the same center. But in general, two isometries do not commute. For example, two reflections commute if and only if the mirror lines of the reflections are perpendicular.

Figure 3 shows an exercise in which students discover that, in general, a translation and a reflection do not commute. But they are asked to go beyond this observation and discover under what circumstances a translation and a reflection *do* commute. They will find that this happens if and only if the translation vector and reflection mirror are parallel.

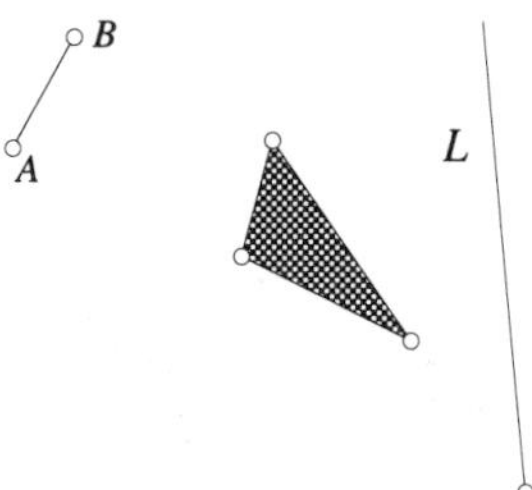

FIGURE 3. Exercise on commutativity of a translation and a reflection. AB is a translation vector, and L is a reflection mirror.

> Instructions to the student: Reflect the triangle (but not its interior) in L, then translate its image by the marked vector AB. Hide the intermediate image. Now select just the interior of the original triangle, and translate it by the marked vector AB, then reflect its image in L. Hide the intermediate image. Do the two final images coincide? What can you say about the commutativity of this translation and this reflection? Now drag point A to change the direction of vector AB. Is there a position of vector AB for which the translation by AB and reflection in L commute (give the same final image for composition of the two isometries in either order)? If so, carefully describe the position and give reasons for your answer. Try the same exercise again, first dragging the endpoint of line L to change its slope.

Generator, Orbits. The Three Points Theorem tells us that there can be a complete identification between the elements of a group of isometries and the images in a pattern generated by the group acting on a single asymmetric motif with three noncollinear points. When the motif is asymmetric, points do not need to be labled, and there can be only one isometry that matches the motif with any of its images in the plane. To illustrate the concepts of generator for a cyclic group and orbit of an object on which a group acts, students are directed to draw a simple asymmetric motif, called the generating motif. They then act repeatedly on the motif and its images by a single transformation (as done in the previous exercise on order of an element) to generate a pattern. The pattern generated should be complete (have all possible images) if the transformation is of finite order, and have a large number of images if the transformation is of infinite order. The symmetry group of the complete pattern is the cyclic group generated by the single transformation. The pattern is also the orbit of the generating motif under the group that is generated by the transformation.

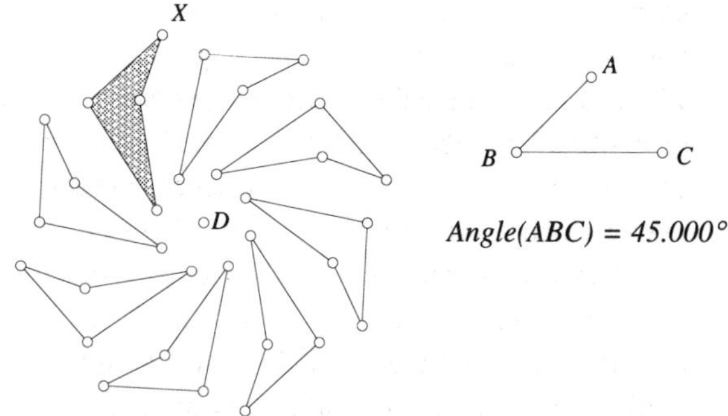

FIGURE 4. Exercise on generating a cyclic group by a rotation and finding subgroups of the cyclic group. D is the center of rotation and angle ABC is the angle of rotation.

Figure 4 shows the setup for a student exercise in which a script generates eight images of a motif by repeatedly rotating it through an angle ABC, initially set at 45°. As the script is played, the shaded interior of the motif rotates 45° like a clockhand about the center D, and at each rotation, fills in one of the blank outlines of the motif. After seven repetitions, the pattern is complete. Orbits of particular points are determined by the students—in general, orbits of points do not all have the same size, and the number of points in the orbit of a single point is determined by its position with respect to fixed centers (or mirrors of the transformation when reflections are involved). By dragging angle ABC to increase its measure, students can discover other generators of the same group. A different rotation will generate this group if and only if exactly the same pattern is generated.

> Instructions to the student: Open the C_8 script and put it at the right side of the window so it doesn't obscure this sketch (as shown in Figure 4). On this sketch, select, in order (hold shift), shaded interior,

points A, B, C, D. Play the script—this shows how one repeated rotation of 45° generates the whole group C_8 (and the whole design from a single motif). Describe the orbit of point X, of point D. Now drag the point A to increase the angle and see what happens to the images of the shaded motif. Find which other angles generate the group. Find which angles generate a subgroup (90°, for instance), and find the order of these subgroups (which is the number of images they produce).

Closure, Subgroup. The same exercises that demonstrate the generation of a cyclic group can also demonstrate the concepts of group closure and subgroups of a cyclic group. In the instructions in Figure 4, students are asked to rotate one side of angle ABC to increase its measure and report what happens. Rotations having angles for which the eight images of the selected motif coincide with the eight images outlined in the C_8 pattern are also generators for the group. Some rotations produce a smaller number of distinct images that all coincide exactly with some of those outlined. This leads students to discover the subgroups of C_8 and generators of these subgroups—for example, a 90° or 270° rotation generates a subgroup of order 4 (shown by four images of the motif) and a 180° rotation generates a subgroup of order 2 (shown by two images of the motif). The importance of group closure is also demonstrated; only for rotations that are multiples of 45° do transformed images of the motif coincide with those outlined.

Different Generators of the Same Group. A dihedral group with two generators is a slightly more complex group than a cyclic group and offers a more interesting example to demonstrate that a group can have different sets of generators. The exercise in Figure 5 builds on the one in Figure 4. The C_8 pattern in Figure 4 is generated by a single rotation. If a reflection whose mirror passes through the center of that rotation acts on the eight images of the motif, a D_8 pattern results. So one rotation and one reflection generate this dihedral group of 16 elements. The exercise shows that the same pattern (and hence the same group) is also generated by two reflections. (See Figure 6 for the completed D_8 pattern.) Students can be instructed to generate this group by two reflections one step at a time themselves, or they can use a script that repeatedly reflects the motif and its images in each of two mirrors. In the exercise, they are also asked to describe orbits of three well-chosen points. (These orbits will have 16 points, eight points, and one point, respectively.) Finally, they are asked to find other sets of generators for this dihedral group.

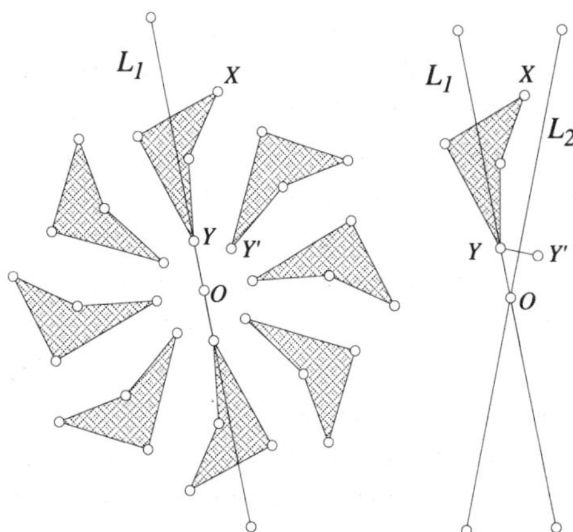

FIGURE 5. Exercise on different generators for D_8. O is the center of rotation; Y' is the image of Y under a $-45°$ rotation about O; L_1 and L_2 are mirrors; L_2 is the perpendicular bisector of YY'.

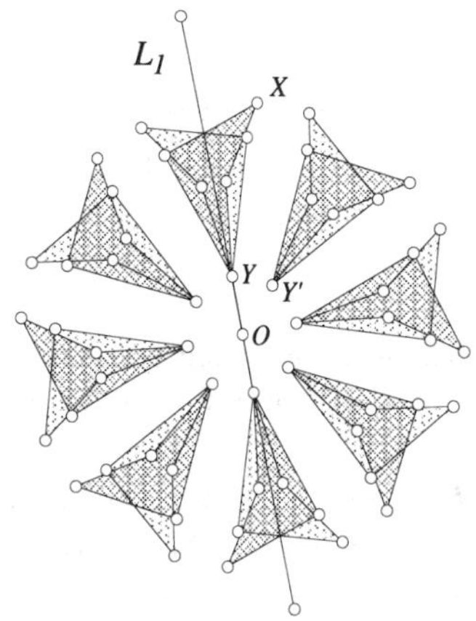

FIGURE 6

Instructions to the student: On the left is the C_8 pattern generated earlier by a clockwise 45° rotation about O, and L_1 is a line through O. On the right is a single motif with L_1 and O in the same position as on the left, and a second line L_2 through O. On the left, mark L_1 as a mirror and then select the whole C_8 design and reflect it in L_1. Now open the D_8 script, and select on the right, in order (hold down the shift key), the shaded interior (only) of the motif, L_2, and L_1. Play the script. It will reflect the motif first in one of the lines L_1, L_2, then the other, then repeat this to reflect the images, and so on, until the design is complete. Compare the design generated on the right with that on the left. Although they were created with different sets of generators, they are the same (check this!). Measure the angle between L_1 and L_2 on the right. How is this related to the rotation that generates C_8? In the completed designs, describe the orbits of each of the points X, Y, O. Find other ways to generate this design, beginning with a single motif (as shown on the right). Can the design be generated with only one transformation?

Group Tables, Closure, Isomorphism. It is a standard exercise to ask students whether certain sets of elements with a given operation form a group. This underscores the need for closure under the operation and also illustrates how combinations of elements may lead to non-obvious new elements outside the set. To prove a finite set is a group, to discover many of its properties, and to decide whether two groups are isomorphic, it is instructive to construct a group table. To construct the group table for S_3 (the symmetry group of the equilateral triangle), I like to have students manipulate cardboard triangles having labeled vertices. However, "hands on" can be tedious for much larger symmetry groups. To make group tables, students need to have facility with compositions of isometries. Depending on the time available, students either can be asked to discover useful characterizing properties of isometries, or they can be given a test to perform that will determine which single isometry equals the composition of two isometries. (If they are merely given the test, it is worthwhile to have them verify it as an exercise with compositions of isometries acting on a triangle.) Figure 7 outlines a very simple test that allows one to give a full description of an unkown isometry that could arise as a composition. One only needs to act on a scalene triangle by the composition of isometries, then connect the vertices of the original triangle to their final images under the composed transformation. The positions of these connecting segments will then allow the complete identification of the transformation.

Translation: The segments that connect points to their images are all parallel, and all have the same length. Each connecting segment represents the translation vector.

Reflection: The segments that connect points to their images are all parallel and not all the same length. The reflection mirror is the perpendicular bisector of each segment.

Rotation: The segments that connect points to their images are not parallel, and their midpoints are not collinear. The center of rotation is the intersection of perpendicular bisectors of the segments (since each point and its image lie on a circle whose center is the center of rotation). The angle of rotation is the directed angle that joins a point of the figure to the center of rotation to the corresponding image point.

Glide-reflection: The segments that connect points to their images are not parallel, and their midpoints are collinear. The glide mirror is the line through the midpoints of the segments. The glide vector is a leg of the right triangle whose hypotenuse connects a point to its image point; the glide vector is parallel to the glide mirror.

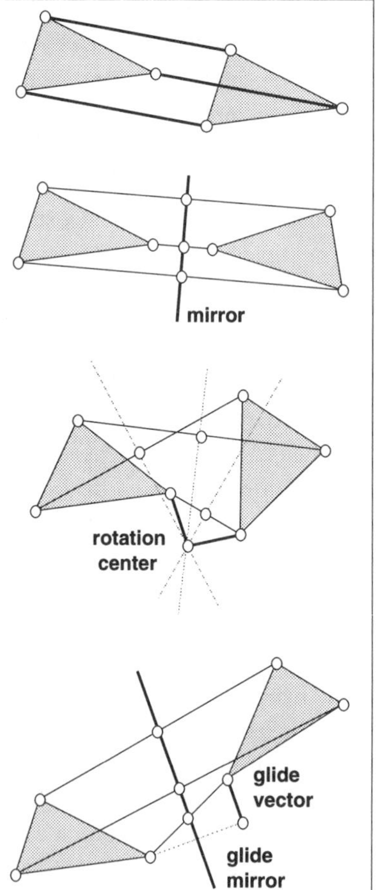

FIGURE 7. Identification of Isometries. An unidentified isometry (that may be a composition of isometries) acts on a scalene triangle, and the vertices of the triangle are joined to their images by line segments. By examining the segments, the isometry may be completely identified.

With this knowledge, students can then make up group tables given the elements of a group or a set of generators for a group. They can answer questions such as: Can a group consist of only reflections and the identity? If so, what can you say about the order of the group? If a group is generated by two rotations what can you say about the group? If a group is generated by two reflections what can you say about the group? What are the subgroups of a given symmetry group?

Students can also show that there are exactly seven symmetry groups of periodic border (frieze) patterns. They will easily identify the isometries that can leave invariant an infinite strip of uniform width. These are the only isometries that can be elements of a symmetry group of a border pattern. A single translation (of minimum period) generates the translation subgroup of a periodic border pattern. To find all possible symmetry groups of such patterns, students need only to consider what kind of elements will be added to the group if one or more different isometries are introduced, and what collections of isometries cannot be a group. For example, a group cannot contain only translations, halfturns, and reflections in mirrors perpendicular to the edge of the strip, since the composition of a halfturn and a reflection is a glide-reflection. The proof that there are just seven symmetry groups of periodic border patterns is one by exhaustion of cases (easily done); it can be found in Washburn and Crowe, pages 278–79.

Although I have included here only a sampling of investigations for students to gain understanding of abstract concepts encountered in the study of groups, there are many more topics that can be given visual illumination. The article by Marjorie Senechal [5] has suggestions for how to illustrate the concepts of stabilizer subgroup, coset, and group products, for example. Many of her suggestions can be adapted to investigations with dynamic geometry software. Whole courses can be devoted to the theory of groups and symmetry—these can be very elementary (see [2]) or more advanced, for math majors or graduate students (see [1] and [4]). At all levels, the use of dynamic geometry software can make the ideas in these courses come alive. Students will not only study and analyze properties, pictures, and proofs, but discover properties, make pictures, and create or illustrate proofs. It is a highly engaging topic in which students can unleash their creativity and literally *see* the mathematics of symmetry and groups.

Endnote

1. It is analogous to the fact that the action of a linear transformation on two linearly independent vectors in the plane completely determines the linear transformation. This is no coincidence; if students have the appropriate background, it is worthwhile to pursue the connection.

References

1. Armstrong, M.A. (1988). *Groups and Symmetry*, Springer Verlag.
2. Farmer, D. (1996). *Groups and Symmetry: A Guide to Discovering Mathematics.* Mathematical World, vol. 5, American Mathematical Society.
3. Gallian, J. (1994). *Contemporary Abstract Algebra*, 3rd ed. D.C. Heath & Co.
4. Gordon, G. (1996). "Using Wallpaper Groups to Motivate Group Theory," *PRIMUS*, vol. IV, no. 4, 355–365.
5. Senechal, M. (1988). "The Algebraic Escher," *Structural Topology*, 15, 31–42.
6. Schattschneider, D. (1986). "In Black and White: How to Create Perfectly Colored Symmetric Patterns," *Computers and Mathematics with Applications*, 12B, nos. 3/4, 673–695.
7. Schattschneider, D. (1978). "The Plane Symmetry Groups: Their Recognition and Notation," *American Mathematical Monthly*, 85, 439–450.
8. Washburn, D. and Crowe, D. (1988). *Symmetries of Culture: Theory and Practice of Plane Pattern Analysis*, University of Washington Press.

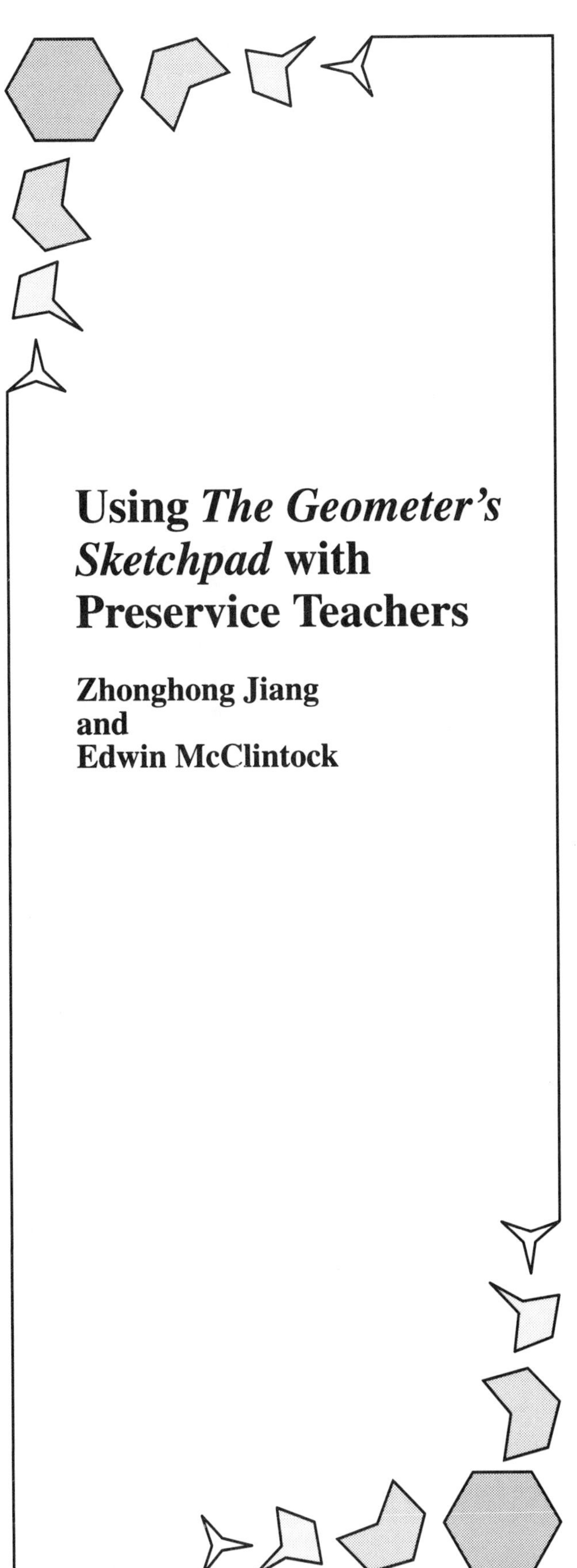

Using *The Geometer's Sketchpad* with Preservice Teachers

Zhonghong Jiang
and
Edwin McClintock

The NCTM *Curriculum and Evaluation Standards* for grades 9–12 advocates a learning "environment that encourages students to explore, formulate and test conjectures, prove generalizations, and discuss and apply the results of their investigations" [3, p. 128]. University undergraduates majoring in mathematics education are expected to become qualified secondary-school teachers who are sophisticated in creating this kind of environment. As their professors, we need first to provide them with this kind of environment. Our way to do this is to help students develop a new learning style—usually starting from initial guesses, then performing investigations and/or experimentations, using the results of their investigations to confirm or modify the initial guesses, formulating more thoughtful conjectures, verifying or proving the correctness of the conjectures, and finally discussing the possible expansions, generalizations, and applications. The effective use of *The Geometer's Sketchpad* [GSP] (Jackiw [1]) has greatly helped us to do this.

We will discuss two examples from our "Computers for Mathematics" class, which provide evidence that, if used effectively, the dynamic geometry software is not only an excellent tool for intuitive teaching and learning but also a facilitator for students' systematic/logical thinking.

Example 1. A Shortest Path Problem

> A road is proposed that will connect two towns A and B on opposite sides of a river. The road will cross the river in a bridge that is perpendicular to the riverbanks. Where should the bridge be placed so as to minimize the total length of the road? (Jennings [2, p. 15])

First, the students were asked to make guesses about the location of the bridge that minimized the road length. One of them said almost immediately, "Because the line segment connecting two points is the shortest distance between the two points, let's connect A and B. Suppose segment AB intersects the riverbanks at C and D, and G is the midpoint of segment CD. Then the bridge should be constructed through G and perpendicular to the riverbanks. In other words, segment AB bisects the bridge." While most of the students seemed to agree with this idea, the others believed that the bridge on the shortest path should lie on the line passing the midpoint of segment AB and perpendicular to the riverbanks. In either case, the "midpoint" intuition was very plausible.

Investigations. We did not comment on the correctness of these initial guesses. Instead, we asked the class to investi-

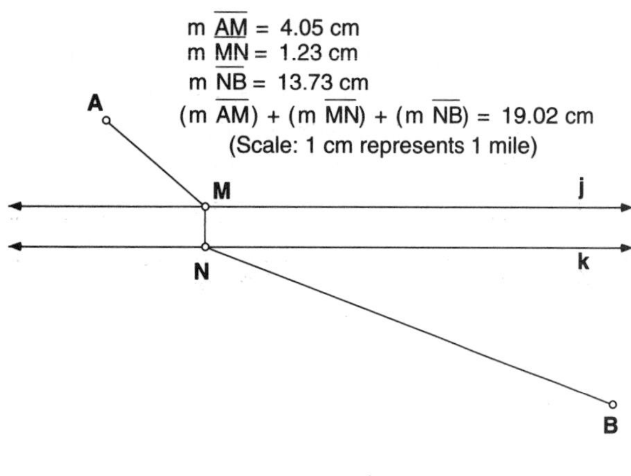

FIGURE 1

gate the problem situation by means of GSP. The students produced a GSP sketch by constructing two points A and B (representing two towns), two parallel lines between the two points (representing a river), a free point M on one of the riverbanks, and a line segment MN perpendicular to the riverbanks, representing a possible (location of) bridge (see Figure 1). With GSP, it was very easy to measure and display the length of each segment and the length of path $AMNB$.

Now came the most interesting part. As the students dragged point M along the riverbank j, changing the location of the bridge, the displayed length of path $AMNB$ was changed accordingly. The students kept dragging point M until they found the minimum length of path $AMNB$. Then the students were asked to check whether or not segment AB bisects the bridge on the shortest path. By constructing the midpoint G of the bridge MN, they immediately saw it did not (see Figure 2). In the meantime, they also found that the midpoint of AB had nothing to do with the shortest path. They were surprised that their intuition could be incorrect and came to realize that they should always test their ideas by appropriate investigations.

Formulating More Thoughtful Conjectures. We now asked the students to find the characteristic(s) of the shortest path $AMNB$ they had discovered. Through observation and checking slopes of the related segments, it was not hard for them to find that AM was parallel to NB. They were then asked to think whether this might help them determine the location of the bridge that would make the shortest path. After thinking and discussing with each other for a few minutes, some students proposed the following idea to the class:

> If we translate point A by vector XX' to point A' (where segment XX' represents the width of the river) and connect A' and N, then quadrilateral $AA'NM$ is a parallelogram since AA' is parallel and congruent to MN. Therefore $A'N$ is parallel to AM. We have already found that AM is parallel to NB, so $A'N$ is parallel to NB. But $A'N$ and NB share point N, hence A', N, and B must be collinear. This implies that if we translate point A by vector MN to point A' and connect A' and B, then N, one endpoint of the bridge which makes the shortest path, can be determined because it is the intersection point of $A'B$ and riverbank k on the side of Town B. With point N determined, the construction process of the shortest path becomes a straightforward procedure: Trans-

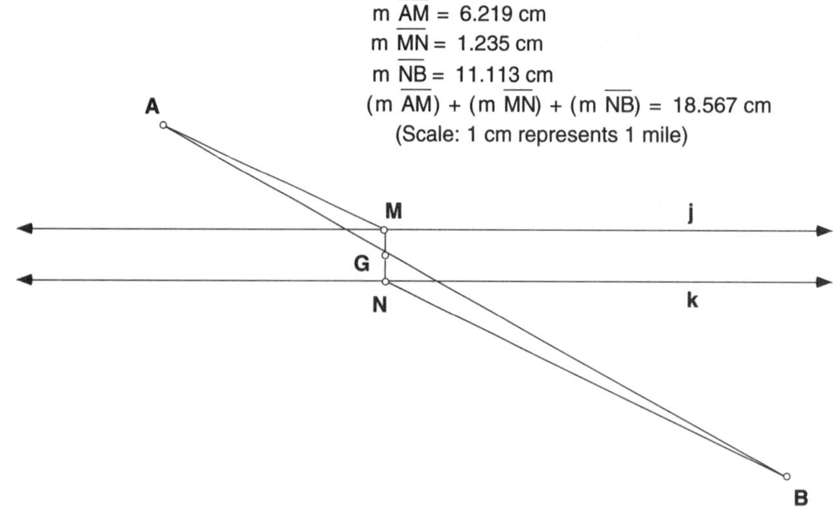

FIGURE 2

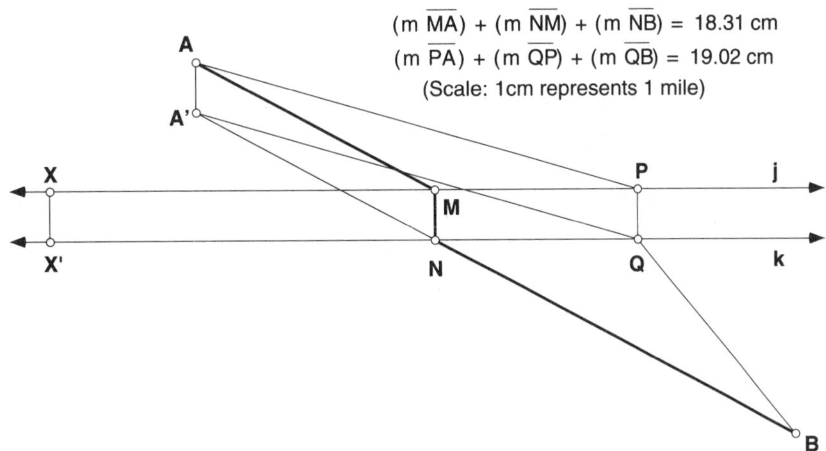

FIGURE 3

late N by vector $X'X$ to M, which must be on riverbank j, then path $AMNB$ is what we seek. See Figure 3 for the construction.

This geometric construction showed that the students had formulated a thoughtful conjecture about the shortest path. It was the dynamic movement and dynamic measurement feature of GSP that helped the students come up with the conjecture.

Verification and Proof. While the students were enjoying their construction, we raised the following question: "Have we finished solving the 1-river problem yet?" The students suggested that the constructed path *AMNB* should be verified. They created an arbitrary path $APQB$ where Q was a free point on riverbank k and PQ represented the bridge (see Figure 3). By continuously dragging point Q, they saw that the length of path $APQB$ was always longer than (or equal to, if both paths coincide) that of path $AMNB$. They were comfortably convinced that $AMNB$ was indeed the shortest path. Even so, however, they still felt that there was a need for the proof to show the correctness of their construction in a logical sense.

The earlier discussions made the proof only a small step further. Some of the students utilized the properties of parallelogram to generate the proof, but most of them felt more interested in the concepts and properties of vector and translation that had been used in their investigation and construction process. The following is a proof made by the students with this consideration:

Let $APQB$ be an arbitrary path as shown in Figure 3. Translate AM and AP by vector XX' to $A'N$ and $A'Q$ respectively. Then, (1) $AM + NB = A'N + NB = A'B$ (since A', N, and B are collinear), which is the length of path $AMNB$ minus the river width; (2) $AP+QB = A'Q+QB$, which is the length of path $APQB$ minus the river width. By the triangle inequality, $A'B \leq A'Q + QB$. Hence $AMNB$ is the shortest path from A to B.

At this point, the problem was completely solved, and the students felt satisfied with their efforts which combined conjecturing, investigating, analyzing, constructing, verifying, and proving.

Generalizations. A natural generalization of the bridge problem is to consider the *n*-river problem in which all conditions are the same except that there are *n* rivers rather than one river between Towns *A* and *B*, where *n* can be any natural number. What is the shortest path from *A* to *B*? Groups of three or four students began their explorations by first considering the 2-river situation. As before, they sketched the situation and found solutions by manipulating the path to be minimized. Several groups proposed the following construction (see Figure 4):

Translate A by vector XX' to A', where segment XX' represents the width of river jk, and translate B by vector $Y'Y$ to B', where segment YY' represents the width of river ih. Connect A' and B', intersecting riverbanks k and i respectively at N and U. Each of the two points represents one endpoint of a bridge. Translate N by vector $X'X$ to M which has to be on riverbank j; translate U by vector YY' to V which has to be on riverbank h, then $AMNUVB$ is the shortest path.

Instead of doing translations on both the A side and the B side, other groups chose to do translations only on the A side

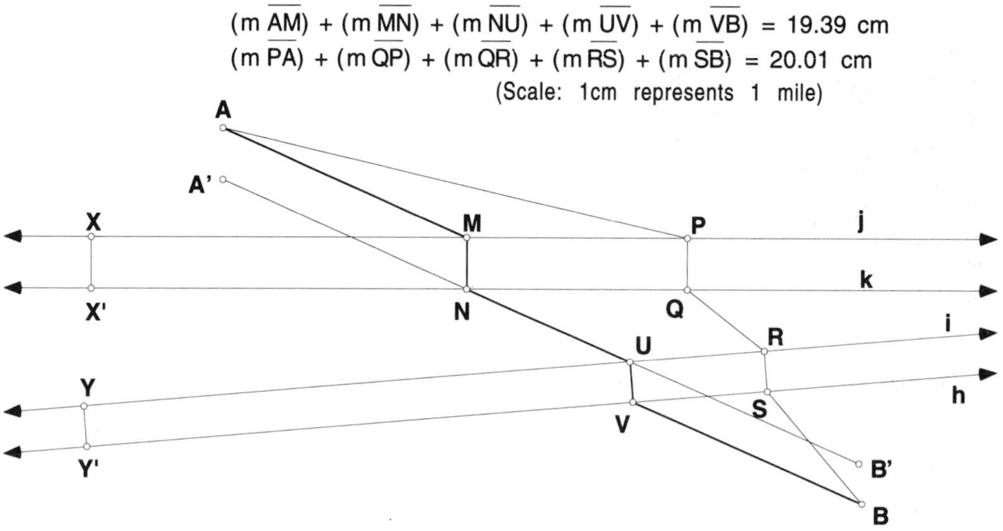

$(m\,\overline{AM}) + (m\,\overline{MN}) + (m\,\overline{NU}) + (m\,\overline{UV}) + (m\,\overline{VB}) = 19.39$ cm
$(m\,\overline{PA}) + (m\,\overline{QP}) + (m\,\overline{QR}) + (m\,\overline{RS}) + (m\,\overline{SB}) = 20.01$ cm
(Scale: 1cm represents 1 mile)

A A' X M P j X' N Q k U R i Y S h V Y' B' B

FIGURE 4

$(m\,\overline{AM}) + (m\,\overline{MN}) + (m\,\overline{NU}) + (m\,\overline{UV}) + (m\,\overline{VB}) = 19.39$ cm
$(m\,\overline{PA}) + (m\,\overline{QP}) + (m\,\overline{QR}) + (m\,\overline{RS}) + (m\,\overline{SB}) = 20.10$ cm
(Scale: 1cm represents 1 mile)

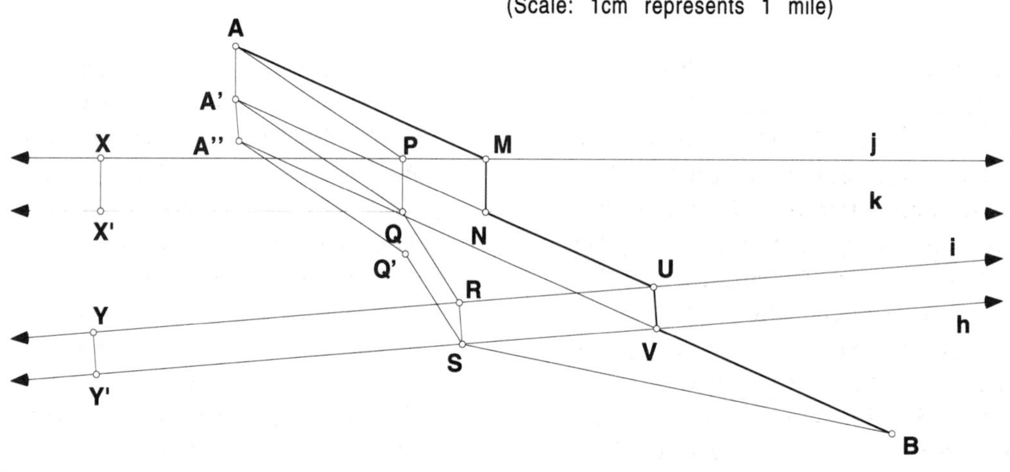

FIGURE 5

(see Figure 5).

> Translate A by vector XX' to A', and then translate A' by vector YY' to A''. Connect A'' and B, intersecting riverbank h at V. Translate V by vector $Y'Y$ to U which has to be on riverbank i. Connect A' and U, intersecting riverbank k at N. Translate N by vector $X'X$ to M which has to be on riverbank j. Then $AMNUVB$ is the shortest path.

We asked the students to compare these two constructions. After serious thinking, they indicated that the two constructions were only slightly different from each other for the 2-river situation, but the latter was better for generalizing to the situation of having three or more rivers, since it would avoid the difficulty of deciding how many translations should be done on each of the A and B sides for the n-river situation. Based on this consideration, they chose to give a proof for the construction shown in Figure 5. Through a serious discussion including repeatedly trying different translations in the sketch, and based on the ideas used in the 1-river proof, each group provided a proof similar to the one described below:

> According to the construction of path $AMNUVB$, $AM + NU + VB = A'N + NU + VB =$ (since

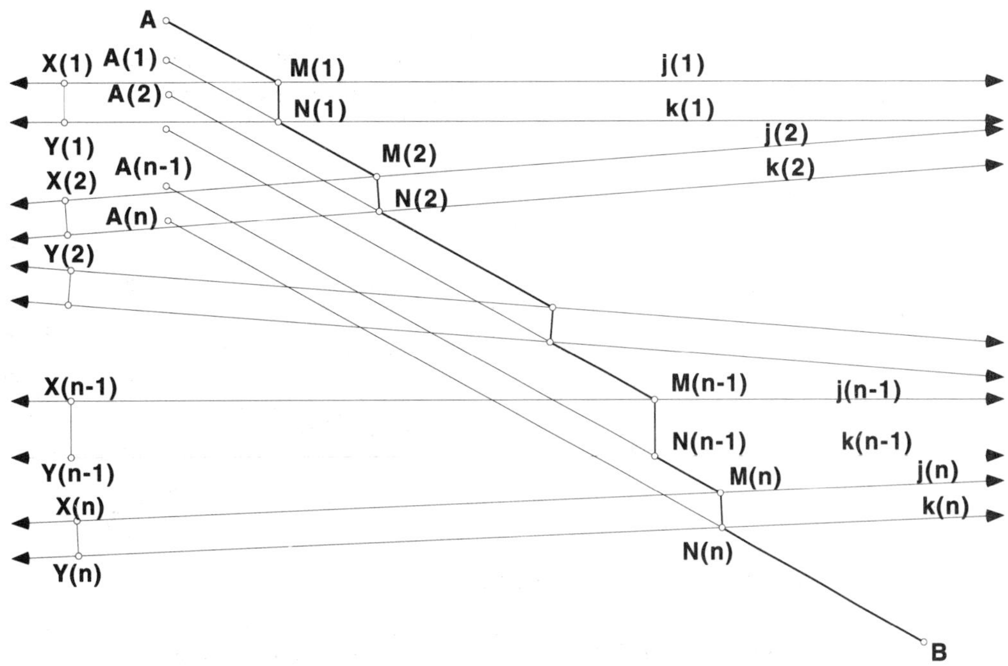

FIGURE 6

A', N, and U are collinear) $A'U + VB = A''V + VB =$ (since A'', V, and B are collinear) $A''B$, which is the length of path $AMNUVB$ minus the sum of the widths of the two rivers. Let $APQRSB$ be an arbitrary path. Translate AP by vector XX' to $A'Q$, and continue to translate $A'Q$ and QR by vector YY' to $A''Q'$ and $Q'S$ respectively. Then $AP + QR + SB = A'Q + QR + SB = A''Q' + Q'S + SB$, which is the length of path $APQRSB$ minus the sum of the widths of the two rivers. Since both $A''B$ and $A''Q' + Q'S + SB$ are paths from A'' to B but $A''B$ is a line segment, $A''B \leq A''Q' + Q'S + SB$. Hence the length of path $AMNUVB \leq$ the length of path $APQRSB$.

The 2-river problem gave the students a deeper understanding of the fundamental difference between the shortest path and any other path. On the shortest path, all component segments (except the bridges) are parallel to each other, so the translation(s) via the "bridge" vector(s) forms a straight line segment to connect A's final image (A'' in the 2-river situation) and B. This made the path the shortest; no other path had this characteristic. The generalization to the n-river situation was natural for the students; they easily formulated and proved the following construction conjecture (see Figure 6):

Let $X(1)Y(1), X(2)Y(2), \ldots, X(n)Y(n)$ represent the widths of river(1), river(2),..., river(n) respectively, with $X(i)$ on riverbank $j(i)$ and $Y(i)$ on riverbank $k(i)$ [i is a natural number between 1 and n inclusive]. Translate A by vector $X(1)Y(1)$ to $A(1)$, then for i increasing from 2 to n, translate $A(i-1)$ by vector $X(i)Y(i)$ to $A(i)$. Connect $A(n)$ and B, intersecting riverbank $k(n)$ at $N(n)$. For i decreasing from n to 2, translate $N(i)$ by vector $Y(i)X(i)$ to $M(i)$ which must be on riverbank $j(i)$, and connect $A(i-1)$ and $M(i)$, intersecting riverbank $k(i-1)$ at $N(i-1)$. Finally, translate $N(1)$ by vector $Y(1)X(1)$ to $M(1)$ which must be on riverbank $j(1)$. Then $AM(1)N(1)M(2)N(2)\ldots M(n)N(n)B$ is the shortest path from A to B.

Most of the students recognized that the idea of the proof in the 2-river situation also worked well here, although the descriptions became somewhat more complicated. Other students tried mathematical induction as a proof strategy.

The Power of GSP. The problem-solving process described above, from the initial guess for the 1-river situation to the proof for the n-river situation, took the students considerable time (several classes) and effort. However, all of them thought their gains were worth it. One group of students wrote: "We feel we had learned a lot of mathematics from this problem because we had to do various types of investigations to find its solution and then to prove it. These investigations involved a variety of concepts, theorems, and problem solving strategies." From the whole process, it is clear that, as a dynamic visualization tool, GSP can provide excellent geometric intuition, effectively helping the students to formulate and test (and hence confirm or modify) conjectures. By its dynamic nature, GSP can reveal the regularity of geometric phenomena, stimulating the students' insight for solution and generalization.

Example 2. A Triangle within a Triangle—Finding Relationships

This example shows the potential of GSP to directly promote students' systematic/logical reasoning, including theoretical proof. The problem is stated and illustrated by Figure 7.

When asked to give initial guesses, two conjectures were made by the students: (1) The two triangles are similar; and (2) $\text{Area}(\triangle PQR) : \text{Area}(\triangle ABC) = 1:9$. While the first guess probably came from rough visualization, the second guess was possibly due to the influence of the 1 : 3 ratios mentioned in the problem statement. However, the students expressed that they were not sure about the guesses. This doubt contrasted with their guess for the 1-river problem, which they were confident was correct.

To find the relationship(s) between these two triangles, the students made GSP sketches and measured the related angles and areas. They found both guesses were wrong. The two triangles were not similar, and $\text{Area}(\triangle PQR) : \text{Area}(\triangle ABC) = 1 : 7$ rather than $1 : 9$, even when they dragged to change the shape of $\triangle ABC$ (see Figure 8).

While the students felt it would be difficult to prove this relationship, we asked them to continue their investigations with GSP. Through further investigations with GSP measurements, the students found the following relationships:

$\text{Area}(\triangle BDP) = \text{Area}(\triangle CEQ) = \text{Area}(\triangle AFR) = \frac{1}{3}\text{Area}(\triangle PQR) = \frac{1}{21}\text{Area}(\triangle ABC)$;

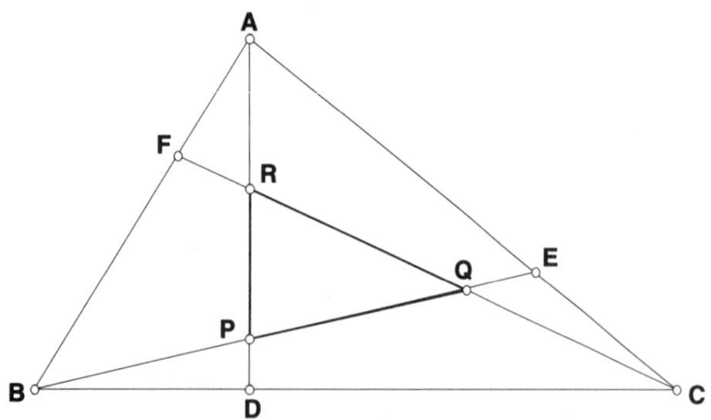

ΔABC is an arbitrary triangle. BD : BC = CE : CA = AF : AB = 1 : 3. ΔPQR is formed by the construction of line segments AD, BE, and CF. What is the relationship between Δ PQR and Δ ABC?

FIGURE 7

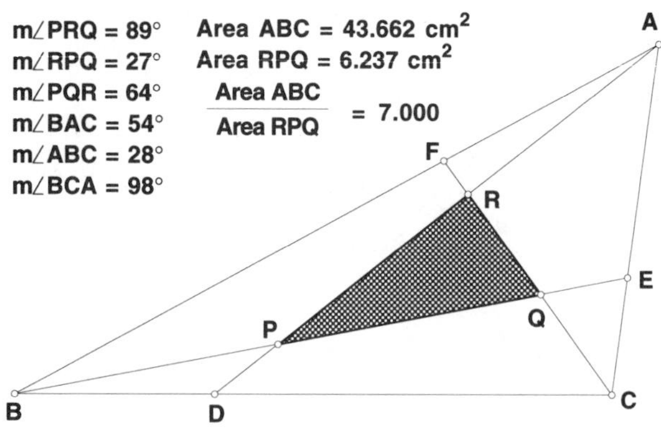

FIGURE 8

$\text{Area (Quad } DCQP) = \text{Area (Quad} EARQ) = \text{Area (Quad } FBPR) = \frac{5}{3}\text{Area}(\triangle PQR) = \frac{5}{21}\text{Area}(\triangle ABC)$;
$BP : PQ : QE = 3 : 3 : 1$;
$CQ : QR : RF = 3 : 3 : 1$; and
$AR : RP : PD = 3 : 3 : 1$.

Stimulated by these new findings, one of the students decided to try to prove the relationship between the smallest triangles (such as $\triangle BDP$) and $\triangle ABC$. He succeeded. Figure 9 gives his proof.

Two other students found similar proofs. All three students felt deeply that it was the investigations with GSP that stimulated their insight for proof. They shared their ideas with the whole class.

Let x be Area(ΔBDP), then Area(ΔCDP)=2x (because of same height and double base). Let y be Area(ΔCEP), then Area(ΔAEP)=2y (same reason).
Hence, we have x + 2x + y = Area(ΔBCE) = (1/3)Area(ΔABC) ... (1)
2x + y + 2y = Area(ΔADC) = (2/3)Area(ΔABC) ... (2)
By simple calculations on (1) and (2), we have x = (1/21)Area(ΔABC).
Using similar method, we can get Area(ΔCEQ) = x = (1/21)Area(ΔABC), and Area(ΔAFR) = x = (1/21)Area(ΔABC). Therefore,
Area(ΔPQR) = Area(ΔADC) - Area(ΔARC) - Area(Quadrilateral DCQP) =
(2/3)Area(ΔABC) - [Area(ΔAFC) - x] - [Area(ΔBCE) - x - x] =
(2/3)Area(ΔABC) - [(1/3)Area(ΔABC) - x] - [(1/3)Area(ΔABC) - 2x] =
x + 2x = 3x = 3*(1/21)Area(ΔABC) = (1/7)Area(ΔABC).

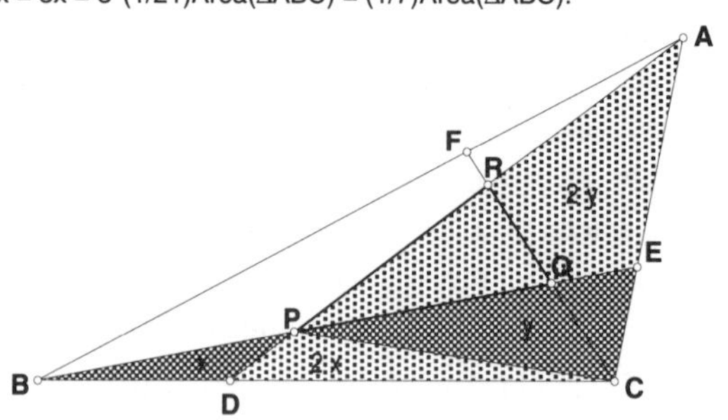

FIGURE 9

$BD:BC = CE:CA =$ $AF:AB =$	$1:3$	$1:4$	$1:5$	$1:6$	$1:7$	$1:n$
$BP:PQ:QE =$	$3:3:1$	$4:8:1$	$5:15:1$	$6:24:1$	$7:35:1$	$n:n(n-2):1$ $= n:(n^2-2n):1$
Area($\triangle BPD$): Area($\triangle ABC$) =	$1:21$	$1:52$	$1:105$	$1:186$	$1:301$	$1:n[n+n(n-2)+1]$ $=1:(n^3-n^2+n)$
Area($\triangle PQR$): Area($\triangle ABC$) =	$1:7$	$4:13$	$9:21$	$16:31$	$25:43$	$(n-2)^2:[n+n(n-2)+1]$ $=(n-2)^2:(n^2-n+1)$

Continued class discussion showed the students that other relationships could be easily proved. For example, the relationship $BP:PQ:QE = 3:3:1$ was proved as follows:

If we connect Q and D, then Area($\triangle DCQ$) $=$ $2 *$ Area($\triangle DBQ$) since they have the same height and $DC = 2 * BD$. But Area($\triangle DCQ$)+ Area($\triangle DQP$) $= y - x + 2x = y + x$. Hence

$$(y+x) - \text{Area}(\triangle DQP) = \text{Area}(\triangle DCQ)$$
$$= 2 * [x + \text{Area}(\triangle DQP)].$$

This implies

$$3 * \text{Area}(\triangle DQP) = y - x$$
$$= \frac{4}{21}\text{Area}\,(\triangle ABC) - \frac{1}{21}\text{Area}\,(\triangle ABC)$$
$$= \frac{3}{21}\text{Area}\,(\triangle ABC),$$

which in turn implies

$$\text{Area}\,(\triangle DQP) = \frac{1}{21}\text{Area}\,(\triangle ABC).$$

Hence Area ($\triangle DQP$) $=$ Area($\triangle BDP$). These triangles have the same height, so base BP $=$ base PQ. Since Area($\triangle CPQ$) $= y - x =$ $\frac{3}{21}$Area ($\triangle ABC$) and Area ($\triangle CEQ$) $=$ $\frac{1}{21}$Area($\triangle ABC$), it follows that Area($\triangle CPQ$) $=$ $3*$Area($\triangle CEQ$). They also have the same height, so base $PQ = 3 *$ baseQE. Therefore

$$BP = PQ = 3 * QE,$$

or

$$BP:PQ:QE = 3:3:1.$$

Based on their investigations so far, we encouraged the students to explore generalizations of this problem. With some hints from us, they changed points D, E, and F in Figure 7 into free points on BC, CA, and AB respectively. Then they dragged each of these points along its corresponding line segment until $BD:BC = CE:CA = AF:AB = 1:4$ (rather than $1:3$), and used the GSP measurements to examine variations in the relationships they had discovered when trying to solve the original problem. After that, they continued similar explorations for the cases that $BD:BC = CE:CA = AF:AB = 1:5$, then $1:6$, and finally, $1:7$. Through a careful study on the data collected during the explorations, they found some non-trivial but interesting patterns. It was a short step for them to make conjectures for the general case of a ratio $1:n$. They also found that their proofs for the particular cases carried over to the general case. Their main results are shown in the table above.

Summary

Our work shows that college mathematics education majors are able to progress quite well with the use of GSP through the guess-investigate-conjecture-verify process, and construct components of this inquiry cycle. A somewhat hidden part of the process is the continued emphasis on producing increasingly more general results. In both of the examples described in this paper, the problems were solved as posed, but afterwards we introduced more general versions for investigation. As students begin to gain confidence in their ability to use the process, we expect them to naturally generalize problems posed for their study. Frequently, in solving a more general version, the need to alter the nature of the investigations arises. This is typically productive, as alternate strategies or solution paths necessary to the generalization process bring greater insight and clarity to the problem situation.

As a significant departure from tradition, proof based on the guess- investigate-conjecture-verify process is viewed as very important and attainable. The verification and analysis leading to ways of building intuition for constructing a proof is a task we will continue to pursue in the future, both for our preservice teachers of mathematics and for their students. Classroom teaching is undergoing great change with innovative dynamic mathematics software packages available, and we mathematics educators have the responsibility to bring their power into full play. A wonderful example of an investigation undertaken in a high-school classroom that is similar to the second problem we have discussed can be found in the article "Morgan's Theorem [4]." We want our future teachers to bring to the classroom the experience, confidence, and enthusiasm generated by the kind of investigations we describe. Perhaps they will also be able to witness some of their students discovering new theorems.

References

1. Jackiw, N. (1991). *The Geometer's Sketchpad* (Software), Key Curriculum Press, Berkeley, CA.
2. Jennings, G.A. (1994). *Modern Geometry with Applications,* Springer-Verlag, New York.
3. NCTM (National Council of Teachers of Mathematics), *Curriculum and Evaluation Standards for School Mathematics,* The Council, Reston, VA. 1989.
4. Watanabe, T., Hanson, R., and Nowosielski, F.D. (1996). "Morgan's Theorem," *Mathematics Teacher,* vol. 89, no. 5 420–423.

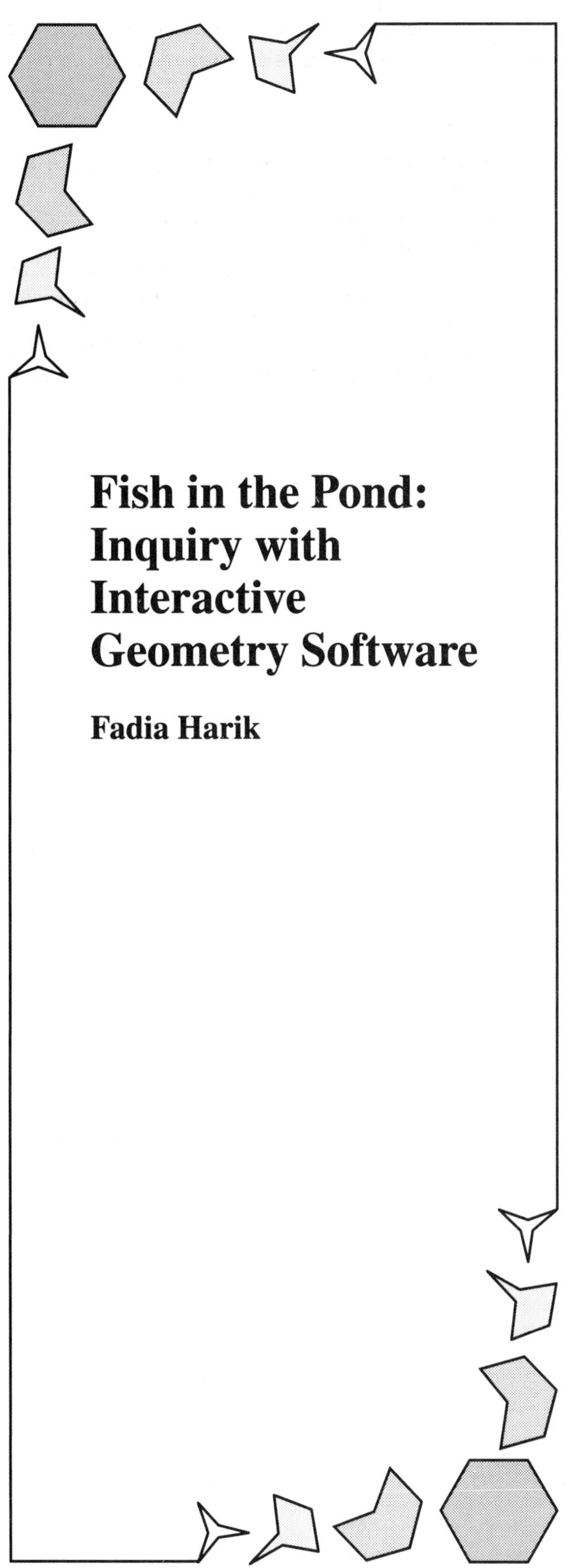

Fish in the Pond: Inquiry with Interactive Geometry Software

Fadia Harik

A colleague described our work best with this analogy: You put a lot of fish in a small pond, you give the learners all sorts of tools—boats, fishing rods, and so on, and you let them loose. Not only will they learn how to fish, but they might tell you a lot about the type of fish in the pond, where you can find them, and, with a little probing, maybe why you can find them.

Our job as instructors becomes that of tuning in to the students' learning processes, providing a little nudge here, an appropriate probe there, and, mostly, acknowledging their achievements. As coaches we become more respectful of the learning we observe and less impulsive about the teaching we are conditioned to give. We become less intrusive and more introspective about what is needed, where, and when. We turn the teaching/learning process into a learning/learning process. They learn the content, we learn the process, and sometimes it works the other way around.

The technology that provides us with dynamic geometry provides us with many ponds, where there are lots of fish. Our challenge is not only to look closely at what we are now teaching and what it is possible to teach, but also at how we can teach it.

In this article my goal is a two-fold theme: to demonstrate pedagogical moves that enhance inquiry-building in the classroom, and to describe computer environments where open-ended, long-term investigations of various mathematical themes are possible. The two components are closely intertwined. Computer environments make possible the design and use of multifaceted mathematical tasks, making small group collaborative work a necessity. The collaborative work in turn becomes the backbone of inquiry-building in a classroom culture.

I will describe the software, the tasks, and the work of middle-school students and teachers in two computer environments: one a physics software simulating the motion of a ball at a table; the other a geometry construction software used to explore some issues related to quilt-making. The pedagogical moves described aim at fostering creativity, enhancing the power of collective group work, and providing for diverse ability levels for mathematical and technological competence.

Bouncing Balls

About the Software. The *Physics Explorer: One Body*[1] simulates a single body moving on the plane of a screen. The body can move at constant or accelerated velocity, in a vacuum or in a viscous medium. The body can move freely or in a walled box where it can collide with the walls in elastic or inelastic ways. You can electrically charge the body

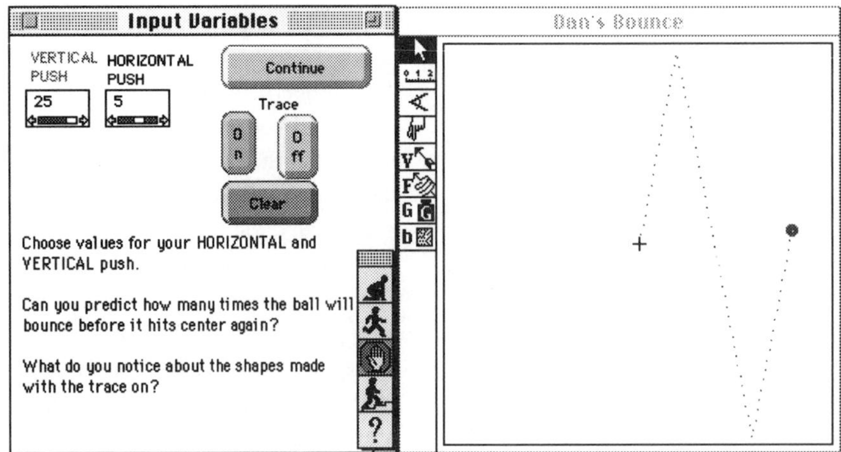

FIGURE 1. A view of the lab screen.

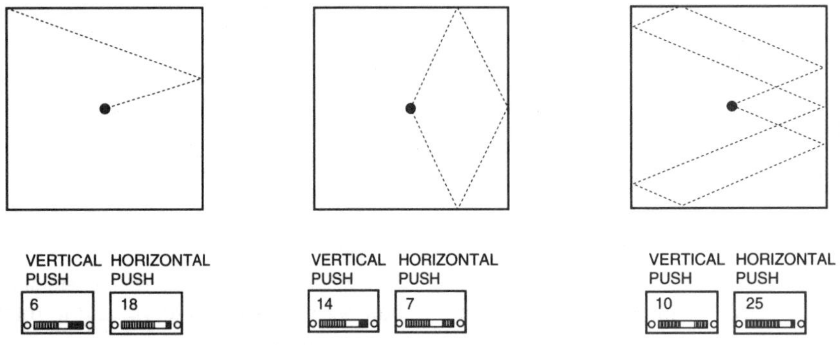

FIGURE 2. Visual patterns made by the paths of a bouncing ball.

and turn on or off a gravitational field, an electric constant field, and a magnetic constant field. The model contains tools for measuring distances and angles. It also contains a spreadsheet, a grapher, and a note taker.

About the Simulation. The options provided by the software are extensive, and most of them are not needed for our purposes; we want to generate some geometry for middle-school kids. So the lab we build has a square-shaped, walled box with inelastic walls, with a body at the center of the square that we call a ball. We use no acceleration, nor any gravitational, electric, or magnetic fields. We call the input boxes for horizontal and vertical velocities *horizontal push* and *vertical push*, place on/off *trace buttons* for tracing the path of the ball when it moves, and limit the velocity, initially, to $+/-30$ distance units per time unit. Finally, we make the ball stop as it passes through the center of the box.

The idealized situation we present to the learners is as follows: you give a ball a push in a horizontal and a vertical direction, by means of number inputs. When you select the RUN button, the ball moves and bounces off the walls, and, when it passes the center again, it stops. If you want to trace the path the ball takes, you select the trace button. See Figure 1.

What Happens. The demonstration of how the simulation works requires little time. We initially limit the field of investigation to a small number of wall bounces and draw attention to the numeric patterns that determine the path of the motion and the visual patterns of the trace as the ball moves (Figure 2). Sixth-graders, after one hour of work on the simulation, report observations such as these:

> There are many ways to create the same pattern.
> If you hit a number and its half, you get three bounces.
> I know how to make it go in one line.
> I can make a diamond-shaped path.
> I can make the path look like many diamonds.
> I could not get an even number of bounces.
> With 5 and 10 it goes this way, and 10 and 5 it goes that way.
> I can make it go in a horizontal line.

It can take several sessions before the students start sifting through the plethora of observations, and they may reach several dead ends, but it won't be long before they can predict the number of bounces, or the visual pattern the ball makes in a number of cases. Soon they can generate a family of pairs of numbers that will give them the "same" pattern such as {(1, 2) (5, 10) (8, 16)} or [{(6, 18) (21, 7) (−6, 2)}.]

Although the students do not have formal training in negative numbers or the coordinate system, they are able to predict the quadrant where the initial move happens. Indeed, their experience with the simulation becomes fertile ground for learning about negative numbers and the coordinate system as well as common factors, relatively prime numbers, equivalent fractions, and slopes of straight lines. To those who question the meaning of "a vertical push of two and a horizontal push of 3," the analogy of two kids pushing an object in one direction and three other kids pushing it in a perpendicular direction suffices. Some kids enjoy and benefit from physically acting out the analogy where various numbers of kids enact each instance.

To organize the patterns they observe, the students begin to classify not just instances but families of instances. They formulate hypotheses and test them (Figure 3) and argue about the value of limiting an investigation to one idea or exploring many ideas at the same time. When asked, they can closely predict the location of consecutive wall hits (an option that can be made available by planning the lab environment to stop at every bounce and to move only after they select the *continue* button). If prompted further, they can measure wall distances and angles and make the shift from what the phenomenon is to why it is so. They learn about the characteristics of reflection both in the present simulation and in the physical context of an air hockey pool table.

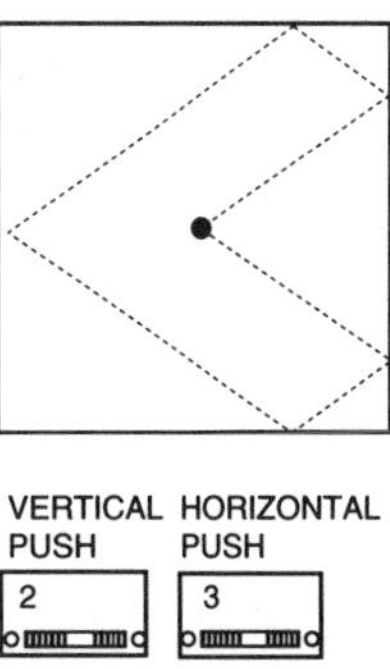

FIGURE 3. "Whatever [the sum of] those [pushes] equal, is how many bounces we have."

Many students are fascinated by the symmetries exhibited in the path patterns and try to associate these symmetries with the families of input numbers. Early in the investigation, some of them start mapping how the different orientations of the same pattern correspond to positive/negative number pairs used as input, and to rotations and reflections of the pattern (Figure 4).

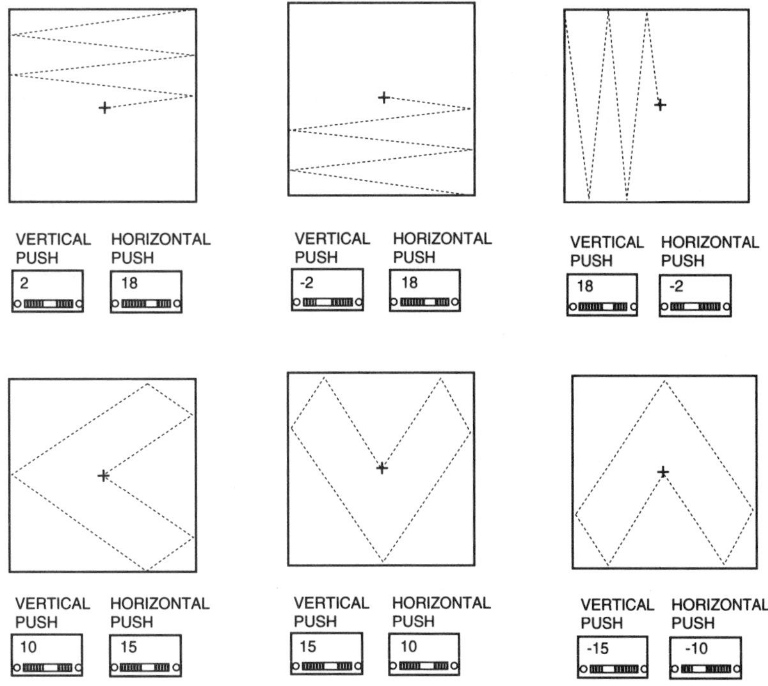

FIGURE 4. Symmetry, orientation, and number patterns.

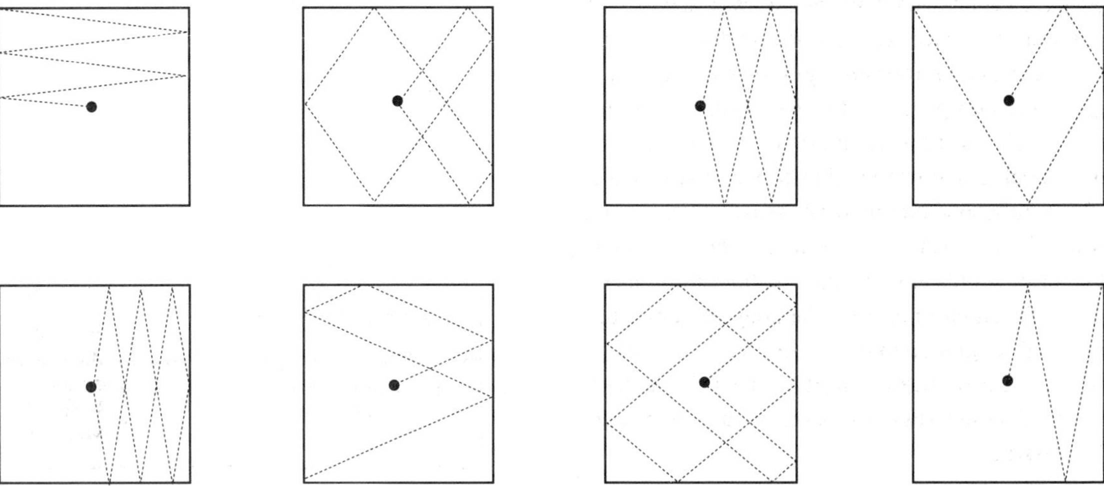

FIGURE 5. Predicting the numeric inputs that produce this collection of patterns accelerates the pace of teachers' inquiries.

In most cases, instructors find themselves responding to the students' discoveries rather than initiating and organizing the investigations for the students. Their roles shift from talkers to listeners. Understanding the thought processes of their students and the directions students take in their investigations becomes as interesting to them as their own ideas.

Teachers who engage in this investigation during our summer workshops respond to the task much as the students do. They recognize patterns faster than students do, but, like the students, they still vary greatly in pathways and styles of investigations.

After getting teachers acquainted with the bouncing ball environment by having them input number pairs and observe the motion and path patterns of the ball, we give teachers a collection of sketches of various paths and ask them to produce the sketches on screen. We find that this reversal process accelerates the pace of their inquiries and provides focus to their explorations (Figure 5).

Like the students, the teachers work in groups of twos and threes, then convene in a large group to discuss their work. In such group discussions, it is easy for a few enthusiasts to spill the beans, so to speak, and give away many results that others have not arrived at, depriving them of the satisfaction of discovering for themselves. To highlight the diversity of paces and pathways of investigation learners take, we ask them not to report actual results but, instead, to report result-related issues such as:

- How their investigation evolved over time
- What phenomena they can already predict
- What directions or hypotheses they still want to pursue.

With such reporting, a large group discussion becomes an avenue for cross-pollination of ideas that will enrich further investigations by the small groups rather than stifle them.

Eventually, what impresses teachers most is the diversity of mathematical topics embedded in the simulation and the relative ease of access to them. Over time, what remains with them is the adventure of investigating and formulating their own mathematics.

Quilt Making

About the Software. The *Geometry Inventor* is a dynamic construction tool where plane geometry can be explored. Its drawing tools provide easy initial access and allow people many redundant but useful ways to draw geometric figures. Among the various tools available are the *polygon* tools, providing ready-made polygons with options to restrict the area or the perimeter; *line segment* tools where parallel and perpendicular lines, among others, can be drawn, a *subdivide line* tool that allows any chosen number of subdivisions of a line segment; and *measuring* tools to be used within the construction process itself (giving Euclid a ruler). The program also provides a spreadsheet, a grapher and a script capability.

About the Tasks. The two tasks described here are part of a quilt-making module designed to be used with middle-school teachers in courses and workshops, with an eye toward having the teachers develop transferable versions of the module for their students. The module is designed to allow a genuine mix between the creative aspects of the craft of quilt making and the embedded mathematics. Users rely on the software as a vehicle for easy and precise drawing, replicating (via script), and performing rigid transformations.

My goal is to convey how embedded ambiguity in the phrasing of a task can provide multiple entries to a task, making it accessible and challenging to many ability levels. Carefully embedded ambiguity can also foster diverse responses to what the task is, how to go about doing it, and what types of results emerge from it. Fostering this diversity has a substantial impact on the body of knowledge co-constructed through collaboration in a classroom environment and the experience of empowerment the learners have as independent investigators inside or outside the confines of the classroom.

The First Task

The first task is: USE THE *GEOMETRY INVENTOR* TO MAKE QUILT SQUARES, EACH COMPOSED OF EIGHT CONGRUENT TRIANGLES, IN AS MANY DIFFERENT WAYS AS YOU CAN THINK OF.

What Happened. Jane and Harvey place on the screen several right triangles from the *polygon* tools menu, and try to move them around with the mouse as if they were moving cutout paper triangles by hand. I make a mental note: yet another way in which the software is used as an extension of our hands and physical experiences.

Susan and Tom start with a right triangle from the *polygon* tools menu and reflect it on its diagonal. Later, as they produce a series of reflections on various available sides, they realize they are not getting a square (Figure 6).

Barry and Kevin place a square from the *polygon* tools menu on the screen and use the *freehand line* tool from the *line segment* tools menu to draw the diagonals of the square (Figure 7).

Vivian and Sonny have already used the *subdivide line* tool to locate the midpoints of each of the sides of a square. They then use the *connect line segment* tool from the *line segment* tools menu to connect the midpoints (Figures 8 and 9).

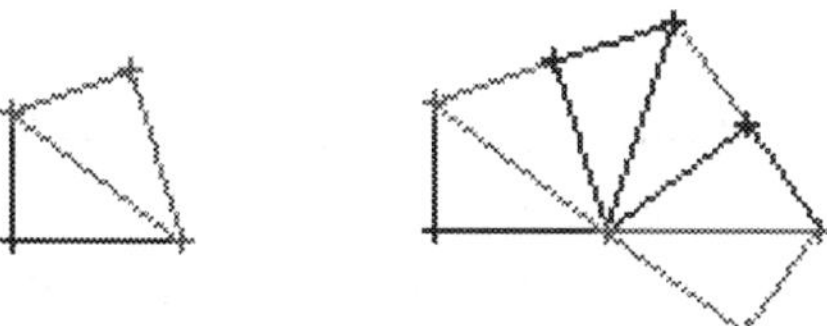

FIGURE 6. Right triangle reflected first along its diagonal.

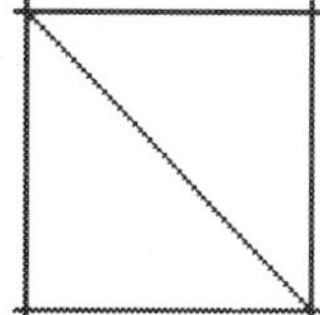

FIGURE 7

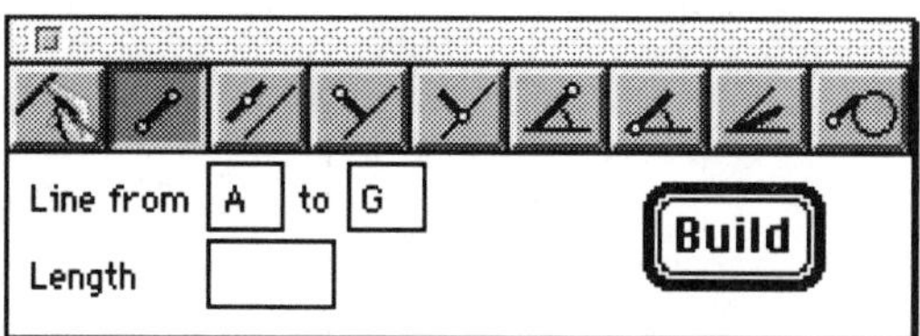

FIGURE 8. Dialogue box for the line segment tool.

FIGURE 9

As the last two groups are making a second quilt square, Harvey from the first group passes by, scouting (a behavior I have encouraged from the beginning of the workshop). He looks intently at what Barry and Kevin are doing, turns to me, and asks: "Is this the way we are supposed to do it? Cut up a square? We are trying to put triangles together." Here, the first intentional ambiguity in the task is surfacing. I tell him that he can do "it" in whatever way he chooses; both ways are fine. Harvey goes back to his table and convinces his partner to shift gears and do "it" the "easy way" according to him.

Barry and Kevin, Sonny and Vivian organize the squares on their screens so they can see all the variations they are constructing (Figure 10).

Susan and Tom experiment with various reflections of the right triangle made available from the polygon tool, still hoping to make a square (Figure 11). Then they shift to rotations of that triangle (Figure 12).

After several trials, it dawns on them that the length of the sides and the angles of the triangle have something to do with why they are not getting a square. This is a lengthy process for them. They then try to stretch one side of the triangle to make the triangle look isosceles but realize

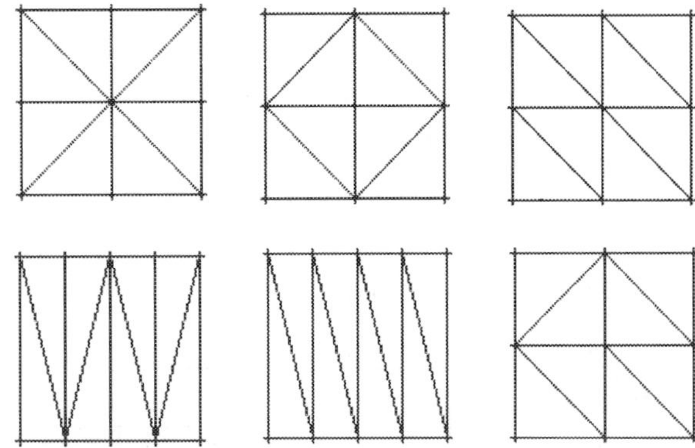

FIGURE 10. Variations on a theme.

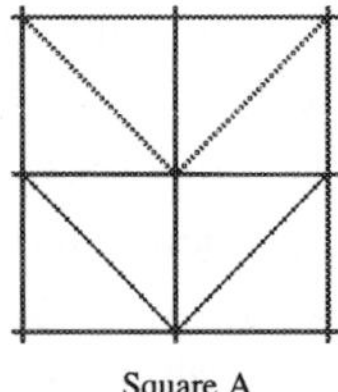

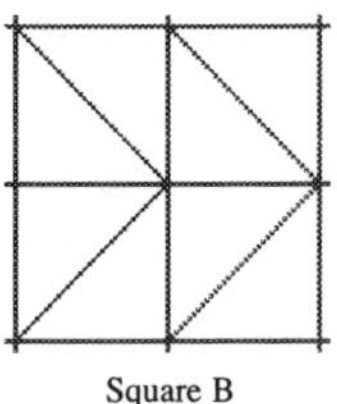

FIGURE 13. Is square A different from square B?

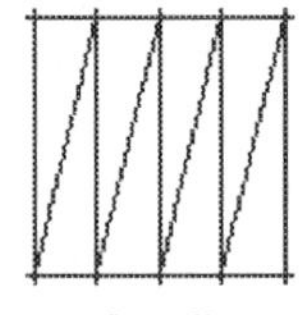

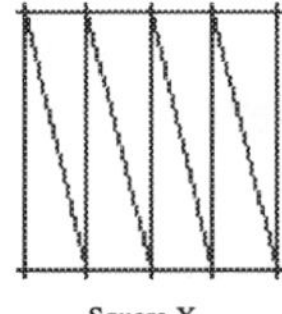

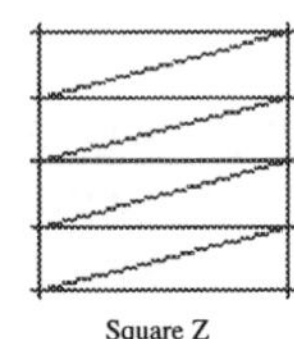

FIGURE 14. To a quilter, squares Y and Z are the same but different from square X.

that the change is not permanent and they cannot make it permanent if they use the right triangle available from the *polygon* tool. They have to build their own triangle with their own specifications first. This is another challenge in their approach that serves them well: They develop a solid understanding of characteristics of right triangles and their dynamic consequences, and they learn how to use the *line segment* tools menu to build a right isosceles triangle.

Vivian, who already has on display several quilt squares, asks me if I would count square A different from square B (see Figure 13). I turn the question back to her. She responds that B is just like A turned around, and do I want them to count the two as same or different? This is another intended ambiguity in the original task.

I tell Vivian and Sonny that they have to determine the criteria for "different" themselves. I sense an unease. Vivian wants me to make the decision so she can move on with the task. Sonny, who is an experienced quilt maker, addresses her dilemma by showing the three quilt squares in Figure 14.

Sonny says a quilter would consider Y and Z the same but not Y and X, because when you flip a quilt square you get the back side of it. Sonny and Vivian proceed to discuss the issue of what criteria they will choose to define "same" for quilt squares. I am no longer the decision-maker in the process.

I ask everyone to go around the room one more time and look at what others have done before we convene as a whole group. Some print their quilt squares to display and compare to the quilt squares of other groups, hoping to find new patterns they did not build.

When the whole group convenes, a host of different issues surface: the various construction techniques and tools, the characteristics of the triangles that had to be used, flipping and turning squares, arguing which are rotations of each other and which are reflections, and the formation of a process of systematization of data to exhaust the possibilities. The self-initiated need to systematize data does not often emerge naturally among teachers who are not mathematicians by training. Yet the simplicity of the task allows for the generation of enough data to call for the economy and feasibility that comes from systematization.

The large group discussion, with its equal emphasis on process and result, gives a chance to those who wrestled with putting the triangles together to offer the wisdom of their experiences for others to address as they take on the challenges presented by those who 'cut up' the square into triangles.

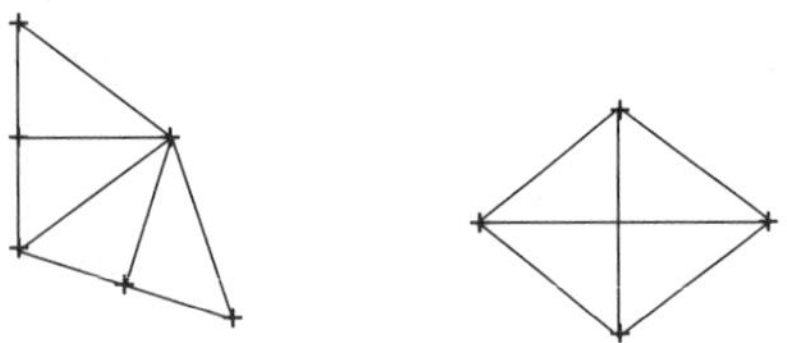

FIGURE 11. Trying to get a square by reflecting the fixed right triangle provided by the program.

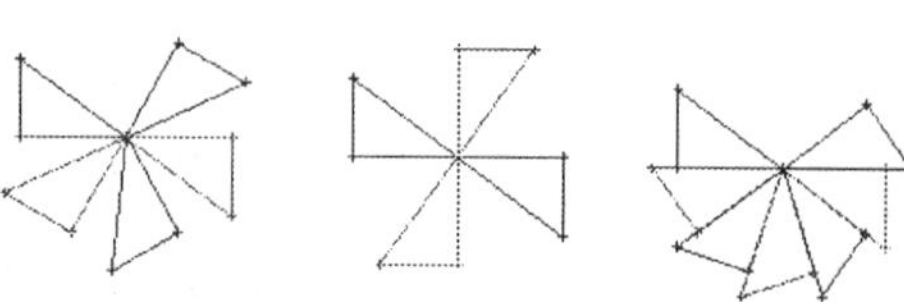

FIGURE 12. Rotating the fixed right triangle provided by the program.

The Second Task

The second task reflects a different level of ambiguity. In the first, the ambiguities were the processes of construction and the criteria of classification. In this task the ambiguity is embedded in the product of the task. They are to make a hexagonal quilt block, and the task is: DO TO A REGULAR HEXAGON WHAT YOU DID TO THE SQUARE.

What Happened. First I ask them to brainstorm about what the task means. They come up with the following:

- 8 congruent triangles
- 6 triangles of equal areas
- 12 right angled triangles
- 2 trapezoids
- 8 trapezoids would fit the analogy
- 4 rectangles and 4 triangles
- 1 rectangle and 2 or 4 triangles

I ask them to choose the ones most analogous to their previous task. They choose six triangles, twelve triangles, two trapezoids, eight trapezoids, and double the number of sides in triangles.

I am intrigued by the trapezoids, because I don't understand where they are coming from, so I ask Ella to elaborate. She says that with the square, she would cut a square into two isosceles triangles, then she would put four squares together. With the hexagon, if she cut it in two, she would get two isosceles trapezoids, so she will put four hexagons together, which will make eight trapezoids. I draw figures (Figure 15), and then I realize they don't exactly match Ella's process, which was a mixture of cutting and building. But it helps me to understand what she is saying.

Ella's interpretation of the analogy is new to me, and I am fascinated by it—so much so that I am hesitant to proceed with the intended analogy, which was the 12 congruent triangles. The validity and creativity of the various interpretations allow us to choose the 12 triangles as a choice among many, because we need a common task to pursue, not because it is necessarily the better analogy or the more creative one.

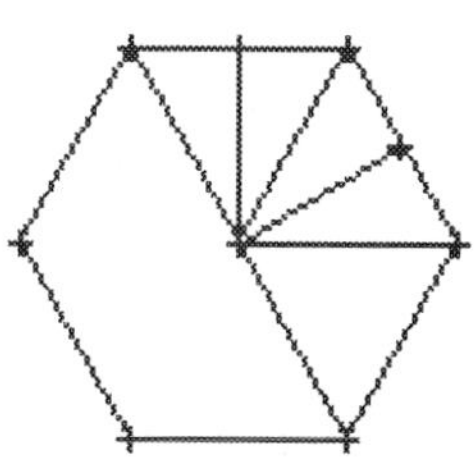

FIGURE 16. Are the triangles congruent?

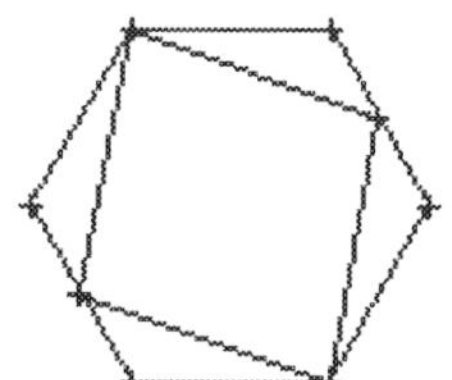

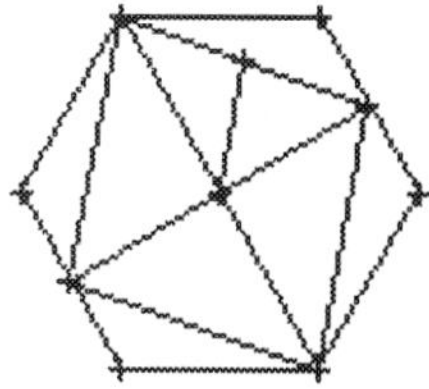

FIGURE 17. Triangles of equal areas?

The teachers change partners and proceed to work on the hexagon problem. Tom and Jane discuss whether they should rotate just one line or a whole triangle to build their hexagon.

Sandy is not sure that the right triangles are congruent (see Figure 16). She says that the fact that they have different orientations makes it hard for her to convince herself visually that they are congruent even though she thinks they are. Vivian and Wendy build a quad inside the hexagon and try to subdivide it. But the attempt is not successful (Figure 17).

As I walk around I observe some completed figures (Figure 18). Then the figures become more creative (Figure 19).

After observing each others' work, there is a whole class discussion. Wendy asks the class if any were able to

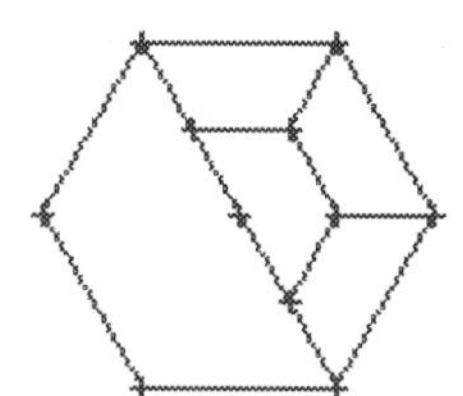

FIGURE 15. A square gives you two isosceles triangles. A hexagon gives you two isosceles trapezoids.

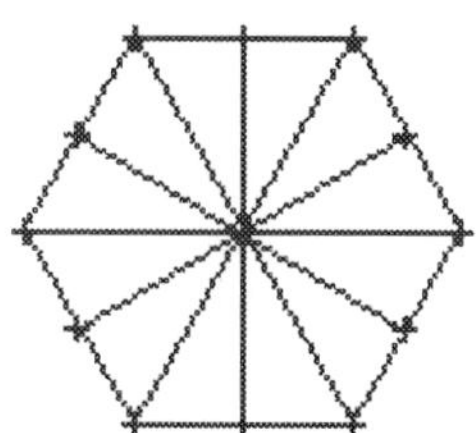

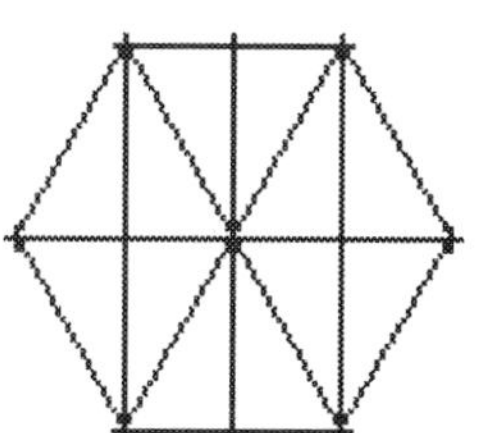

FIGURE 18. Completed figures.

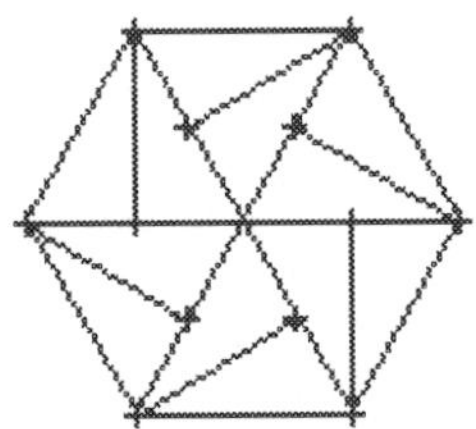

FIGURE 19. Getting more creative.

subdivide the hexagon into triangles that are not of the 30-60-90 type. None had. I suggest that, if they start by subdividing the hexagon into rhombi, they might find other triangles. They quickly disperse to their computer stations. In a few minutes I hear the AH! coming from different corners of the room (Figure 20). After the AH! come hushed and puzzled voices (see Figure 21):

> How can we tell the triangles are congruent?
> The three-dimensional perspective is confusing me.
> What I think I am seeing I am not sure is true.
> We have to measure the angles.
> Wait, if this angle is 120, then that one should be less.

The next phase has started; verifying congruence beyond the visual and discerning the difference between the "measuring" argument and the reasoning argument.

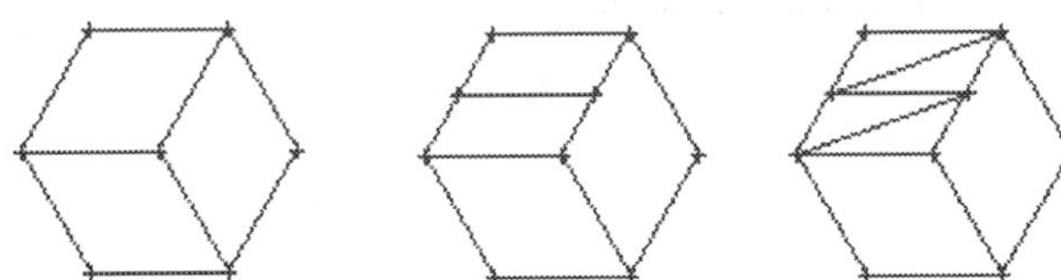

FIGURE 20. The AH!s emerging.

FIGURE 21. Which have congruent triangles?

Conclusion

The learners in these workshops are provided with open-ended and accessible tasks in a non-structured and trusting environment where performance pressures are not pronounced. What we find is that the learners become investigators without fear, creating and re-creating their own problems within a task. They witness and participate in making collaboration a powerful and inspiring avenue for learning.

One teacher writes in her daily journal:

> *[I learnt today] that I could be totally immersed in an exploration that 3 days ago I wouldn't have known how to even embark upon, that I could get stuck and bail myself out, that I could get "software" stuck and ask for help without getting "exploration" stuck.*

A major challenge facing the mathematics and mathematics education communities is the use of the powerful tools that technology provides, not only to revise the mathematics we are to teach, but also to transform our understanding of the teaching and learning processes and to empower the lay mathematician within every learner.

Acknowledgements. The work reported in this article is supported by the National Science Foundation under grant number PPE-9153760. All opinions expressed are solely the responsibility of the author.

Special thanks to Dan-Klemmer, Claire Groden, and Laurie Pattison-Gordon for their contributions to this work.

Endnote

1. Stand-alone software for the bouncing ball simulation is now available through the author.

Bibliography

Physics Explorer: One Body, Logal Educational Software 1994, Cambridge, MA.

Geometry Inventor, Logal Educational Software 1994, Cambridge, MA.

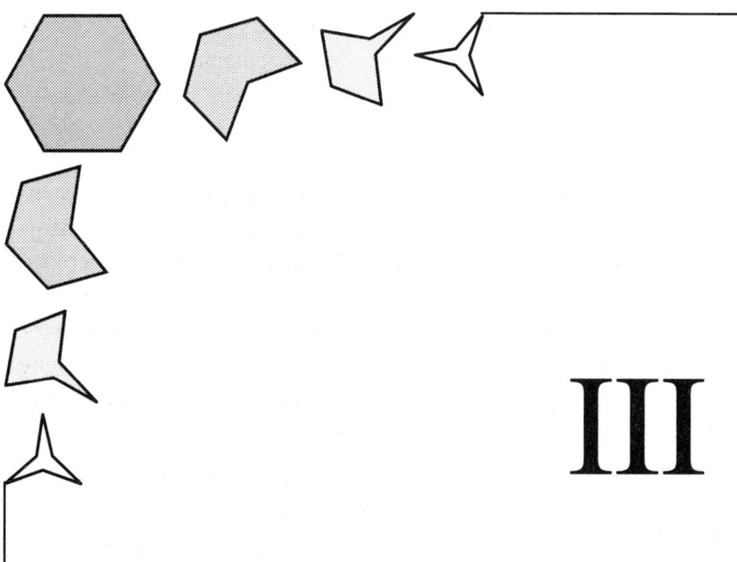

III

Dynamic Visualization in History, Perception, Optics, and Aerodynamics

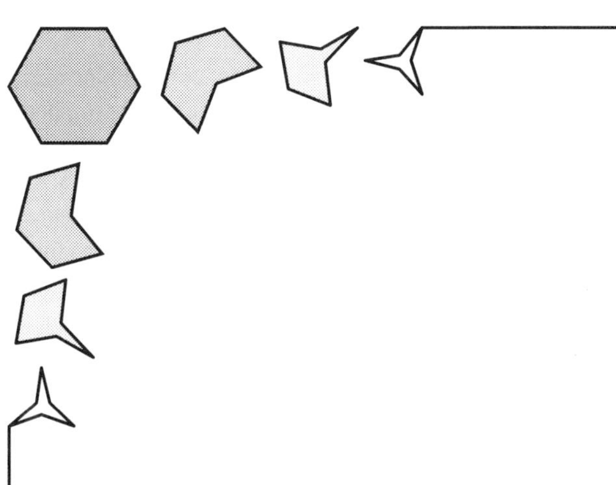

Drawing Logarithmic Curves with *Geometer's Sketchpad:* A Method Inspired by Historical Sources

David Dennis
and
Jere Confrey

In this article we first describe a mechanical linkage device from Descartes's *Geometry* (1637), for finding any number of geometric means between any two lengths. This linkage device draws a family of curves which we will discuss briefly. We then employ Descartes's device as part of a construction that will find any number of points, as densely as desired, on any logarithmic or exponential curve. This second construction uses the dynamic geometry software, *Geometer's Sketchpad*, in a modern adaptation of several historical ideas, combined in ways that provide useful and enlightening curricular suggestions, including an elementary approach to tangent slopes. Several important educational and epistemological questions are raised by our approach, and we conclude by discussing these.

Descartes's Construction of Geometric Means

In the *Geometry* (1637), Descartes considered the problem of finding n mean proportionals (i.e., geometric means) between any two lengths a and b (with $a < b$). To do this, one must find a sequence of lengths beginning with a and ending with b such that the ratio of any two consecutive terms is constant. In modern algebraic language, this means finding a sequence $x_0, x_1, \ldots, x_{n+1}$ such that for some fixed ratio r, $x_k = a \cdot r^k$ and $x_{n+1} = b$. Hence the terms of the sequence have a constant ratio of r, and form a geometric sequence beginning with a and ending with b.

Descartes began, as always, with a geometric construction. He imagined a series of rulers with square ends, sliding along and pushing each other, creating a series of similar right triangles (see Figure 1, which is reproduced from an original 1637 edition of the *Geometry*). Let Y be the origin with A and B on a circle of radius a centered at Y. The tangent to the circle at B intersects the line ZY (the x-axis) at the point C. As angle XYZ increases, C moves farther to the right on the x-axis. The vertical from C then intersects the line XY at D which is still farther from the origin. The triangles YBC, YCD, YDE, YEF, etc., are all similar, since they are all right triangles that contain the angle XYZ. Hence we have $\frac{YB}{YC} = \frac{YC}{YD} = \frac{YD}{YE} = \frac{YE}{YF} = \cdots$. Therefore, the sequence of lengths $a = YB, YC, YD, YE, YF, \ldots$ form a geometric sequence.

If we let $a = 1$ and angle $XYZ = 60°$, we form the sequence: $1, 2, 4, 8, 16, 32, \ldots$. If angle $XYZ = 45°$, the sequence of lengths is: $1, \sqrt{2}, 2, 2\sqrt{2}, 4, 4\sqrt{2}, 8, \ldots$ which is a refinement of the previous sequence. As angle XYZ decreases, we obtain increasingly dense geometric sequences. In modern terms, the relationship between the constant ratio r and angle XYZ is given by the equation: $\sec(\text{angle } XYZ) = r$. This relationship is never mentioned

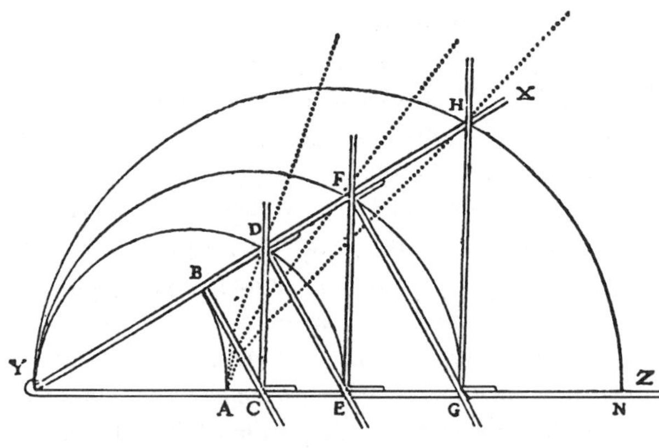

FIGURE 1

in the *Geometry*. Descartes instead emphasized the curves traced by the points D, F, and H as the angle XYZ is varied (shown in Figure 1 by the dotted lines). These curves all have algebraic equations, as opposed to the secant which can only be computed with some infinite process.

To solve the original problem of finding n mean proportionals between a and b, Descartes suggested using the curves drawn by the device. For example, if two mean proportionals are sought between a and b, mark off length $b = YE$ on the line YZ (recall $a = YA$). Next construct the circle having diameter YE, and find its intersection (D) with the first of these curves (see Figure 1). Then drop the vertical line from that point (D) to YZ, to locate the point C. YC and YD will then be the desired mean proportionals. This method uses the curve drawn by D to determine the appropriate angle of the device so that the point E will fall on any specified length.

The equations of the curves traced by $D, F, H, \ldots$ can all be found by successively substituting into the similarity relations upon which the device was built. To find these equations one can proceed as follows. Let $Y = (0, 0)$, $D = (x, y)$, and let $YD = z$. Now $z^2 = x^2 + y^2$, but one also knows that $\frac{z}{x} = \frac{x}{a}$. Hence $z = \frac{x^2}{a}$, and therefore by substitution one obtains, for the path of point D, the equation: $\boldsymbol{x^4 = a^2(x^2 + y^2)}$.

Now let $F = (x, y)$, and let $YF = z$. Now $\frac{z}{x} = \frac{x}{YD}$, hence $YD = \frac{x^2}{z}$. One also knows that $\frac{x}{YD} = \frac{YD}{YC}$; so substituting and solving for YC one gets: $YC = \frac{x^3}{z^2}$. Lastly, one knows that $\frac{YD}{YC} = \frac{YC}{a}$; hence: $\frac{ax^2}{z} = \frac{x^6}{z^4}$. Solving for z, one obtains: $z = \sqrt[3]{\frac{x^4}{a}}$. As before, $z^2 = x^2 + y^2$, so $\sqrt[3]{\frac{x^8}{a^2}} = x^2 + y^2$. Cubing both sides, one obtains, for the path of point F, the equation: $\boldsymbol{x^8 = a^2(x^2 + y^2)^3}$.

In a similar fashion, one can find that an equation of the curve traced by the point H is: $\boldsymbol{x^{12} = a^2(x^2 + y^2)^5}$. Note that all of these curves pass through the point $A = (a, 0)$, and that, as one moves from one of these curves to the next, the degree of the equation always increases by four (on both sides of the equation). In the *Geometry*, Descartes proposed a system which classified curves according to *pairs* of algebraic degree: i.e., lines and conics form the first "genre" of curves; those with third and fourth degree equations form the second "genre" of curves; and so on.[1] In many examples, Descartes found that iterating various forms of mechanical linkages tended to jack up by twos the degrees of the equations of the curves. The curve traced by D is of his second *genre*; the curve traced by F is of the fourth *genre*; etc. To get from any one of these curves to the next one involves two perpendicular projections, each of which raises by one the Cartesian *genre* of the curve.

Descartes, after stating that, "there is, I believe, no easier method of finding any number of mean proportionals, nor one whose demonstration is clearer," (1637, p. 155) goes on to criticize his own construction for using curves of a higher *genre* than is necessary. Finding two mean proportionals, for example, is equivalent to solving a cubic equation and can be accomplished by using only conic sections (first *genre*) while the curve traced by D is of the second *genre*. The solution of cubics by intersecting conics had been achieved in the thirteenth century by Omar Khayyam and was well known in seventeenth-century Europe (Joseph, 1991). Descartes spent much of the latter part of the *Geometry* discussing the issue of finding curves of minimal *genre* that will solve various geometry problems (1637).

Descartes expounded an epistemological theory which sought a universal structural science of measure which he called "mathesis universalis" (Lenoir, 1979). Fundamental to his program was his classification of curves in geometry. He wanted to expand the repertoire of curves that were allowed in geometry beyond the classical restrictions to lines and circles, but he wanted to include only curves whose construction he considered to be "clear and distinct" (Descartes, 1637). For him this meant curves which could be drawn by linkages (i.e., devices employing only hinged rods and pivots). Such curves could all be classified by his system according to pairs of algebraic degrees, since the class of curves that can be drawn by linkages is exactly those which have algebraic equations (Artobolevski, 1964). Algebraic curves were called "geometrical" by Descartes because he wanted to expand the constructions allowed in geometry to include those curves. All other curves he called "mechanical."

This distinction is equivalent to what Leibniz would latter call "algebraic" and "transcendental" curves. Descartes viewed "mechanical" (i.e., transcendental) curves as involving some combination of incommensurable actions.

Examples that he specifically mentioned are the spiral, quadratrix, and cycloid. These curves all involve a combination of rotation and linear motion that can not be connected and regulated by some linkage. The drawing of such curves involves rolling a wheel or the unwinding of string from a circle. Descartes was aware that such curves could not be classified by his system. This is not to say that Descartes did not address himself to problems concerning transcendental curves (see Dennis, 1995, for some of Descartes's thoughts on the cycloid).

A Dynamic Construction of Logarithmic Curves

In order to construct logarithmic curves, we must first define the term "subtangent." Given a smooth curve and an axis line, for each point P on the curve let T be the intersection of the axis with the tangent line to the curve at P, and let O be the foot of the perpendicular from P to the axis (i.e., PO is the ordinate, see Figure 2). For each point P the subtangent is then defined as the line segment TO. Throughout the seventeenth century, such geometric entities were studied for symmetry and invariance as the point P moved along the curve. Such investigations played a very important role both in the study of curves and in the development of the concept of functions (Arnol'd, 1990; Dennis and Confrey, 1995).

Two years after the publication of the *Geometry* Descartes addressed a problem that was sent to him by De Beaune which asked for the construction of a curve in a skewed coordinate system where the ratios of the subtangents to the ordinates are everywhere equal to the ratio of the ordinates to a fixed segment, i.e., a type of logarithmic curve, the requirement being equivalent to a first-order differential equation (Lenoir, 1979). Descartes generated a method for pointwise approximation of this curve and also provided a detailed study of how the curve could be drawn by a combination of motions with particular progressions of speeds. He then stated:

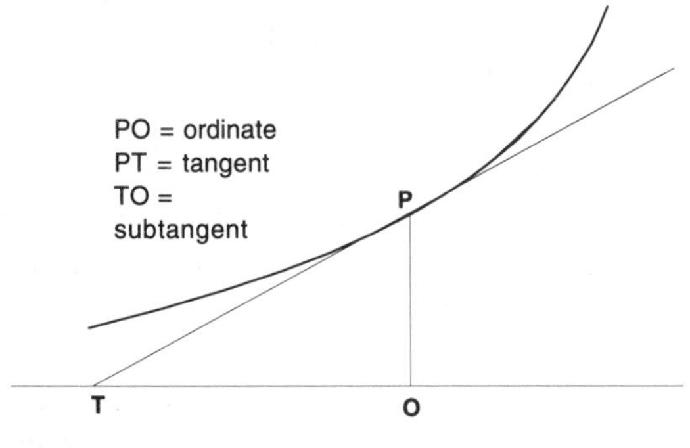

FIGURE 2

> I suspect that these two movements are incommensurable to such an extent that it will never be possible for one to regulate the other exactly, and thus this curve is one of those which I excluded from my *Geometry* as being mechanical; hence I am not surprised that I have not been able to solve the problem in any way other than I have given here, for it is not a geometrical line. (Descartes, quoted in Lenoir, 1979, p. 362)[2]

We will now proceed to construct pointwise approximations of logarithmic curves, but we will not follow the particular example discussed in Descartes's letters to De Beaune. That example turns out to have been a transformation of a logarithm added to a linear function. We will instead construct standard logarithmic curves using Descartes's device shown in Figure 1, together with the original conception of logarithms by Napier as pairings of geometric and arithmetic sequences (Smith and Confrey, 1994; Edwards, 1979).

This construction connects curve drawing with a covariational view of functions. Covariation is essentially a viewing of tables of data that looks for methods to simultaneously extend or interpolate values in both columns separately, rather than looking for a rule which relates values in the first column to those in the second. This approach to functions was central in the thinking of Leibniz (Leibniz, 1712) and has been shown to be important in the thinking of students (Rizzuti, 1991; Confrey and Smith, 1995).

Napier and others in the early seventeenth century made tables of logarithms by placing arithmetic sequences alongside geometric sequences. They devised various ways to make these tables dense (Edwards, 1979). These early approaches to logarithms were entirely tabular and calculational and did not involve curves or graphs. When Descartes constructed a curve as a solution to De Beaune's problem, he did not view the curve as a "logarithm." A fully flexible view that could go back and forth between curves, tables, and equations did not evolve until the end of the seventeenth century, especially with respect to transcendental curves whose general coordinates could be found only by using series expansions (Dennis & Confrey, 1993).

Our aim here is to provide modern students with a hands-on way to build logarithmic and exponential curves through a series of simple geometric constructions using *Geometer's Sketchpad.* From the standpoint of covariation there is little difference between exponentials and logs. The pair of actions which builds one also builds the other. We have constructed the following curves as logarithms, but the same constructions could be viewed as exponentials, by simply repositioning the constructive actions.

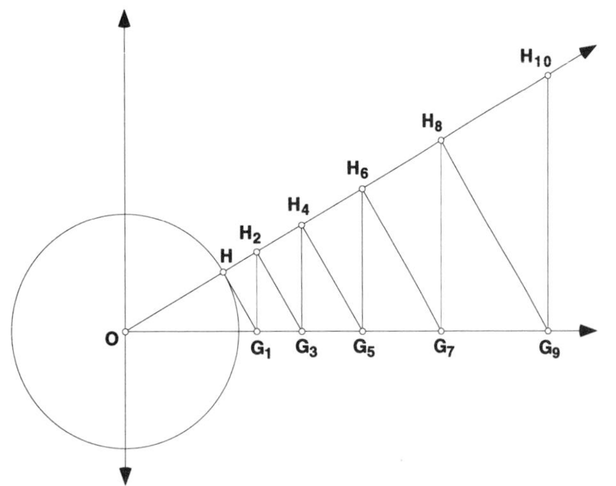

FIGURE 3

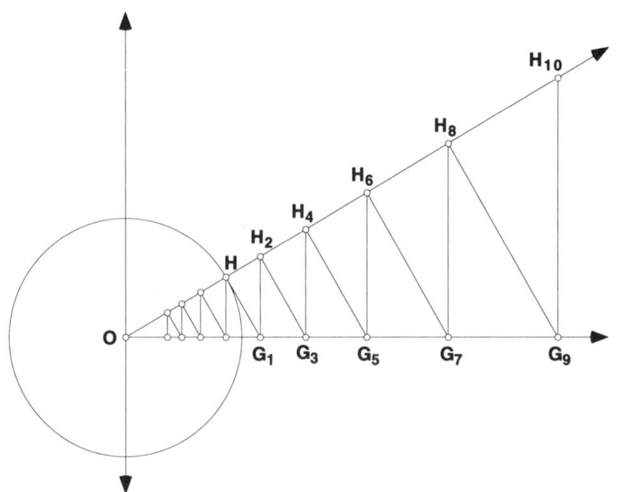

FIGURE 4

We start by building a computer simulation of Descartes' device for the construction of geometric sequences with $a = 1$ (see Figure 3). Let O be the origin and let H be any point on the unit circle. By moving H around the circle, the distances of the labeled points from the origin will form geometric sequences with any common ratio. That is, if $r = OG_1$, then $r^2 = OH_2$, $r^3 = OG_3$, $r^4 = OH_4$, etc.

This construction can also be extended to the interior of the unit circle to obtain segments whose lengths are the negative powers of r. Once again, as with the preceding construction, the odd powers of r are on the horizontal while the even powers of r are on the line OH. This can be seen, as before, by considering the series of similar triangles with common vertex O (see Figure 4).

In order to get the entire geometric sequence on one line, we will now tranfer the lengths marked on the line OH onto the x-axis by using circles centered at O (see Figure 5). Thus we now have a geometric sequence laid out on the x-axis whose common ratio, or density, can be varied as the point H is rotated. The point where the circle through H_2 intersects the x-axis, we will name G_2, likewise for H_4, H_6, etc. The points on the x-axis that are inside the unit circle we will call G_{-1}, G_{-2}, ... etc., where the subscripts correspond to the powers of r that represent their distances from O.

In order to construct logarithmic curves we must now construct an arithmetic sequence $\{A_i\}$ on the y-axis with a variable common difference. This can be achieved in a variety of ways (e.g., create OA_1, and then translate repeatedly along the y-axis by OA_1). The common difference (d) in the arithmetic sequence can be adjusted here by moving A_1 along the y-axis. We next vertically translate each of the points G_i in the geometric sequence by lengths OA_i

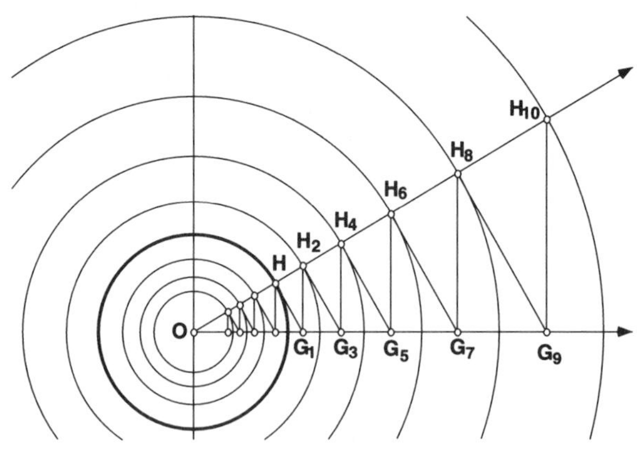

FIGURE 5

corresponding to consecutive points in the arithmetic sequence, thus creating the points (G_i, A_i). These translated points will then lie on the graph of a logarithm. We can connect the new points by line segments to approximate a log curve (see Figure 6). Using the measuring tools available in *Geometer's Sketchpad*, one can continuously monitor the lengths in both the sequences and, hence, the coordinates of the points on the log curve. Some of the construction lines in Figure 6 have been hidden for greater visual clarity.

We now have an adjustable curve. By moving H around the unit circle or A_1 along the y-axis, one can map any geometric sequence against any arithmetic sequence. In Figure 6 the point H is adjusted so that $OG_4 = 2$, and the point A_1 is adjusted so that $OA_4 = 1$. Hence this curve is a graph of the log base 2.

By readjusting A_1 so that $OA_8 = 1$, the curve shifts dynamically to become a graph of the log base 4 (see Figure 7).

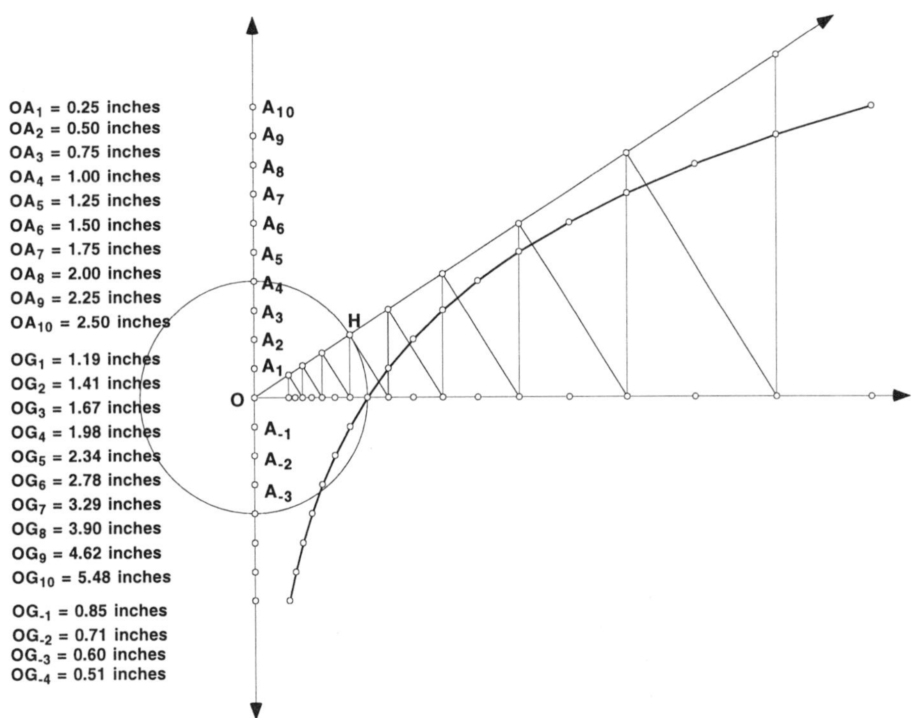

FIGURE 6

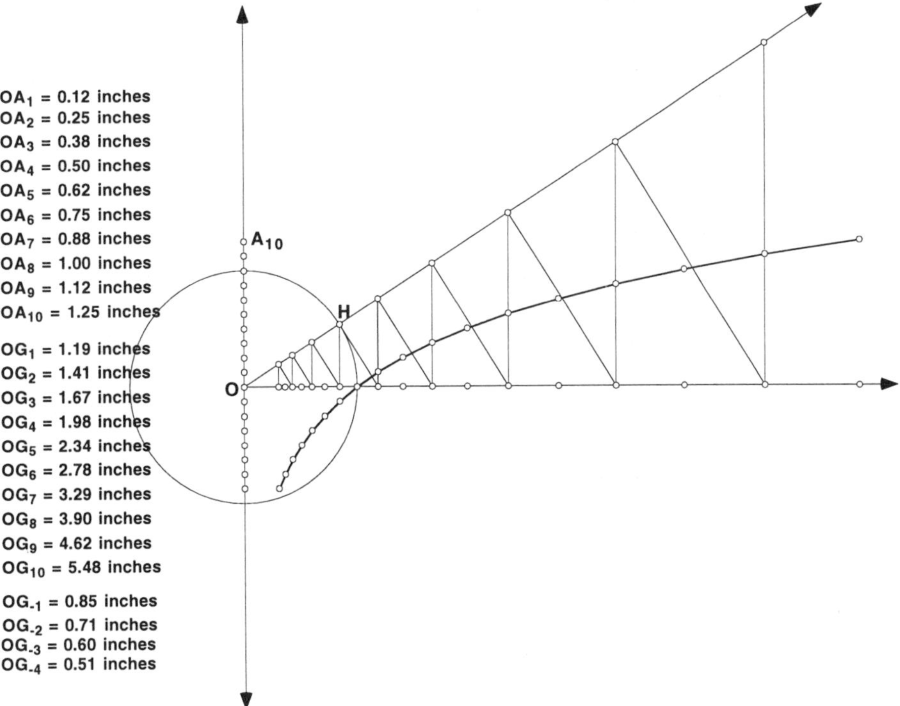

FIGURE 7

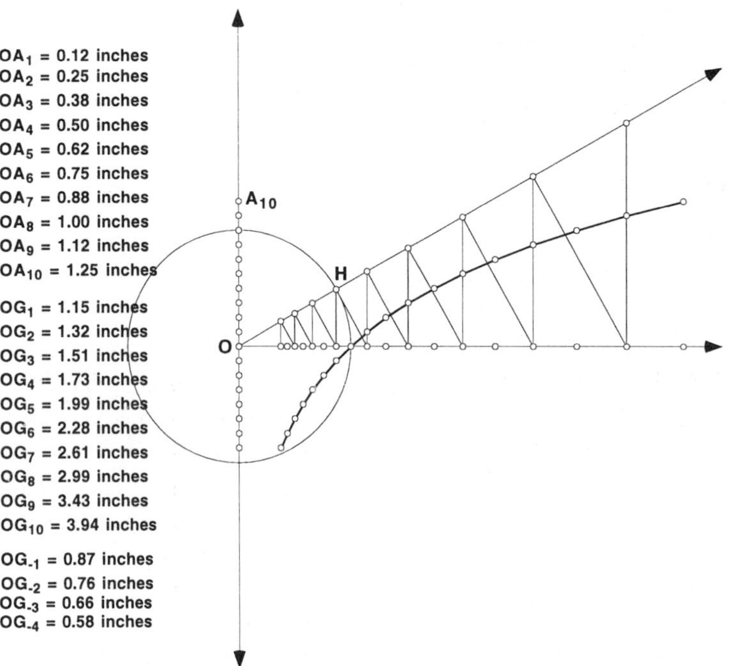

FIGURE 8

By readjusting the point H so that $OG_8 = 3$, one obtains a graph of the log base 3 (see Figure 8). Since the monitors are measuring to the hundredth of an inch, which is smaller than a pixel, it is not always possible to get exactly the desired numbers by simple direct actions (e.g., OG_8 reads 2.99 in Figure 8). One can get around the pixel problem by using the appropriate rescaling window commands, but for a first experience this would decrease the sense of a direct physical approach, which we feel is important for students.

It is fascinating to watch these curves flex and bend as the arithmetic and geometric sequences are manipulated. Even when the points are quite broadly spaced, as in Figure 6, the graphs look very smoothly curved, though they are all made up of line segments. When the angle of H is increased, the geometric sequence spreads out rapidly off the screen. By scanning far to the right, it is instructive to see just how incredibly flat log curves become.

When the arithmetic and geometric sequences are both spread out, the graphs can eventually become "chunky" since the points are being connected with line segments. However, by manipulating both sequences it is possible to increase the density of points on any particular log graph without changing the base. For example, we could create another graph of the log base 2 by setting $OA_8 = 1$, and $OG_8 = 2$ (see Figure 9). This is the same curve as the one in Figure 6, but with a much higher density of constructed points.

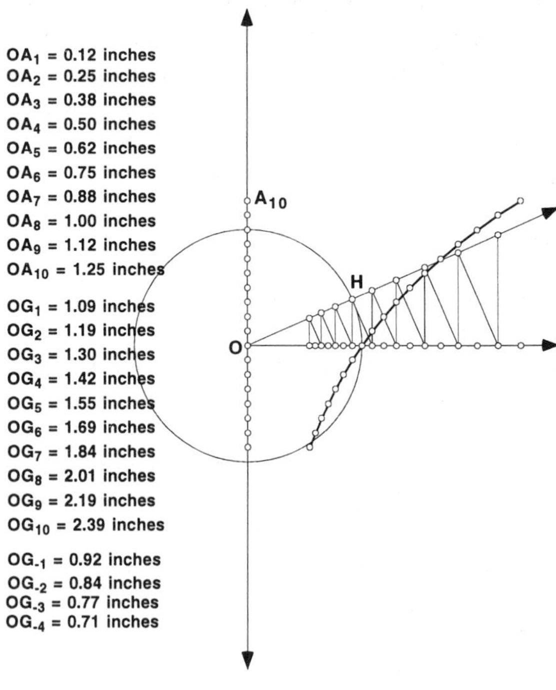

FIGURE 9

Descartes's device allows us to carry out geometrically the calculational aims of Napier and other seventeenth-century table makers. Geometric sequences can be built as densely as one desires, and paired against any arithmetic sequence.

An Investigation of the Slopes of Logarithmic Curves

After looking at these log curves shift and bend dynamically, one can begin to look carefully at the slopes of segments that join points on the curves. Several interesting patterns come to light. If the slopes of segments that join consecutive constructed points are used to approximate the tangent slope at a point, say, for example, at $(1, 0)$, it is visually apparent that this calculation is not the best thing to use. The slope between the nearest points to the right and left of a point gives a better approximation of the tangent slope at that point. This is true for most curves, not just the logarithm.[3] Here, at (1, 0), the best appoach to the tangent slope is to calculate the secant slope between G_{-1}, and G_1. Letting r equal the common ratio of the geometric sequence, and d equal the common difference of the arithmetic sequence, we calculate this slope as:

$$\text{tangent slope at } (1, 0) \approx \frac{2d}{r - \frac{1}{r}} = \frac{2rd}{r^2 - 1}$$

We will use k for this approximate slope at $(1, 0)$. Suppose we now approximate in the same way the slope at any other point on the constructed curve. The approximate tangent slope at $(G_n, A_n) = (r^n, nd)$ is found by computing the secant slope between G_{n-1} and G_{n+1}. The calculation yields:

$$\text{tangent slope at } (G_n, A_n) \approx \frac{2d}{r^{n+1} - r^{n-1}} = \frac{1}{r^n} \cdot \frac{2rd}{r^2 - 1} = \frac{k}{r^n}$$

Here one has the approximate tangent slope at a point on a logarithmic curve written as the function $\frac{1}{x}$ times a constant k that is the slope of the curve at $(1, 0)$. Of course these slopes are all approximations, but, once the slope at $(1, 0)$ is approximated, it can be divided by the x-coordinate at any other point to get the corresponding slope approximation at that point. When we make the constructed points on the curve denser, the approximations all improve by the same factor. Thus the essential derivative property of logarithms is revealed without recourse to the usual formalisms of calculus. In fact, even more is being displayed here than the usual derivative of a logarithm. One sees that all the slope approximations converge uniformly, as the density of the constructed points is increased.

This constant k (= the slope at $(1, 0)$) can be seen geometrically in another way. If we view these curves and tangent constructions using the vertical axis (i.e., as exponential functions), then we find that the subtangent is constant for all points along the curve, and is always equal to k. This can be established algebraically from the previous discussion, but it is nice to see it geometrically on the curve, and verify it using the measurement capability of *Geometer's Sketchpad*. This is shown in Figure 10 for two different points on the log base three curve. The tangent lines and slope are approximated by using the points adjacent to the one under consideration, and the accuracy is quite good (a calculator gives $k = .910$).

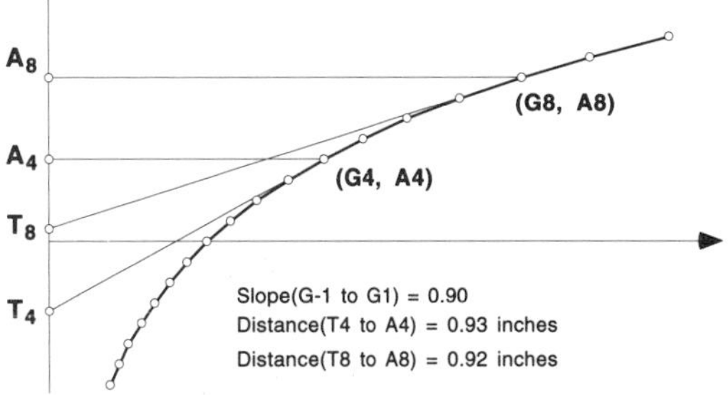

FIGURE 10

This constant subtangent property was at the heart of Descartes's discussion of De Beaune's curve. The constant subtangent was the hallmark by which logarithmic and exponential curves were recognized during the seventeenth century (Lenoir, 1979; Arnol'd 1990). One way to think of this property is to imagine using Newton's method to search for a root of an exponential curve. The method will march off to infinity at a constant arithmetical rate, where the size of the steps will be the constant k.

In order to construct the natural logarithm, we want the slope k at $(1, 0)$ to be equal to 1. This is the property from which Euler first derived the number e (Euler, 1748). Returning to the construction, with a measure that monitors the secant slope k between G_{-1} and G_1, we now rotate H until the slope measurement reads as close to 1 as possible. We have now constructed a close approximation to the graph of the natural logarithm. The approximate slope at any point on the curve is the inverse of its x-coordinate. Note that, since $OA_5 = 1$, the value of G_5 is approximately the number e (see Figure 11).

This geometric construction of points on log curves achieves the goals set out by Napier. It allows one to construct logarithms (and also exponents) as densely as one desires. Of course, Napier achieved these goals through interpolation schemes (Edwards, 1979), and throughout the seventeenth century increasingly subtle methods of table interpolation were developed, e.g., those of John Wallis (Dennis and Confrey, 1996). The story of these calculational

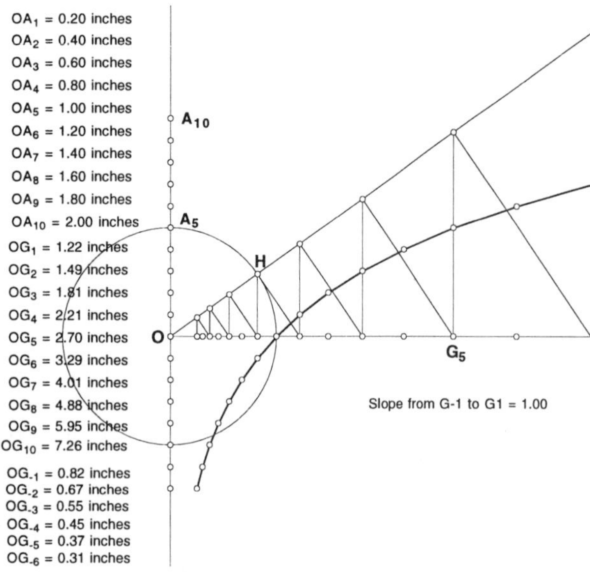

FIGURE 11

techniques is a very important one leading eventually to Newton's development of binomial expansions for fractional powers. Euler routinely used Newton's binomial expansion techniques to calculate log tables to over 20 decimal places (Euler, 1748). Theoretically the geometric construction has unlimited accuracy, and this can be achieved through appropriate rescalings, but directly using the measuring capability of *Geometer's Sketchpad*, one is limited to at most three decimal places.

Conclusions

Our approach raises several general educational and epistemological questions:

1) How are tools and their actions related to mathematical language and symbols?
2) What notion of functions arises from constructions made with particular tools?
3) What are the educational and epistemic roles of mathematical history?

The first question is addressed in theoretical detail by Confrey (1993). She suggests that effective education in mathematics must be approached as a *balanced* dialogue between "grounded activity" and "systematic inquiry." New technological tools, such as dynamic geometry software, help to populate the dialogue between physical investigations and symbolic language, allowing it to flow with greater ease. Confrey stresses the need for balance in this dialogue. All too often in our mathematics curriculum, symbolic language is given a preeminent role, de-emphasizing visual and physical activity, especially at more advanced levels. The examples discussed in this paper demonstrate the value of striking a balance.

The most common notion of a function that is taught in mathematics is to see it as a rule or process which computes or predicts one quantity from another, i.e., a correspondence notion. These quantities are most often real numbers, and functions can then be represented by graphing these quantities in the plane. Our investigations employed two different conceptions of functions, both of which have played important historical roles. The first conception involved creating a curve as a primary object through some kind of physical or geometric action, and only afterwards analyzing it by means of the geometric properties of particular magnitudes, such as abscissas, ordinates, and subtangents (Dennis, 1995). Algebra and equations then become secondary representations of curves, rather than primary generators. The second conception of functions that came into play in our constructions is "covariation" (Confrey and Smith, 1995). This view sees functions as a pair of independent operations on columns in a table, where one finds ways to simultaneously extend or interpolate values on both sides of a table, thus generating a correspondence when the columns are paired in a given position.

Both of these conceptions of functions have played important roles in the genesis of analytic geometry and calculus. Neither one is entirely reducible to the current "definition" of a function. For example, in Descartes's *Geometry*, never once is a curve created by plotting points from an equation. Descartes always began with a physical or geometric way of generating a curve, and then, by analyzing the actions which produced the curve, he obtained equations. Curves were primary and gave rise to equations only as a secondary representation. Equations allowed him to create a taxonomy of curves (Lenoir, 1979). Furthermore, Leibniz originally created his calculus notation from his many experiments with tables, where a covariational approach was his fundamental form of generation (Leibniz, 1712).

These alternative conceptions of functions are not just awkward phases of historical development which should be abandoned in light of more modern developments. Quite the contrary, they provide ways of working which give mathematicians a powerful flexibility. They can help to balance the dialogue between the physical world and our attempts to represent it symbolically, especially when combined with newly available tools, such as *Geometer's Sketchpad*.

We ask the reader to compare this investigation of logarithms with the approaches more frequently taken in classrooms. Many students are introduced to logarithms in a formal algebraic way, with no references to geometry or to table construction. Such students often have no method for geometrically or numerically constructing even a square root. Such an approach leads, at best, to only a superficial understanding of the grammar of logarithmic notation. There is no dialogue at all between geometrical, numerical, and algebraic experience.

Another approach that is frequently taken is to see logarithms as the accumulated area under a hyperbola (usually $y = 1/x$). This approach can provide many fascinating insights that connect logarithms to both geometry and to the numerical construction of tables. The study of hyperbolic area accumulation was fundamental in the early work of Newton, but was always linked in his work to extensions of the table interpolations of John Wallis. It was in this setting that Newton created his first infinite binomial expansions (Dennis and Confrey, 1993; Edwards, 1979). Although the hyperbolic area approach can create a fascinating and balanced dialogue, it is not usually taken with students until they are already involved with calculus. The fundamental theorem of calculus, for example, is usually invoked to show that the hyperbolic area function must have a derivative of $1/x$.

The approach that we have described here is strictly pre-calculus. It involves only a systematic use of similar triangles, in a hands-on setting that is both visual, physical, and geometric. It provides a specific form of grounded activity that allows students to manipulate, extend, and interpolate both logarithms and continuous exponents. Rather than using calculus to create a balanced dialogue, this approach uses the dialogue to achieve some of the results of calculus in a very simple setting. It highlights the power of iterated geometric similarity (Confrey, 1994).

Reading this paper can not truly convey the feeling one gets while physically manipulating the curves. The investigation of the slopes of log curves depends logically only on the properties of a table which maps a geometric sequence against an arithmetic sequence, but we did not notice this piece of algebra until many fluctuating examples of log curves had appeared on the screen. Dynamic geometry can heighten the intuition so that fruitful conjectures emerge. The power of suggestion should not be underestimated. The association of rotation around the unit circle with the building of logarithms is a wonderful foreshadowing of the connections between these functions and the trigonometric functions, when extended to the complex numbers (Euler, 1748).

Acknowledgments. The authors wish to thank David Henderson and Paul Pedersen for their probing questions and comments during the conduct of this research.

This research was funded by a grant from the National Science Foundation. (Grant #9053590). The views and opinions expressed are those of the authors and not necessarily those of the Foundation.

Endnotes

1. This same classification by pairs of degrees is used in modern topology in the definition of the "genus" of a surface. Classification by pairs of algebraic degrees often makes more sense geometrically.
2. For a fascinating social and philosophical analysis of why Descartes would adopt such an attitude see the article by Lenoir (1979). It certainly had nothing to do with his ability to contend with such problems.
3. It is strange that, when the derivative is developed in calculus classes, it is defined using secant slopes from the point in question, rather than around the point. It would seem that nobody is directly interested in secant slope approximations, except as an algebraic device from which to define a limit. The practical geometry of using secant slopes is ignored.

Bibliography

Arnol'd, V.I. (1990). *Huygens & Barrow, Newton & Hooke*. Boston: Birkhäuser Verlag.

Artobolevskii, I.I. (1964). *Mechanisms for the Generation of Plane Curves*. New York: Macmillan Co.

Confrey, J. (1993). The role of technology in reconceptualizing functions and algebra. In Joanne Rossi Becker and Barbara J. Pence (eds.) *Proceedings of the Fifteenth Annual Meeting of the North American Chapter of the International Group for the Psychology of Mathematics Education*, Pacific Grove, CA, Oct. 17–20. Vol. 1, pp. 47–74. San José, CA: The Center for Mathematics and Computer Science Education at San José State University.

——. (1994). Splitting, Similarity, and Rate of Change: New Approaches to Multiplication and Exponential Functions. In G. Harel, and J. Confrey, eds., *The Development of Multiplicative Reasoning in the Learning of Mathematics*. Albany NY: State University of New York Press, pp. 293–332.

Confrey, J. and Smith, E. (1995). Splitting, covariation, and their role in the development of exponential functions. *Journal for Research in Mathematics Education*. Vol. 26, No. 1.

Dennis, D. (1995). *Historical Perspectives for the Reform of Mathematics Curriculum: Geometric Curve Drawing Devices and their Role in the Transition to an Algebraic Description of Functions*. Unpublished Doctoral Dissertation, Cornell University, Ithaca, New York.

Dennis, D. and Confrey, J. (1993). The Creation of Binomial Series: A Study of the Methods and Epistemology of Wallis, Newton, and Euler. Presented at the *Joint Mathematics Meetings* (AMS–CMS–MAA) Vancouver, Aug. 1993. Manuscript available from the authors.

——. (1995). Functions of a curve: Leibniz's original notion of functions and its meaning for the parabola. *The College Mathematics Journal*. Vol. 26, No. 2, March 1995, pp. 124–130.

——. (1996). The Creation of Continuous Exponents: A Study of the Methods and Epistemology of Alhazen and Wallis. In J. Kaput and E. Dubinsky (eds.), *Research in Undergraduate Mathamatics Education II*, CBMS, Vol. 6, Providence, RI: American Mathematical Society, pp. 33–60.

Descartes, R. (1637). *The Geometry*. From a facsimile edition with translation by D.E. Smith and M.L. Latham. (1952) LaSalle, Ill. Open Court.

Edwards, C.H. (1979). *The Historical Development of Calculus*. New York: Springer-Verlag.

Euler, L. (1748). *Introduction to Analysis of the Infinite Book One*. Translaion by J.D. Blanton, (1988). New York: Springer-Verlag.

Jackiw, N. (1994). *The Geometer's Sketchpad* (version 2.1), Berkeley, CA.: Key Curriculum Press.

Joseph, G.G. (1991). *The Crest of the Peacock, Non-Euopean Roots of Mathematics*. New York: Penguin Books.

Leibniz, G.W. (1712). History and origin of the differential calculus. In J.M. Child, ed. 1920. *The Early Mathematical Manuscripts of Leibniz*. Chicago: Open Court.

Lenoir, T. (1979). Descartes and the geometrization of thought: The methodological background of Descartes's geometry. *Historia Mathematica*, 6, pp. 355–379.

Rizzuti, J. (1991). *Students' Conceptualizations of Mathematical Functions: The Effects of a Pedagogical Approach Involving Multiple Representations*. Unpublished Doctoral Dissertation, Cornell University, Ithaca, New York.

Smith, E. and Confrey, J. (1994). *Multiplicative Structures and the Development of Logarithms: What was Lost by the Invention of Functions?*. In G. Harel and J. Confrey, eds., *The Development of Multiplicative Reasoning in the Learning of Mathematics*. Albany NY: State University of New York Press, pp. 331–360.

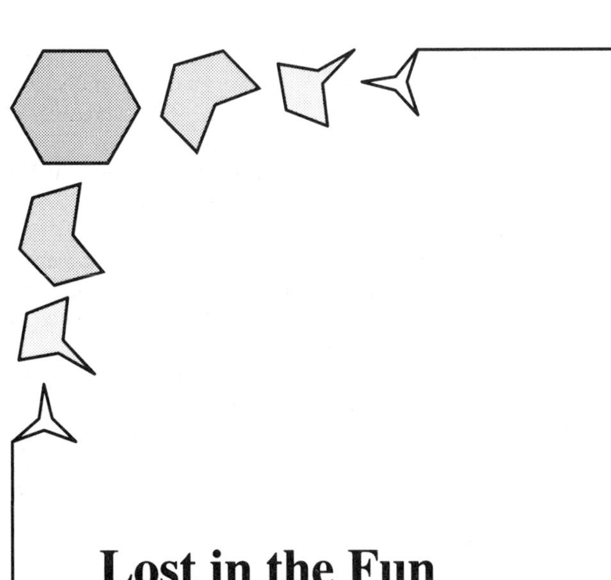

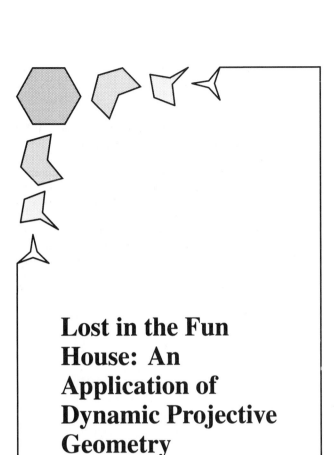

Lost in the Fun House: An Application of Dynamic Projective Geometry

Susan Addington and Stuart Levy

Ames rooms were invented by the perceptual psychologist Adelbert Ames to test the mind's reactions to ambiguous visual cues. Today, Ames rooms can be found in science museums and fun houses. An observer who looks into an Ames room from a peephole at one end sees what appears to be an ordinary rectangular room. Ames rooms are typically decorated with checkered floors, windows with panes, pictures on the walls: props which reinforce the supposed rectangularity of the room. In fact, the room has no right angles at all. The only clue that the room is not as expected is the apparent size of a person standing in the room: a person in one far corner seems to be a giant; in the other, a midget.

Ames rooms illustrate several ideas in projective geometry. We have written an interactive computer demonstration in which users can change the shape and viewpoint of an Ames room, thus gaining a concrete understanding of projective transformations. The program uses *Geomview*, a package for viewing and manipulating 3-dimensional objects.

The program opens three windows with different views of the same Ames room.

The first window (see Figure 1) shows an Ames room from the "correct" viewpoint: the unique point from which the room appears rectangular. Chess pieces represent humans in the back corners of the room. Although the bishop on the left appears half as tall as the bishop on the right, the two bishops are actually the same size. A perspective projection is used for this window; this is essential to the whole illusion.

The second window (see Figure 2) shows another view of the Ames room. Here it is clear that the Ames room is not rectangular; the corresponding normal room is outlined in gray. Lines from the Ames viewpoint E through the back corners of the Ames room also pass through the corresponding corners of the normal room. This shows why the Ames room and the normal room have the same image from the Ames viewpoint. A user can inspect the scene by flying around and zooming in or out using *Geomview*'s controls.

The third window (see Figure 3), in contrast to the first two, uses a parallel projection rather than perspective. It shows the floor plans of both the Ames room and the normal room, with the bishops in the back corners of the Ames room. In this window, it is clear that the left-hand bishop appears smaller because it is farther away. The lines of sight that connect the corresponding corners of the normal room and the Ames room are also clearer in this window.

If a user clicks the mouse in the third window, a new Ames room will be drawn in all three windows. The point

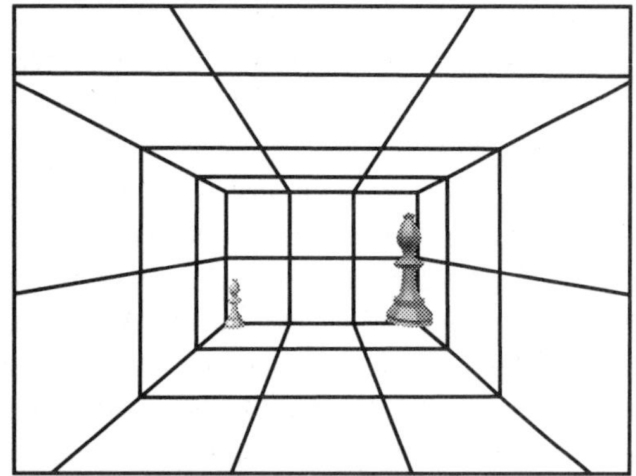

FIGURE 1. "Correct" view of the Ames room.

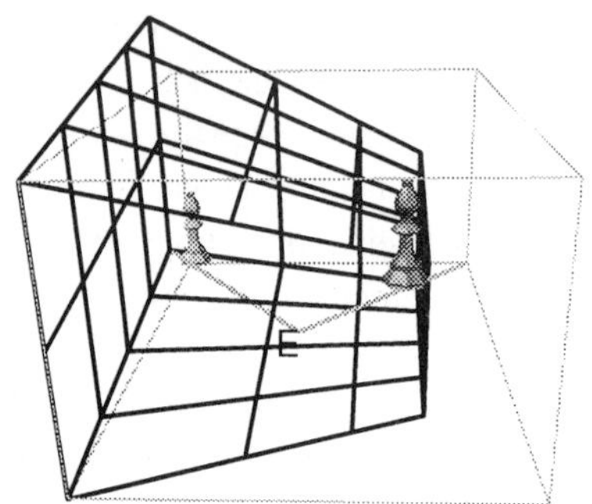

FIGURE 2. Another view of the Ames room.

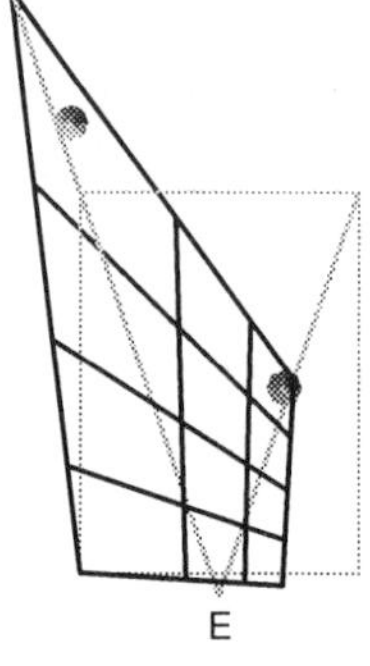

FIGURE 3. Floor plan of the Ames room.

at which the mouse is clicked determines the two free parameters that control the shape of the Ames room. When a user touches the space bar, an animation begins in which the bishops walk towards each other and exchange places. At the midpoint of their common path, the bishops coincide, showing that they really are the same size.

A Simple Model of Visual Perception

Artists and mathematicians since the Renaissance have used the analogy of the picture plane to explain how to produce a realistic image of a scene [1]. The artist is assumed to view

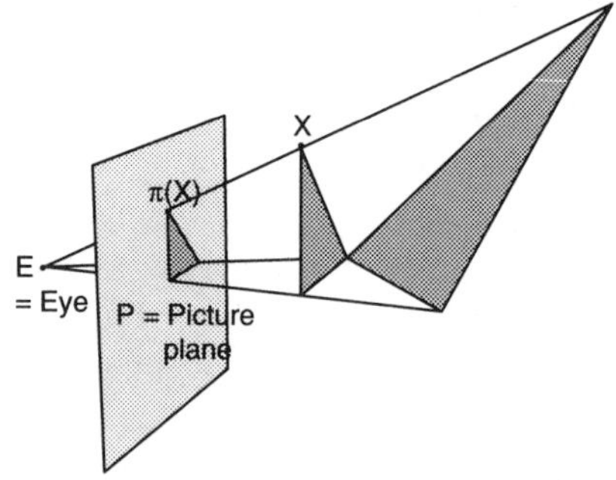

FIGURE 4. Two triangles with the same image.

the scene with one eye, which remains at a fixed position, E. An imaginary glass window, P, is between the scene and the eye. For each light ray XE from a point X in the scene to the eye, the point where that ray intersects the window is the image, or projection, of X.

Figure 4 shows a scene consisting of two triangles, the eye E, picture plane P, a point X in the scene, and its image, $\pi(X)$, in P. The triangles, which are neither congruent nor similar, are placed so that they have the same image.

The effectiveness of this model of vision is shown by *trompe l'oeil* paintings, painted stage and movie backdrops, and, more recently, the visual part of virtual reality experiences.

The projection π (almost) defines a function from $\mathbf{R}^3$ to P:

$$\pi(X) = EX \cap P$$

Note that $\pi(X)$ is undefined if the line EX is parallel to P.

One of the hallmarks of perspective drawings is that the images of parallel lines often converge. How can parallel lines have non-parallel images? Suppose two parallel lines, L_1 and L_2, are in the scene, and neither contains E (see Figure 5). Then $\pi(L_i)$ is $P_i \cap P$, where P_i is the plane

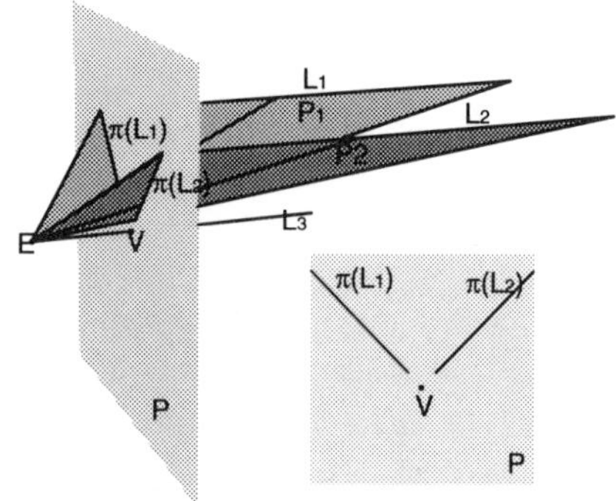

FIGURE 5. The image of two lines perpendicular to P.

determined by L_i and E. Hence, if $\pi(L_i)$ is defined (if E is not on L_i and P_i is not parallel to P), it is a line in P.

Since P_1 and P_2 intersect (at E), they intersect in a line, L_3, and $V = L_3 \cap P$ is the intersection of $\pi(L_1)$ and $\pi(L_2)$. The point V, called a vanishing point, is the point of intersection of the images of parallel lines. In fact, if $f_i(t)$ is a parametrization of L_i, then $V = \lim_{t\to\pm\infty} \pi(f_i(t))$. Hence V is not in the image of L_i under the function π.

Projective Geometry

To unify the treatment of concurrent and parallel lines, define a point at infinity for each family of parallel lines, and declare each point at infinity to lie on each of the parallel lines that define it. (For details, at least in two dimensions, see [2], for example.)

Now parallel lines meet at infinity, and vanishing points are the images of points at infinity. For points $Y \neq E$ in the plane parallel to P through E (for which π was not previously defined), $\pi(Y)$ is a point at infinity.

Definition 1. *Three-dimensional projective space*, $\mathbf{P}^3$, is $\mathbf{R}^3$ together with all points at infinity.

Projective space can be coordinatized, not by triples of real numbers, but by quadruples, called the homogeneous coordinates of a point. Homogeneous coordinates are unique only up to multiplication by a nonzero scalar; that is, $(x, y, z, w) \sim (ax, ay, az, aw)$ if $a \neq 0$. A point (x, y, z) in $\mathbf{R}^3$ is called a finite point, and corresponds to the 4-tuples equivalent to $(x, y, z, 1)$. Points at infinity correspond to $(x, y, z, 0)$, where (x, y, z) is a direction vector for the lines in $\mathbf{R}^3$ defining the point at infinity.

Definition 2. A *projective transformation* is a bijection $T\colon \mathbf{P}^3 \to \mathbf{P}^3$ that takes lines to lines.

Theorem. *If T is a projective transformation, then $T(X) = MX$, where M is an invertible 4×4 matrix, and X is a 4-vector of homogeneous coordinates. The matrix M is unique up to a nonzero scalar multiple.*

Construction of an Ames Room

Projective transformations preserve all the geometric cues we use to recover the shape of an object S from its image $\pi(S)$. A projective transformation can take a finite point to a point at infinity and vice versa. This is realistic, since we can see points at infinity: for example, the point on the horizon where the railroad tracks meet. In real life, parallel lines are distinguishable from concurrent lines only because of additional cues: the blurring effect of the atmosphere on distant objects, and previous experience: the rails on a train track stay a fixed distance apart.

To construct an Ames room A that looks like a normal room R, we need to find a projective transformation T such that $A = T(R)$ and $\pi(A) = \pi(R)$ for a fixed eye point E and picture plane P. For this, it is sufficient to require that: (a) the three families of parallels determined by the edges of the normal room are transformed into parallel or concurrent families; and (b) each corner of the normal room, the corresponding corner of the Ames room, and the eye are collinear.

For reasons of visual believability, we impose two more conditions: (c) We want verticals to remain vertical under T: if $P_{\infty 1}$ is the point at infinity in the vertical direction, then $T(P_{\infty 1}) = P_{\infty 1}$. (d) Let $P_{\infty 2}$ be the intersection of the left–right horizontals of the normal room; we want $T(P_{\infty 2})$ to be the finite point $E + (v, 0, 0)$ for some $v > 0$. That is, the point at infinity to the right of the eye is transformed to a finite point to the right of the eye.

Choose a coordinate system such that the x axis in $\mathbf{R}^3$ points to the right, the y axis points up, and the z axis is pointing into the room from the eye. The z-coordinate of the rear left corners of the Ames room and v can be chosen freely. These are the two parameters in the floor plan for the Ames room.

Assume that the dimensions of the normal room are L, W, D, and that its front lower left corner is at the origin. Let the coordinates of the eye be $E = (e_1, e_2, e_3)$, and the picture plane P be parallel to the xy plane. Then the Ames room A is a projective transformation of the normal room R, and the projective transformation is given by

$$M = \begin{pmatrix} 1 + \frac{e_1}{v} & e_1 \frac{1-L}{DL} & 0 & 0 \\ \frac{e_2}{v} & 1 + e_2 \frac{1-L}{DL} & 0 & 0 \\ \frac{e_3}{v} & e_3 \frac{1-L}{DL} & 1 & 0 \\ \frac{1}{v} & \frac{1-L}{DL} & 0 & 1 \end{pmatrix}$$

The Program

After Addington wrote a version of this program in *Mathematica*, Levy wrote the present version as an external module driving *Geomview*. *Geomview*'s graphics capabilities are far better than *Mathematica*'s, and it is easier to build in interactivity.

Geomview is an interactive package for displaying geometric objects in 3-D. It was written at the Geometry Center by Levy, Tamara Munzner, Mark Phillips, et al. *Geomview* is free software; versions for a variety of UNIX workstations are available by anonymous FTP to `geom.umn.edu` under `/pub/software/geomview`, or from

`http://www.geom.umn.edu/software/download/geomview.html`

The Ames program, written in perl, runs as an external module for *Geomview*, sending *Geomview* Command Language commands to create and move objects, and responding to requests to reshape the Ames room or animate the chess pieces. Since *Geomview* uses homogeneous coordinates, it constructs the Ames room simply by asking *Geomview* to transform the normal room by the matrix M above. Other interactive manipulation of the scenes is handled by *Geomview* alone. The Ames program is available by anonymous FTP to `geom.umn.edu` under `/priv/slevy/ames.shar`.

References

1. Alberti, Leon Battista. (1991). *On Painting*; tr. C. Grayson; Dover.
2. Cederberg, Judith. (1989). *A Course in Modern Geometries*; Springer.

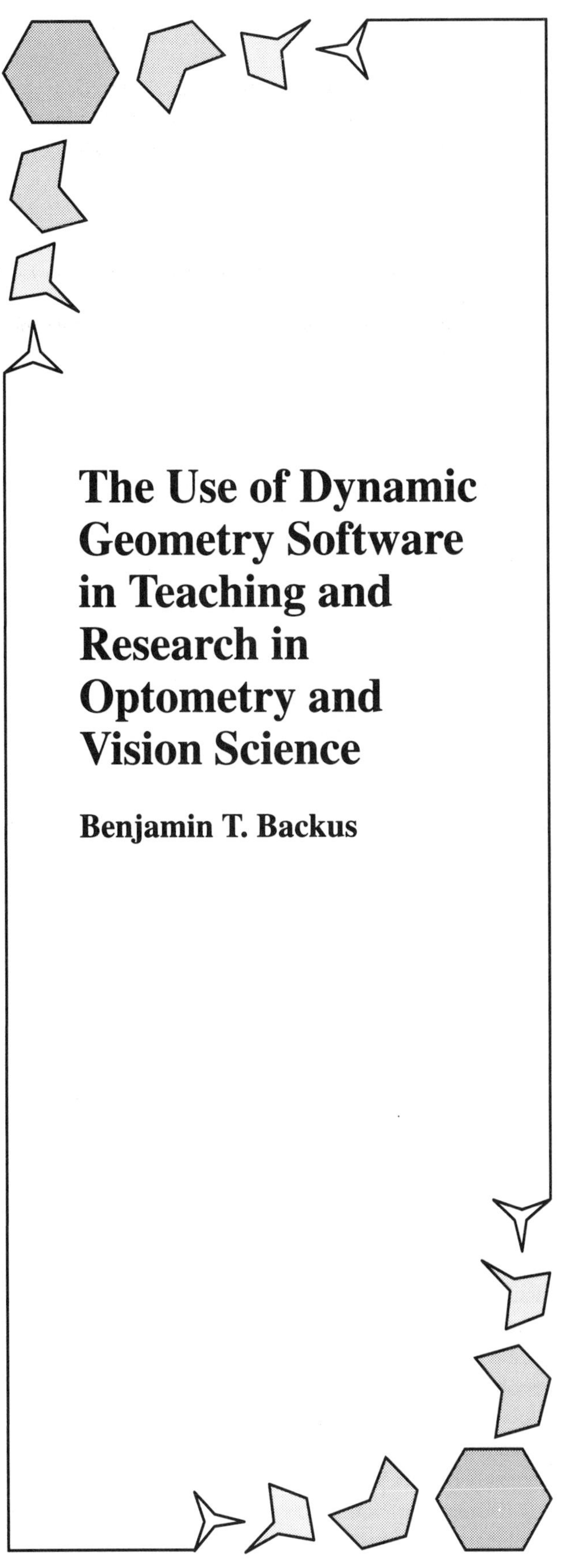

The Use of Dynamic Geometry Software in Teaching and Research in Optometry and Vision Science

Benjamin T. Backus

Light travels in straight lines. It is this property that makes vision possible. It is also the property that makes light a good candidate for modeling with dynamic geometry software (DGS) tools. At the University of California School of Optometry in Berkeley, we use *The Geometer's Sketchpad* (GSP) both for teaching and for research. Dynamic geometric sketches are used for teaching both during lectures and during student laboratory exercises. A geometric optical bench can be built using GSP which allows students to manipulate the focal lengths of lenses, the distances between lenses, and the locations of objects and images. In this use, it is especially helpful that changes in the drawing take place instantaneously. Students can watch an image change in size and position as the focal length of a lens is "dragged" through a series of values, or as the object is moved about.

We also find the software useful as a research tool. Although not designed for these purposes, it serves as a geometric calculator, as a conceptual tool when our intuitions fail us, and as drafting board for the design of equipment.

It is possible to do limited kinds of three-dimensional (3D) work with GSP, and several techniques for doing so are mentioned in this paper. The development of a truly 3D DGS should be a priority; such a tool can be expected to receive wide use in education, research, and industry.

I. Use of DGS in Teaching Optics

The value of dynamic geometry software for teaching geometry is well established. It is also the case that many systems that seem unintuitive when first encountered behave in essentially geometric ways, and lend themselves to modeling with DGS. For students of such systems, the goal is not to understand the laws of geometry per se, but rather to develop an intuition for the laws that pertain to the particular domain of interest. For example, a student of linear algebra might want to gain an intuition for the behavior of affine transformations; a landscape architect might want to gain an intuition about how the relationship between an area and its border length changes with the shape of the area.

DGS is especially well suited to developing an intuition for optics, because light travels in straight lines, and interacts with lenses and mirrors in geometrically well-defined ways. Figure 1 shows a sketch in which a student can watch what happens to the optical image formed by a "thin" lens as the object is moved or as the focal power of the lens changes. Esoteric concepts such as optical infinity can be grasped when the student sees the image depart in one direction and come back from the other. The thin lens equation is $V = P + U$, where U is the vergence of the entering wave front, V

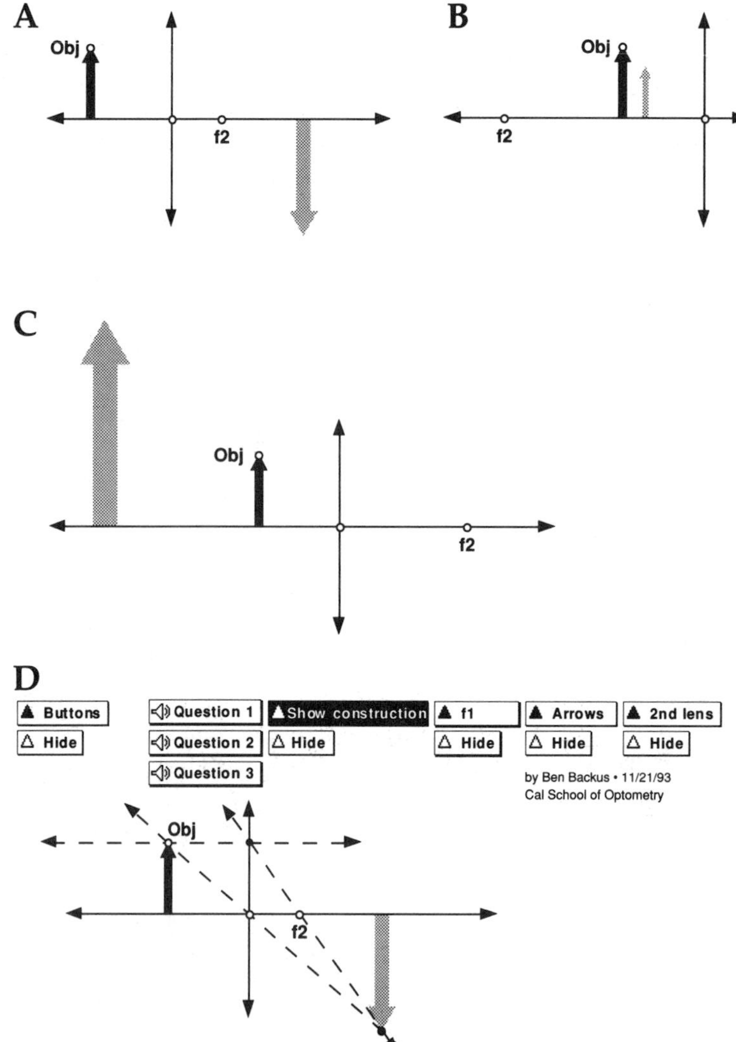

FIGURE 1. Sketch of a thin lens. A. Converging lens, real image. The vertical line represents the lens, the horizontal line represents the optical axis, the dark arrow represents the object, and the light arrow represents the image. The object (**Obj**), focal point (**f2**), and the lens itself can be moved by dragging. B. The point **f2** has been dragged to the other side of the lens, making it a diverging lens with a virtual image. C. Point **f2** has been dragged back again: It is now a converging lens with a virtual image. D. Same as A, but with buttons and authorship note unhidden. The button that shows how the image was constructed is pressed. The "Question" buttons, when pressed, ask verbal questions such as "Have you tried moving the object to the other side of the lens yet?" The "2nd lens" button unhides a second thin lens and the image it forms, using the image from the first lens as the object.

is the vergence of the exiting wave front, and P is the dioptric power of the lens. (Vergence is the reciprocal of the radius of curvature of a spherical or cylindrical wave front, usually expressed in m^{-1}, and dioptric power is the reciprocal of focal length.) The thin lens equation can be verified by the student for an infinite number of combinations of lenses and object distances.

Other lenses besides thin lenses can be modeled. This requires the use of Snell's Law:

$$n_1 \sin\theta_1 = n_2 \sin\theta_2$$

where n_1 and n_2 are the indices of refraction of air and the lens material, respectively, and θ_1 and θ_2 are the angles of incidence formed by a ray of light with respect to the surface normal on either side of the air-glass boundary. There is a geometric method for finding the value of any one of these variables, given the other three. The sketch in Figure 2 allows one to watch what happens as the index of refraction of the lens material changes, and as the angle of incidence of the ray of light is varied.

Distance-correcting eyeglass lenses are made by the combination of two spherical refracting interfaces (Figure 3).

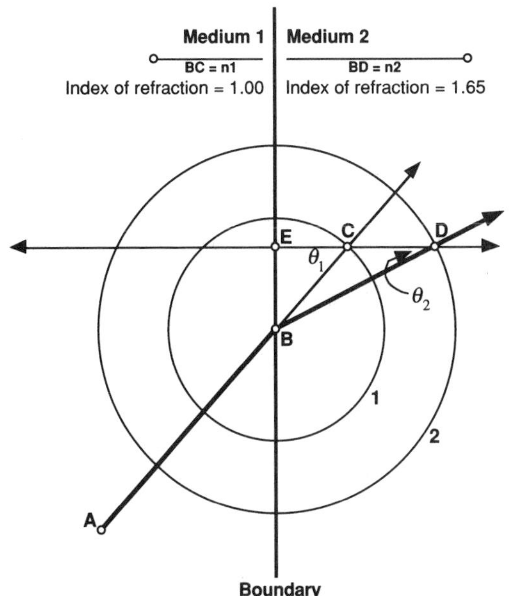

FIGURE 2. Snell's Law for describing the bending of light at the boundary between two media. By construction, circles 1 and 2 have their common center on the boundary at B, and have radii that are proportional to the indices of refraction in media 1 and 2, respectively. The indices can be set using the sliders at the top of the sketch. C is the intersection of the incident ray, continued into medium 2, and circle 1. Line EC is perpendicular to the boundary. It can be seen that BE is equal both to $BC \sin\theta_1$ and to $BD \sin\theta_2$, so that $n_1 \sin\theta_1 = n_2 \sin\theta_2$. The indices of refraction shown are typical for air and glass.

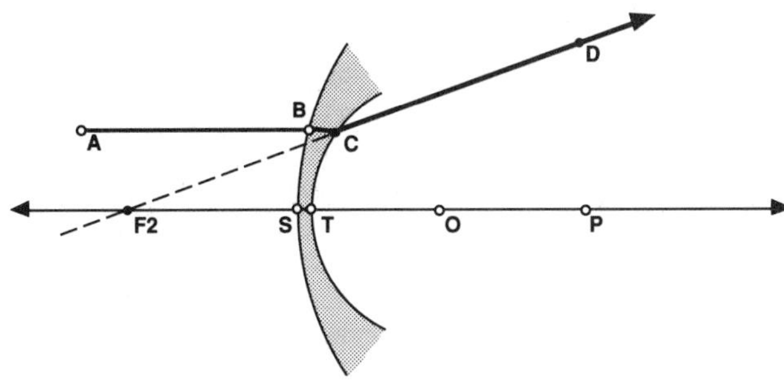

FIGURE 3. Two spherical refracting interfaces, as are commonly used to make the lenses in eyeglasses. All of the open circles can be dragged. Points O and P are circle centers, and they control the curvatures of the lens surfaces. B is confined to the larger circle, and determines the angle of incidence of ray AB upon the lens; B is positioned here so that ray AB is parallel to the optical axis. The *back vertex power* of an optical system is the vergence of the light leaving the system when plane waves are incident on the front. Back vertex power is the quantity used to specify an ophthalmic correction. It can be found here by tracing ray DC back to its intersection with the optic axis to the *secondary focal point*, $F2$, and then taking the reciprocal of the signed distance from T to $F2$. This is a diverging lens, so the back vertex power is negative.

The path of a ray of light through the lens can be constructed by applying the method of Figure 2 to each interface in turn. Spherical aberration can be studied by varying where the light strikes the lens, the angle of incidence, and the focal power of the lens.

Banks (1994), for an optometry course at Berkeley in binocular vision, made a sketch containing two circular eyes, each containing a projection center (single point in its middle) and a "retina" (the back of the circle). Several additional points were placed "in front" of the eyes. Each eye "sees" a given point as the intersection of the retina with the ray that runs from the point through the eye's projection center. The points in front of the eyes can be dragged about. To the side is an anaglyph (red-green stereogram) derived from the retinal images of the free points. When viewed through red-green glasses, the anaglyph shows a 3D image of vertical bars at the same depths relative to the observer as was specified by the locations of the points relative to the eyes in the sketch. (See *Anaglyphs and stereograms*, below.)

A laundry list of other optical phenomena could be examined using DGS: angular magnification, lateral magnification, prisms, telescopes, blur circles, curved and flat mirrors, contact lenses (which have air only on one side), depth of field, aperture stops and entrance pupils, Fresnel lenses, nodal points of Gauss systems, image inversion and reflection, lenses with cylindrical components, and more. We anticipate adding to our small library of pedagogically useful sketches over the course of the next few years.

II. Use of DGS in Research

Three classes of use have been identified for DGS in our research: the modeling of the geometry of vision; geometric calculation; and the design of scientific equipment. Assertions made in the literature about stereoscopic vision can frequently be represented geometrically, and doing so gives one an intuition for the behavior of the quantities involved and for the scope of applicability of a result. GSP is also useful for getting quick answers to the questions that come up during scientific rumination. Figure 4 shows an early sketch used by the author to understand what an observer should see when a given pair of two-point images is presented one to each eye.

We also use GSP as a geometric calculator. We made sketches, for example, to determine how large to cut the holes in the black frames that hide the borders of our CRT monitors, and to check that the glass plate that sits in front of a monitor during an anti-distortion calibration procedure would not cause significant errors in the apparent positions of dots once the plate was removed.

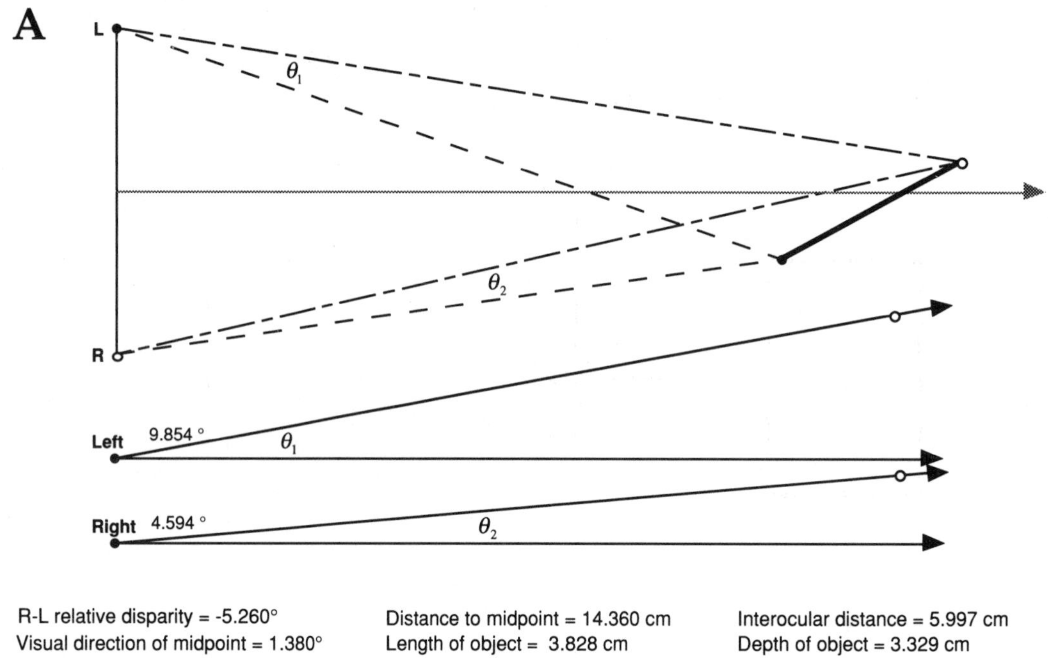

R-L relative disparity = -5.260°
Visual direction of midpoint = 1.380°
Distance to midpoint = 14.360 cm
Length of object = 3.828 cm
Interocular distance = 5.997 cm
Depth of object = 3.329 cm

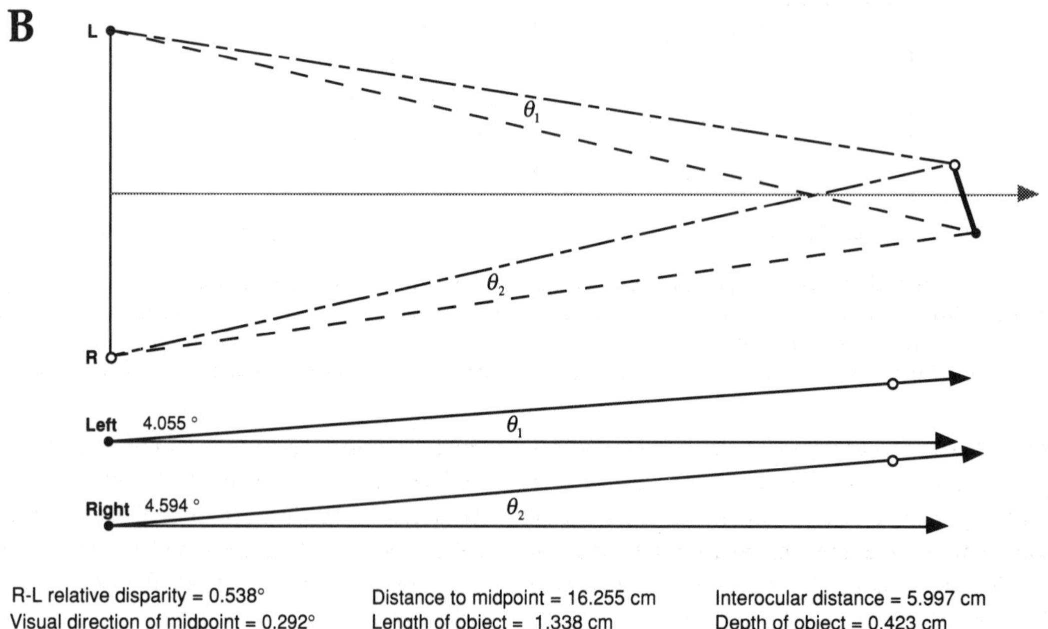

R-L relative disparity = 0.538°
Visual direction of midpoint = 0.292°
Distance to midpoint = 16.255 cm
Length of object = 1.338 cm
Interocular distance = 5.997 cm
Depth of object = 0.423 cm

FIGURE 4. Projection of a two-point object into space as a function of the location in space of one of the points, and the angle subtended by the points in each eye. The thick segment predicts the perceived length and orientation of the segment that connects the points in space. The open-circle endpoint of the segment is free; the filled-circle endpoint is determined. The angles subtended in either eye are set at the bottom of the sketch. The eyes (**L, R**) can be moved to see the effect of interocular distance. A. The points subtend a larger angle in the left eye than in the right eye. B. The points subtend a larger angle in the right eye.

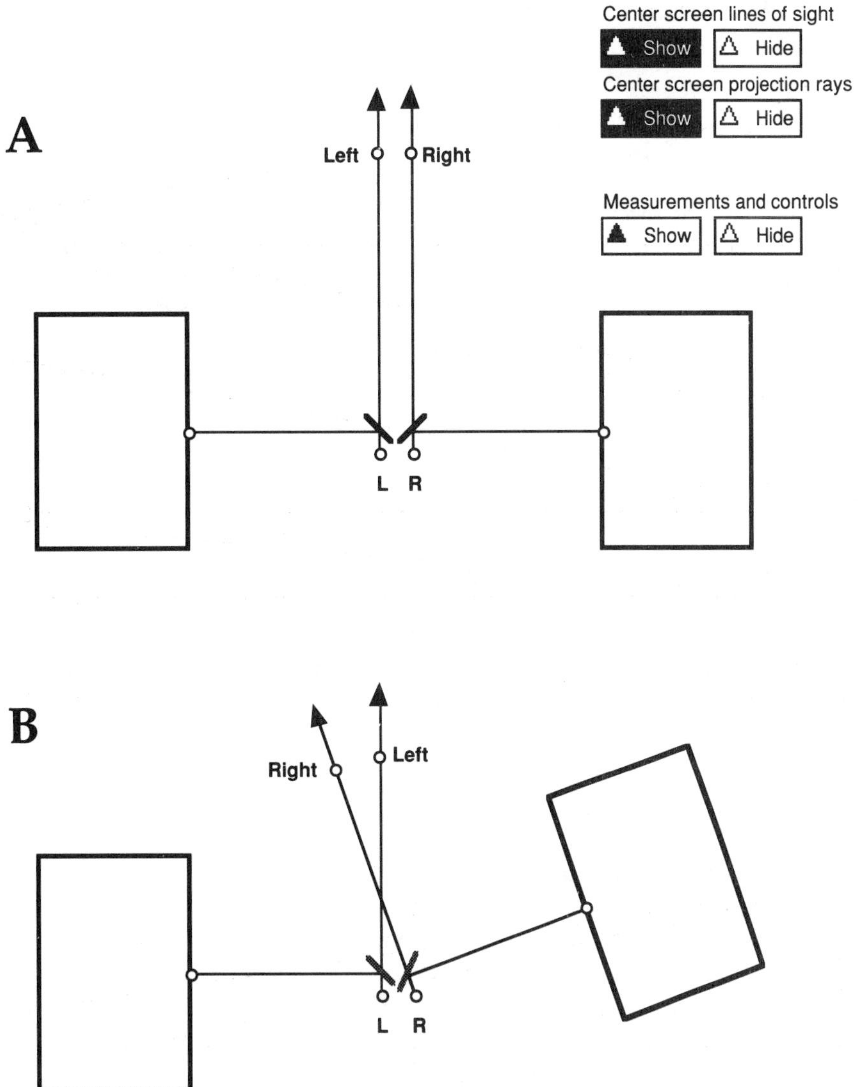

FIGURE 5. Haploscope design. The short thick, angled segments are mirrors. Points **L** and **R** represent eye positions. The large blocks are CRT monitors. Both the monitors and mirrors rotate about the eyes, so that the lengths of the visual paths, and hence the sizes of the displays, remain constant. The various control sliders are not shown. A. A pair of points at the centers of the CRT screens appear to be a point at infinity. B. With the monitors converged, the point appears to be located at a finite distance in front of the observer.

GSP has also proved useful in the design of our scientific equipment. GSP is well-suited to the design of devices that have moving parts, so long as the movements consist of translations parallel to a single plane and rotations about axes perpendicular to that plane. We used GSP to make a model of a haploscope, a device for showing separate images to the two eyes. The basic design is shown in Figure 5. The physical pieces of the device and the optical path are included in the same sketch, and move together when the parts are moved. The arms, with CRT monitors attached, rotate about axes that go through the centers of rotation of the eyes. The advantages that the model in GSP has over traditional drawings are that (a) one can easily fiddle with various parameters, such as monitor arm length, viewing distance, and the distance between the observer's eyes; and (b) after such fiddling, the monitor arms can still be rotated (to check, for example, whether the mirror size is still big enough).

Despite the joys and benefits of using GSP, version 2.0 was not optimal for use either as a geometric calculator or as a design tool. The restrictions that made the program simple and elegant for learning geometry often became a hindrance during practical applications. We made some recommendations for an industrial version; some of these have been incorporated in version 3.0. We encourage further enhancements to GSP that will improve its performance for these special uses.

III. 3D Techniques

The use of one- and two-point linear perspective as a way of rendering 3D shape is fairly straightforward in GSP, at least in some cases (see for example, [2], p. 110). In this section, two new techniques for incorporating a third dimension into sketches are discussed.

Anaglyphs and stereograms. The left and right eyes see the world from slightly different locations. This difference creates disparities between the images formed on the two retinas, and these disparities are used by the brain to help reconstruct the 3D shapes and locations of objects. The most straightforward method for creating a pair of images that could actually have arisen from a real scene is to project a set of 3D points onto a pair of properly oriented image planes or spheres in three-space. However, acceptable stereograms can be made by trial and error, or by using the rule of thumb that a difference in viewpoint of about six degrees will give a suitable range of disparities. Using DGS, objects can be animated so as to move in depth (Figure 6).

An anaglyph can be made from a stereogram by coloring the objects to be seen by one eye green, and those to be seen by the other eye, red. A pair of red-green glasses filter the image so that only the properly colored set of objects can be seen by each eye. This technique is practical with color monitors, and does not require the observer to "free-fuse" separate images, a skill with which many students have difficulty.

Simultaneous representation of three orthogonal dimensions within a plane. Another technique is to represent more than one dimension simultaneously within the plane of a sketch. These multiple representations can then be hidden from the viewer, leaving only the two-dimensional image of a 3D object. This technique requires a thorough understanding of how the two-dimensional projections of the object will behave as the various degrees of freedom in the sketch are exercised. Figure 7 shows the two-dimensional rendering of a globe that was used by the author to become familiar with the four different coordinate systems commonly used to describe eye position. It is an orthographic projection of the original 3D object, based on what happens to the two-dimensional projections of circles when they are tilted in three-space.

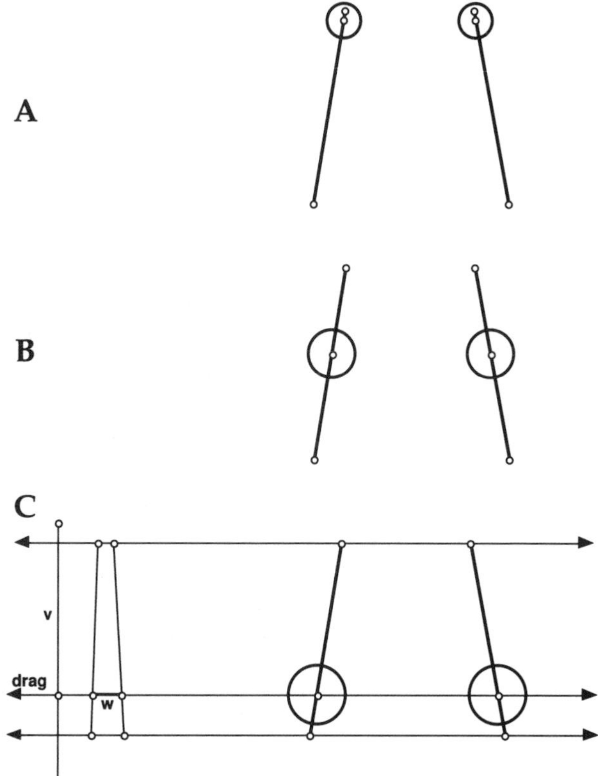

FIGURE 6. Dynamic stereogram. Shown here as a pair of cross-fusible pictures, the stereogram can be implemented as an anaglyph if colors are available. Part C shows how the stereogram is constructed. The radius of the circles is **w**, and the point **drag** can be animated on segment **v**, which causes the appearance of a circle sliding in and out of the computer screen, changing image size as it goes.

Figure 8 shows a sketch based on the planar perspective projection of a vertically-oriented rectangle that is free to rotate in three-space about a vertical axis through its center. On the left side of the figure, the viewing point and the top of the rectangle are visible. On the right side of the figure is the image of the rectangle as seen from the viewpoint. The distance in the image from the center to the vertical sides comes from a straightforward one-point perspective projection of the top (horizontal) edge of the rectangle. The vertical sides of the rectangle had to be rotated into the plane of the sketch before the projection could proceed. In the original rectangle, the viewing geometry was chosen so that each vertical side would be perpendicular to the line between its midpoint and the viewpoint. This same relationship was maintained when the vertical sides were redrawn in the plane of the sketch. Each of the vertical sides has to have its own image axis, from the viewpoint not to the center of the rectangle, but to the midpoint of the side.

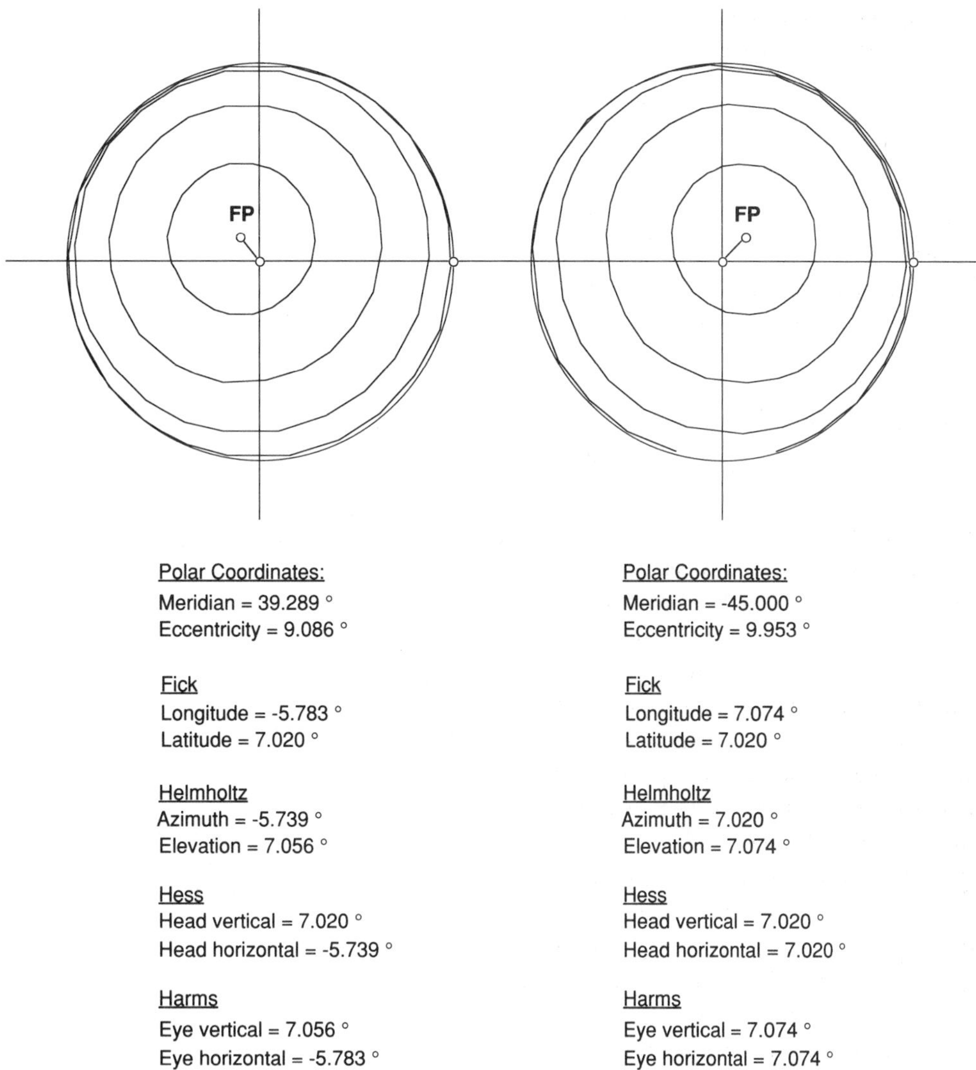

FIGURE 7. Model for understanding spherical coordinate systems. The two different positions of the globe shown can be cross-fused to give an idea of the depth one sees from motion cues when dragging the "fixation point," **FP**. For the polar coordinates shown, the meridian coordinate is 0° on the vertical line, and the eccentricity coordinate is the angular deviation from a vector coming straight out of the page.

These examples are quite complicated, despite the fact that the objects being portrayed are themselves very simple. Orthographic images are in general easier to construct than are planar perspective images. The author does not know whether the limitations of this type of modeling are theoretical limitations, or simply that the sketches become very complex.

The desirability of true 3D dynamic geometry. As an engineering tool, a 3D analog of GSP should be extraordinarily valuable. Present day computer-aided design programs are good for modeling 3D objects, but this author is unaware of any implementation in which strongly constraining relationships between objects are held constant, while, simultaneously, the values of a number of free parameters are manipulated. GSP sketches do this almost instantaneously.

With such software, an engineer could model a physical device with moving parts, and see, for example, whether the parts still clear each other after some of them are modified. The animation of motorized objects would be greatly simplified. Perhaps most useful, the ability to easily manipulate, visualize, and compute with objects in three dimensions would make many tasks much easier. As an example, consider the problem of developing an intuition for how retinal

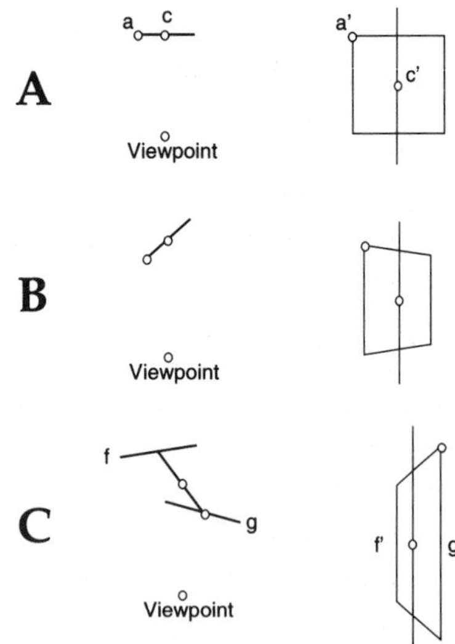

FIGURE 8. Planar perspective projection of a rotating rectangular object in three-space. A. The single horizontal bar represents a view of the rectangle from above. The rectangle rotates about its center on a vertical axis. The figures on the right show how the rectangle would look in planar perspective projection, as seen straight on from **viewpoint**. B. The rectangle has been rotated, bringing point **a** closer to **viewpoint**. C. The vertical sides of the rectangle are drawn in the same plane as the horizontal sides of the rectangle. Segments **f** and **g** are each oriented perpendicular to the lines of sight to their respective midpoints, as is the case in the three-dimensional model. (The rectangle in C is slightly larger than in A and B).

images change as one moves through the environment. With 3D DGS, one could simply draw lines from points in the environment to the eyes' projection centers, and see where they intersect the retinas as the eyes are dragged through the environment.

Conclusion

DGS is used at the University of California School of Optometry to help teach optics, to answer research-related questions, and in the design of experimental equipment. There is a need for a DGS application that is tailored to use in engineering and industrial design. Although there are 3D modeling techniques that can be used in two dimensions, they are of limited utility. A three-dimensional version of this software would be highly desirable, and would likely find the wide acceptance in education, research, and industry that would justify its development.

References

1. Banks, M.S. (1994). Personal communication. December. Berkeley, California.
2. Key Curriculum Press. (1994). *The Geometer's Sketchpad:* User guide and reference manual, Version 2.1. Key Curriculum Press, Berkeley, California.

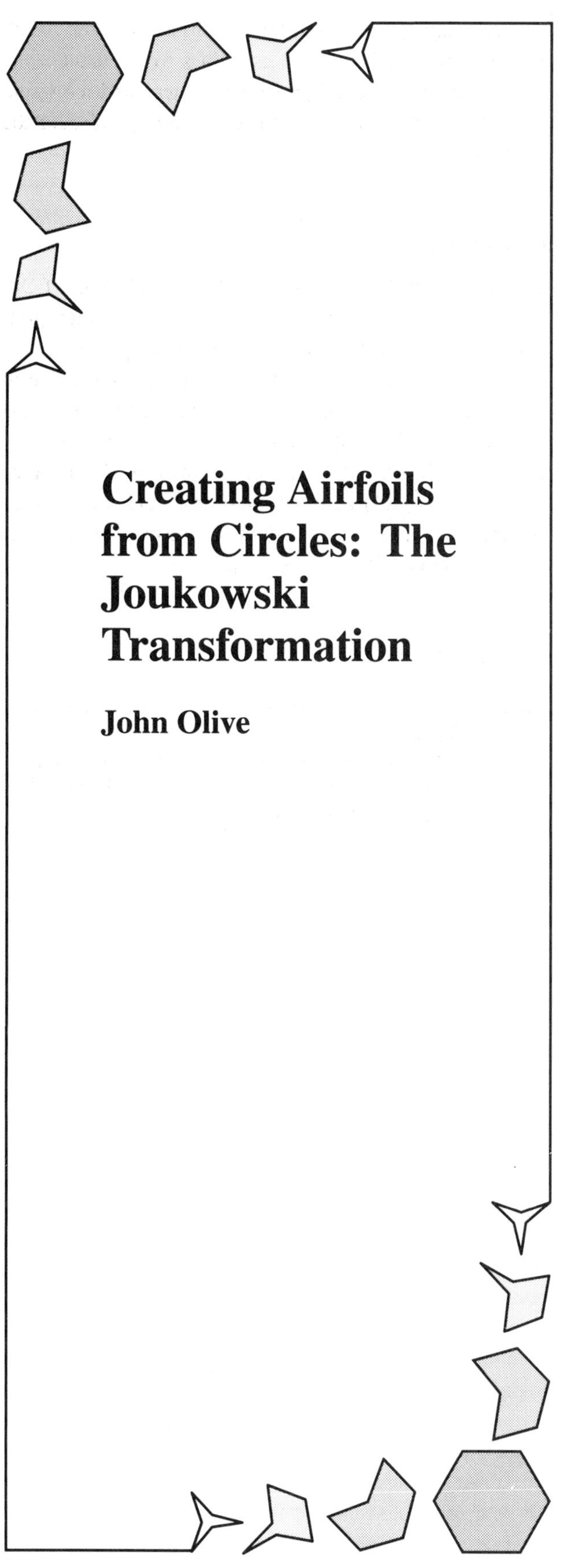

Creating Airfoils from Circles: The Joukowski Transformation

John Olive

Nikolai Joukowski (1847–1921) was a Russian mathematician who did research in aerodynamics (James and James, 1992). He used circle inversion in the complex number plane to study airfoil shapes. By applying Joukowski's transformation to the mathematical models of airflow over a cylinder (or a circle in two-space), aeronautical engineers could predict the amount of lift and drag that certain airfoil shapes would produce in an aircraft wing built using the airfoil shape as its cross-section. (For an explanation of the aerodynamics of airplane flight, see Foote, 1952 and Ludington, 1943.) With the aid of dynamic geometry software, such as *The Geometer's Sketchpad*, it is now possible for students to use Joukowski's transformation to create and explore their own airfoil shapes. In so doing, the students gain an appreciation for the important application of complex numbers, their representation as points in two-space, and the geometry of circle inversion. The recursive scripting facility in *Sketchpad* also provides students with an opportunity to investigate results of recursive applications of the Joukowski transformation that lead to new, aesthetically interesting curves and seemingly chaotic behavior. These recursive extensions can help students make connections between art and mathematics, as well as involve them in investigations of the new mathematics of chaos through dynamic visualization of iterations of functions in the complex plane.

I begin with a brief review of complex numbers and their representation as points in the plane. The Joukowski transformation involves the vector addition of a point and its reciprocal in the complex plane. I explain the reciprocal of a point in the complex plane as a composite of two geometric transformations in the plane: inversion in the unit circle and reflection about the x-axis. I present two different ways of constructing the circle inversion using *Sketchpad*. The Joukowski transformation is then achieved using *Sketchpad*'s dynamic vector composition.

Having constructed the Joukowski transformation of a free point in the plane, students can investigate the locus of the Joukowski point as the free point moves on a circle. The resulting locus can be regarded as the image of the animation circle under the Joukowski transformation in the complex plane. Special relations between the animation circle and the unit circle produce airfoil shapes. Version three of *Sketchpad* provides a facility to construct the locus of the Joukowski point as a dynamic curve that will change as the relations between the animation circle and unit circle are changed.

The recursive extensions of this investigation are achieved by simply applying the Joukowski transformation to the constructed Joukowski point, and constructing the locus of the resulting point as the original free point moves on the animation circle.

An Introduction to Complex Numbers and the Complex Plane

A complex number can be represented as $z = x + iy$, where i is $\sqrt{-1}$ and x and y are real numbers. The complex number plane is the set of points (x, y) in the plane such that $z = x + iy$. The unit circle in the complex plane is the set of points (x, y) such that $|z| = \sqrt{x^2 + y^2} = 1$. The reciprocal of any complex number z in the plane is that number whose complex product with z gives 1; it is represented as $\frac{1}{z}$. Now $\frac{(x+iy)(x-iy)}{(x^2+y^2)} = 1$; thus $\frac{1}{z} = \frac{(x-iy)}{(x^2+y^2)}$, which is the same as $\frac{(x-iy)}{|z|^2}$. Geometrically, $\frac{1}{z}$ is the image of z under the composite of an inversion in the unit circle and a reflection in the x-axis.

Inversion in a Circle. The *inverse* of a point in a given circle is the point on the ray from the center of the circle through the original point such that the ratio of the distance of the original point to the center and the radius of the circle is equal to the ratio of the radius and the distance of the inverse point to the center. In Figure 1 this identity can be expressed in the following way: $\frac{AC}{\text{radius}} = \frac{\text{radius}}{AG}$ or, alternatively, as: $AC \cdot AG = \text{radius}^2$. A is the center of the circle, C the original point, and G the inverse point.

The inverse of a free point in a circle can be constructed with *The Geometer's Sketchpad* in several ways (see King, 1995, pp. 162–164). The above proportion can be constructed geometrically using similar triangles (see Figure 1) or by using the dynamic dilation facility built into the *Sketchpad.*

In Figure 1, point G is the inverse of point C in the circle with center A and radius AB. To construct G for an arbitrary point C, point D is placed on the circle of inversion and the segment CD constructed. The line AC and the ray AD are also constructed. In the triangle ACD, AD is the radius of the circle. Point E is the intersection of AC with the circle; thus, AE is also the radius of the circle. By constructing a line through E parallel to CD, a triangle AEF similar to ACD is produced by the intersection (point F) of this new line with the ray AD. This pair of similar triangles provides us with the necessary ratios to produce the inverse point of C in the circle AB: $\frac{AC}{AE} = \frac{AD}{AF}$, as AE and AD are both radii of the circle. The inverse point, however, needs to be on a line connecting C to the center of the circle. The intersection of the circle (center A) through F with the ray AC gives us the inverse point G, as $AG = AF$.

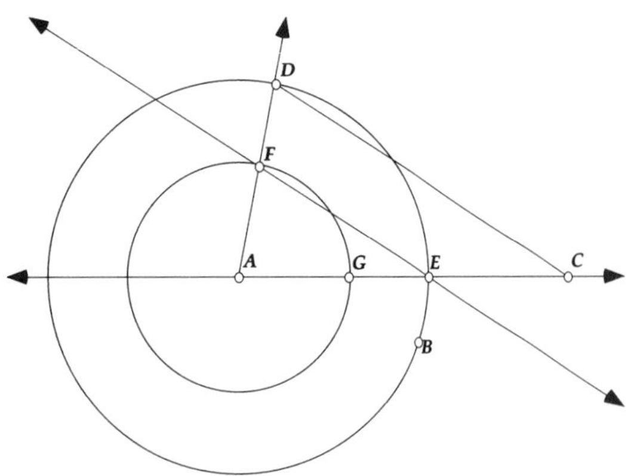

FIGURE 1

As point C was a free point in the sketch, it can be moved around (dragged) in order to investigate the relation of C to its inverse point G.

A more efficient way to obtain the necessary relation between point C and its inverse G is to use *dynamic dilation.* Point G can be obtained through the dilation of point E about the center A using the *directed* ratio $\frac{AE}{AC}$ as the dilation ratio. To do this in *Sketchpad*, you need only to construct the ray through A and C and the intersection point E of this line with the circle of inversion. Then select A as the *center of dilation* and mark the ratio $\frac{AE}{AC}$ by selecting the points A, C, and E **in that order**, and, while holding the *option* key down (on a Mac) or Shift key (Windows), select **Mark Ratio** $\frac{AE}{AC}$ from the **Transform** menu. You then select the point E and dilate it by the marked ratio. I leave it to the reader to verify that the dilated image of point E about the center A of the circle by the ratio $\frac{AE}{AC}$ is G, the inverse of point C.

The Geometric Construction of the Joukowski Transformation

I now return to the construction in the complex plane of the point representing the reciprocal $\frac{1}{z}$, from the point (x, y) representing $z = x + iy$. Figure 2 illustrates the two-step construction of $\frac{1}{z}$ from the point z. The point z' is the geometric inverse of point z in the **unit** circle: thus $|z||z'| = \text{radius}^2 = 1$; the point $\frac{1}{z}$ is the reflection of z' in the horizontal line through the center of the unit circle (the x-axis).

The *Joukowski transformation* of the complex plane is defined as $\mathbf{w} = z + \frac{1}{z}$. This transformation has the property of mapping both z and $\frac{1}{z}$ to the same point, since addition of complex numbers is commutative: $z \to z + \frac{1}{z}$, and $\frac{1}{z} \to \frac{1}{z} + z = z + \frac{1}{z}$.

Addition of complex numbers can be represented as vector addition in the complex plane. The complex number $z = x + iy$ is represented by the point (x, y), which also can be thought of as the vector from the origin $(0, 0)$ to the point (x, y). The sum of two complex numbers represented

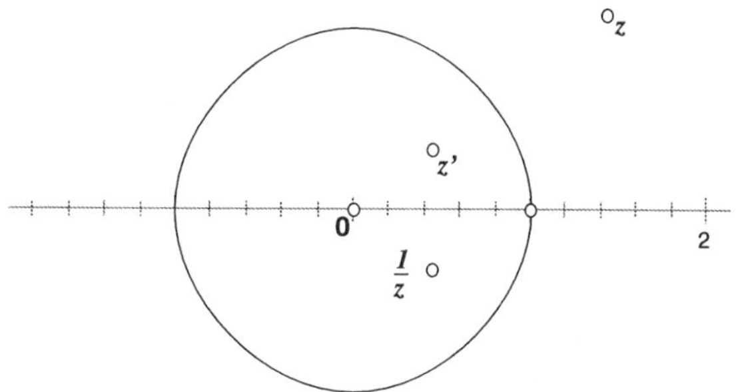

FIGURE 2. Construction of $\frac{1}{z}$.

by the points (x, y) and (x', y') is $(x + iy) + (x' + iy') = (x + x') + i(y + y')$. This sum is represented by the point $((x + x'), (y + y'))$, which is the vector sum of (x, y) and (x', y'). Geometrically, the point $((x + x'), (y + y'))$ is constructed by translating the point (x, y) by the vector (x', y'). Figure 3 shows the construction of w, the Joukowski transformation of the point z. The construction was accomplished in *Sketchpad* by selecting the points 0 and $\frac{1}{z}$ as a **Marked Vector** and then translating the point z by this marked vector.

The Joukowski airfoil profile is produced by the locus of w as z moves on a circle that passes through the point $z = -1$ (the left-hand intersection of the unit circle with the x-axis), and has the point $z = +1$ in its interior. The circle also needs to be offset slightly above the x-axis (see Figure 4).

The locus can be drawn by tracing point w as z is animated around the circle, or, with version 3 of *The Geometer's Sketchpad*, users can construct the locus of the airfoil profile as a continuous curve (rather than the trace of w) by selecting the point w, the free point z, and the circle with center G (in that order), and, while pressing the Option key (Mac) or Shift key (Windows), selecting **Locus** from the **Construct** menu. Students can use this sketch to investigate the unique properties of the Joukowski airfoil by changing the size and position of the circle (center G) that is transformed into the airfoil profile. (I shall refer to this circle as circle 2.)

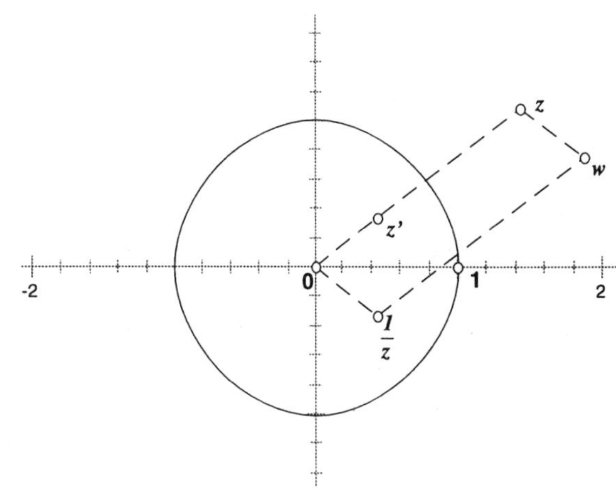

FIGURE 3. $w = z + \frac{1}{z}$.

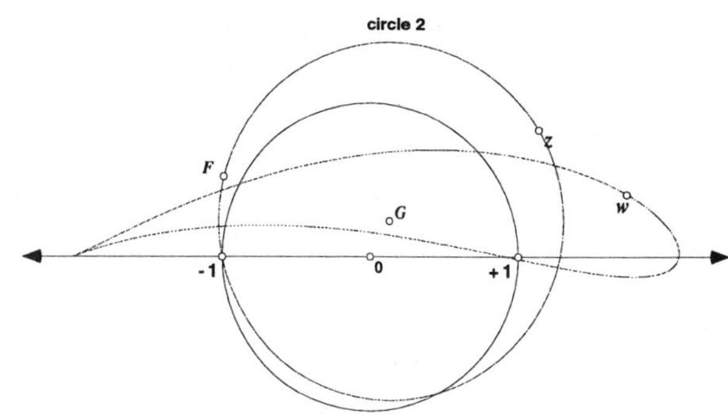

FIGURE 4. Joukowski Airfoil Profile.

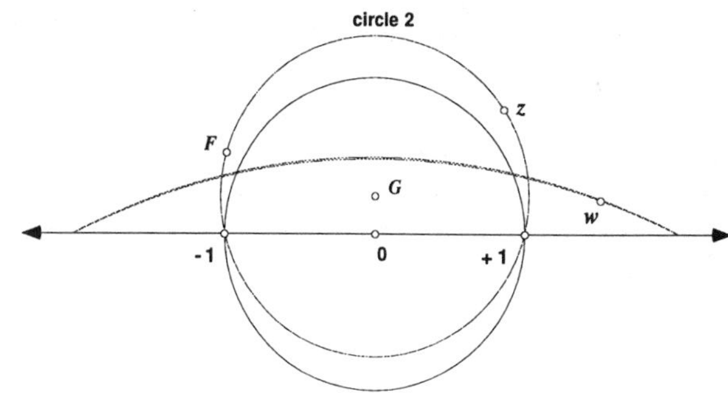

FIGURE 5. Joukowski profile with $z = +1$ intersecting circle 2.

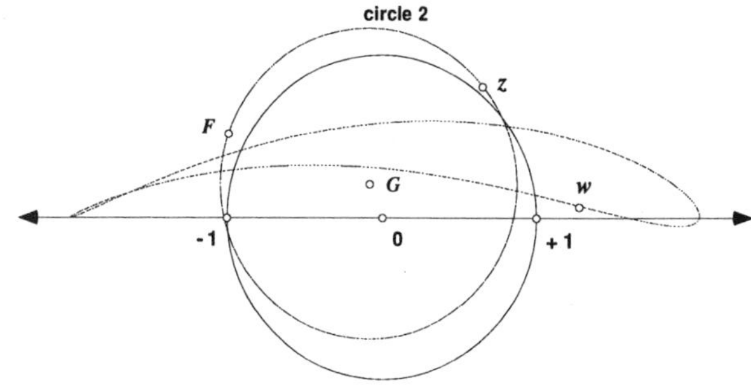

FIGURE 6. An airfoil profile with $z = +1$ exterior to circle 2.

Investigating Airfoils. James and James (1992) state that the point $z = +1$ must be in the *interior* of the transformed circle in order to produce an airfoil profile. Students can test this property by reducing the size of circle 2 slowly until the point $z = +1$ is outside the circle. Figure 5 illustrates what happens to the airfoil profile as $z = +1$ intersects circle 2, but still passes through $z = -1$.

Further reduction of the size of circle 2 (Figure 6) produces a profile very similar to the airfoil shape in Figure 4 with $z = +1$ outside circle 2!

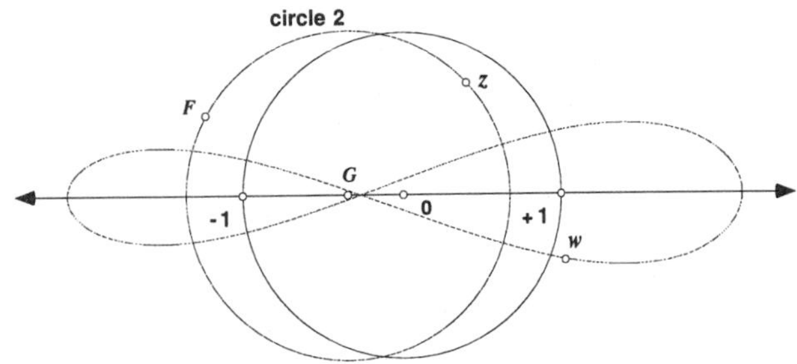

FIGURE 7. A non-airfoil profile.

Students can find out how different parts of circle 2 are mapped to the airfoil profile by animating z on circle 2 slowly and following the traced paths of z and $\boldsymbol{w}$. Of particular interest are the parts of the Joukowski profile that correspond to the arcs of circle 2 that are interior to the unit circle, those that are exterior to the unit circle, those that are above the x-axis, and those arcs that are below the x-axis.

Students can investigate the critical roles of the points $+1$ and -1 with respect to circle 2 in determining the shape of the Joukowski profile. Figure 7 illustrates a resulting profile when circle 2 does not pass through either point.

In Figure 7, the center of circle 2 (point G) is on the x-axis. Students can investigate what happens when the center G is *below* the x-axis.

Students should be encouraged to find as many different kinds of profiles as possible and to form ways of categorizing them. In our investigations of the airfoil profile, we discussed the relative merits of different airfoil shapes for the design of aircraft wings. With G very close to the x-axis, slim, aerodynamic airfoils can be produced. Such airfoils would give minimal lift to an aircraft but would also generate minimum drag. In contrast, as the center of circle 2 moves farther above the x-axis, highly curved airfoils are created. Such airfoils would create greater lift for an airplane wing but would also increase the drag on the wing. It became more clear to us why it is that large jet aircraft change the shape of their wings to a profile that is more curved for takeoff and landing (when maximum lift is required), while at cruising speeds (when minimum drag and lift are required) the shape of the aircraft wing is more like the slim, aerodynamic airfoil.

An Extension to the Iterated Joukowski Transformation

What will be the result if the point $\boldsymbol{w}$ is transformed by the Joukowski transformation to create the point $\boldsymbol{w}' = \boldsymbol{w} + \frac{1}{\boldsymbol{w}}$? By investigating the iterated application of the Joukowski transformation, students can create curves suggestive of aquatic creatures, helmets, and Greek letters.

Students can create a script in *Sketchpad* to produce $\boldsymbol{w}$ from z by following the constructions in Figures 2 and 3. The script should require three given points: the center and radius points of the unit circle and the free point z. Students should apply their script to the same center and radius points that define the unit circle, but they should select $\boldsymbol{w}$ instead of z as the third given.

The locus of $\boldsymbol{w}'$ as z moves on a free circle passing through the point $z = -1$ can be constructed in the same way that the locus of $\boldsymbol{w}$ was constructed (see Figure 4 above). Figure 8 shows one possible result. I find this shape aesthetically pleasing and suggestive of some kind of water creature, dinosaur, or bird. I would like to call it a *RAT* profile for Recursive Airfoil Transformation.

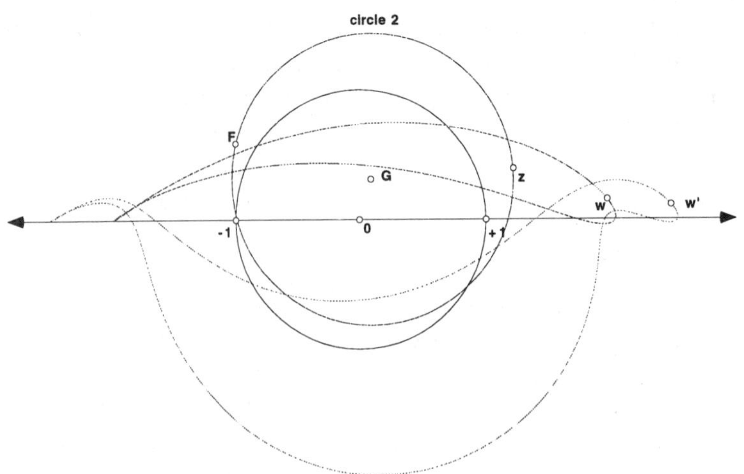

FIGURE 8. Joukowski Airfoil and its *RAT*.

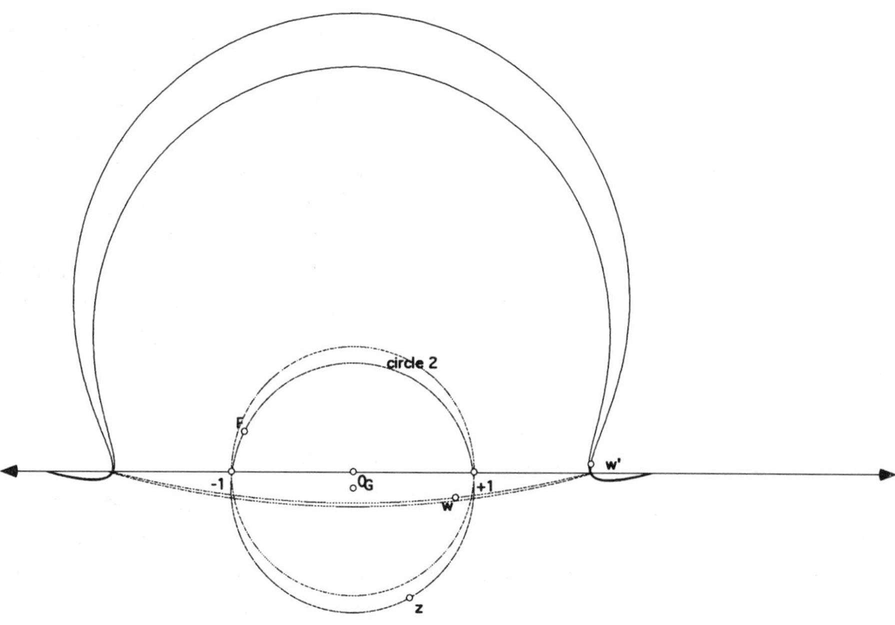

FIGURE 9. An "Omega" *RAT*.

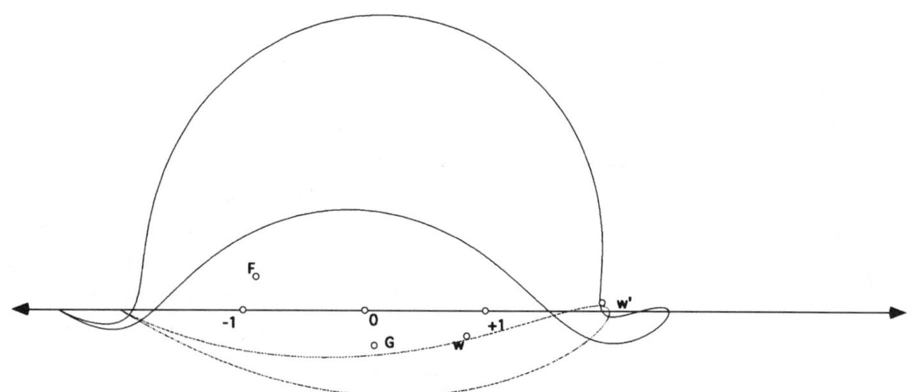

FIGURE 10. An upside-down airfoil with its "helmet" *RAT*.

Students can investigate the *RAT* by repositioning and changing the size of the transformed circle. By gradually moving the circle so that it is offset *below* the x-axis, the point w creates an upside-down airfoil. Some examples of what happens to the associated *RAT* are demonstrated in Figures 9 and 10.

Exploring Iterated Images of a Point

The Geometer's Sketchpad provides the capability to create recursive scripts that will automatically replay the script on the new objects created by the script. The user determines the depth of recursion when playing the script. A *Recursive Joukowski script* can be used to make further iterations of the Joukowski transformation. Figure 11 shows the result of playing a recursive Joukowski script to a depth of 24 recursions. It shows an apparent hyperbolic sequence of transformed points for a particular z just below the center of the unit circle. The first transformed point is D'''. The pattern appears to show subsequent transformations approaching the x-axis from above. Further iterations of the transformation verify this trend.

Following are some of the questions that were raised by students during this investigation. Are there points in the z-plane for which the iterations appear to diverge rather than converge? Are there regions of the complex plane that appear to converge to the real line (the x-axis)? Or to the imaginary line (the y-axis)? Why might this be so? Are there points around which the transformed, points appear to act chaotically? All of the above questions were answered in the affirmative. Of special interest was the transformation of the unit circle. Under the first Joukowski transformation,

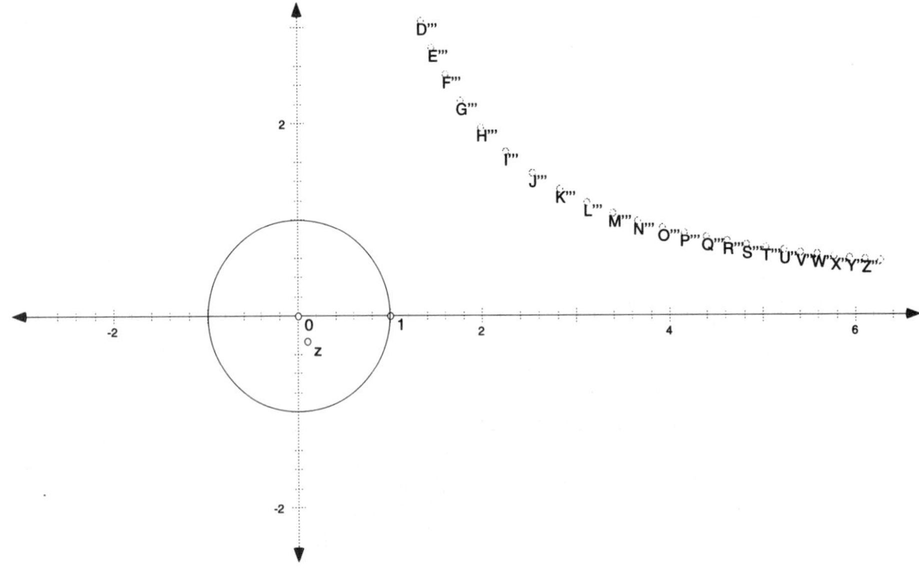

FIGURE 11. 24 iterations of the Joukowski transformation.

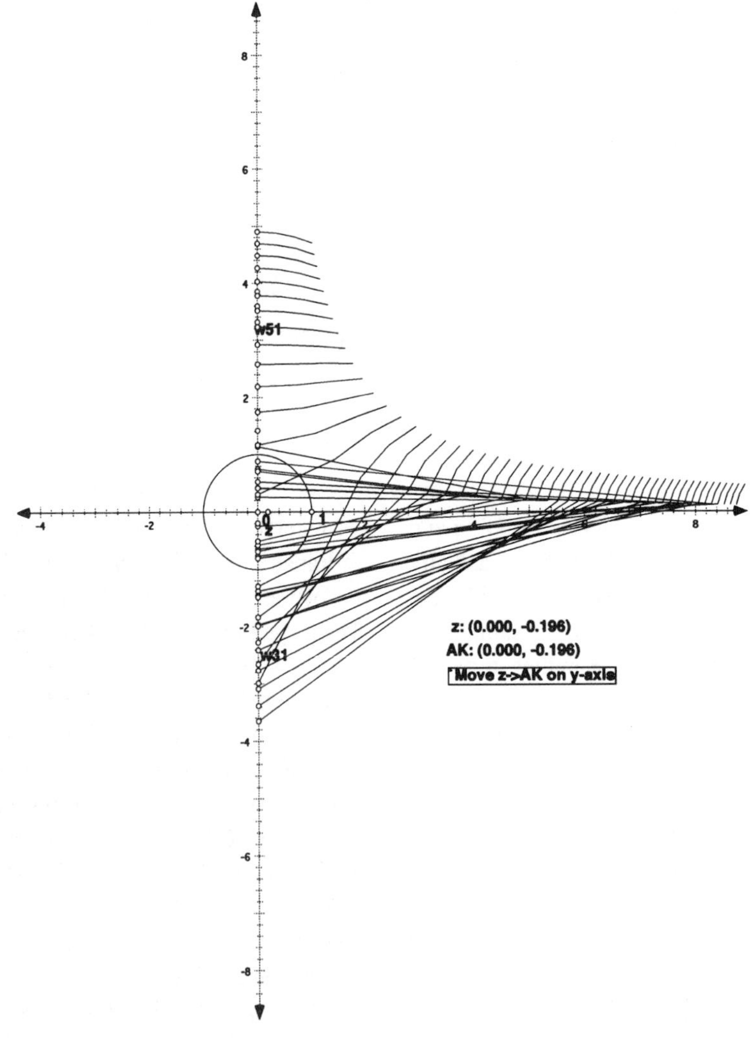

FIGURE 12. Trace of movement of the transformed points.

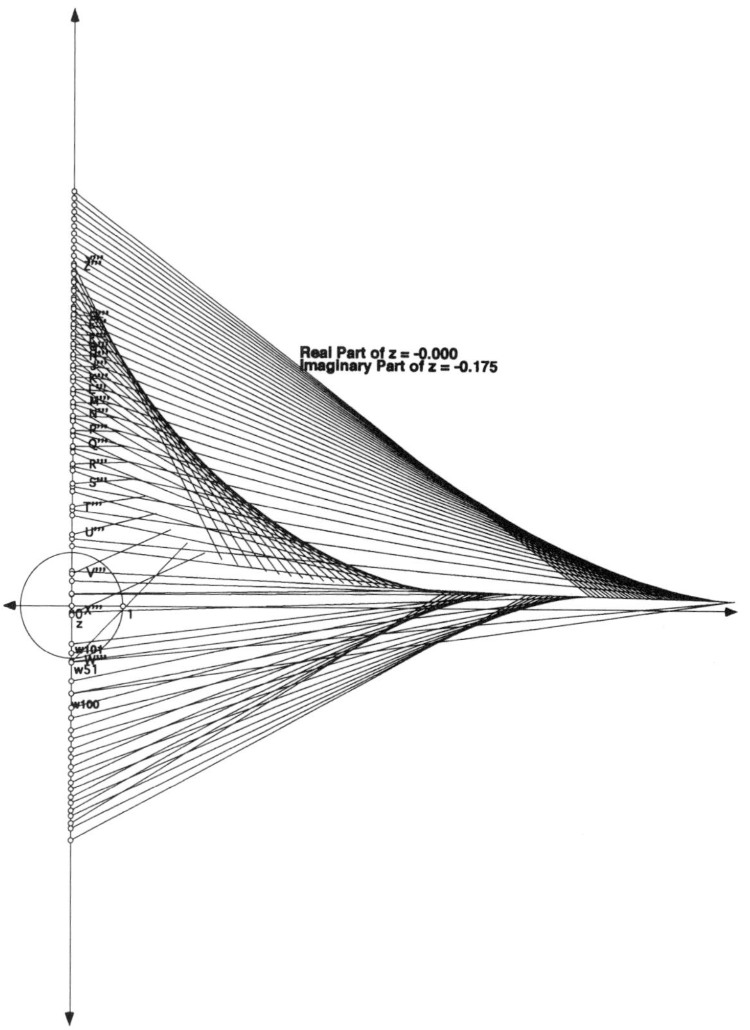

FIGURE 13. Real part of z is zero.

the unit circle is mapped to the segment $[-2, 2]$ on the real line. Subsequent iterations appear to oscillate between either side of zero on the real line, but move farther and farther away from zero. With 31 iterations, the traced paths reached approximately as far as $+135$ and -45.

Apparently chaotic behavior was observed as z approached the imaginary axis from points not on the unit circle or the real line. Figure 12 shows the dramatic result of a tiny change in the position of z (from the point $(0.123, -0.196)$ to the point $(0.000, -0.196)$) so that z ends up directly below the center of the unit circle. All points are now mapped to the y-axis instead of approaching the x-axis! (The iteration was extended to 51 points.)

Figures 13 illustrates even more dramatic results when the transformation is iterated to a depth of 100 recursive calls (giving 101 iterations of the w transformation).

This seemingly chaotic behavior in the iterated function occurs only when z is very close to the imaginary axis. The dramatic change in the location of the iterated points can be partially explained by the apparent convergence of the iteration to the real axis for any z with a non-zero real part. When the real part of z becomes zero, the iterations must stay on the imaginary axis. What is not explained by this necessary shift from the real axis to the imaginary axis the seemingly unpredictable locations on the imaginary for each iterated point. The relative order in the i points as they converge towards the real axis is c lost when they shift to the imaginary axis. W this lack of relative order becomes apparent o point representing z moves towards the imagin slowly. The above traces were obtained in *Th Sketchpad* by creating a point on the imagin using a **Movement** button to move the point z point on the imaginary axis.

In order to check whether or not the beha seeing was a result of mathematical inexac

Sketchpad program, we investigated the chaotic behavior using the *Excel* spreadsheet, calculating values out to ten decimal places. The spreadsheet values supported the wild oscillations visible in Figure 13. The chaotic behavior of the iterated Joukowski function near $x = 0$ does open up possibilities for rich mathematical investigations. The trace function in *The Geometer's Sketchpad* provides an interactive way of visualizing this chaotic phenomenon.

Bibliography

Foote, D.K. (1952). *Aerodynamics for Model Airplanes; How and Why a Model Airplane Flies.* New York: A.S. Barnes.

James, R.C. and James, G. (1992). *Mathematics Dictionary*, 5th ed. New York: Van Nostrand Reinhold.

King, J. (1996). *Geometry Through the Circle with The Geometer's Sketchpad.* Berkeley, CA: Key Curriculum Press.

Ludington, C.T. (1943). *Smoke Streams; Visualized Air Flow.* New York: Coward-McCann, Inc.

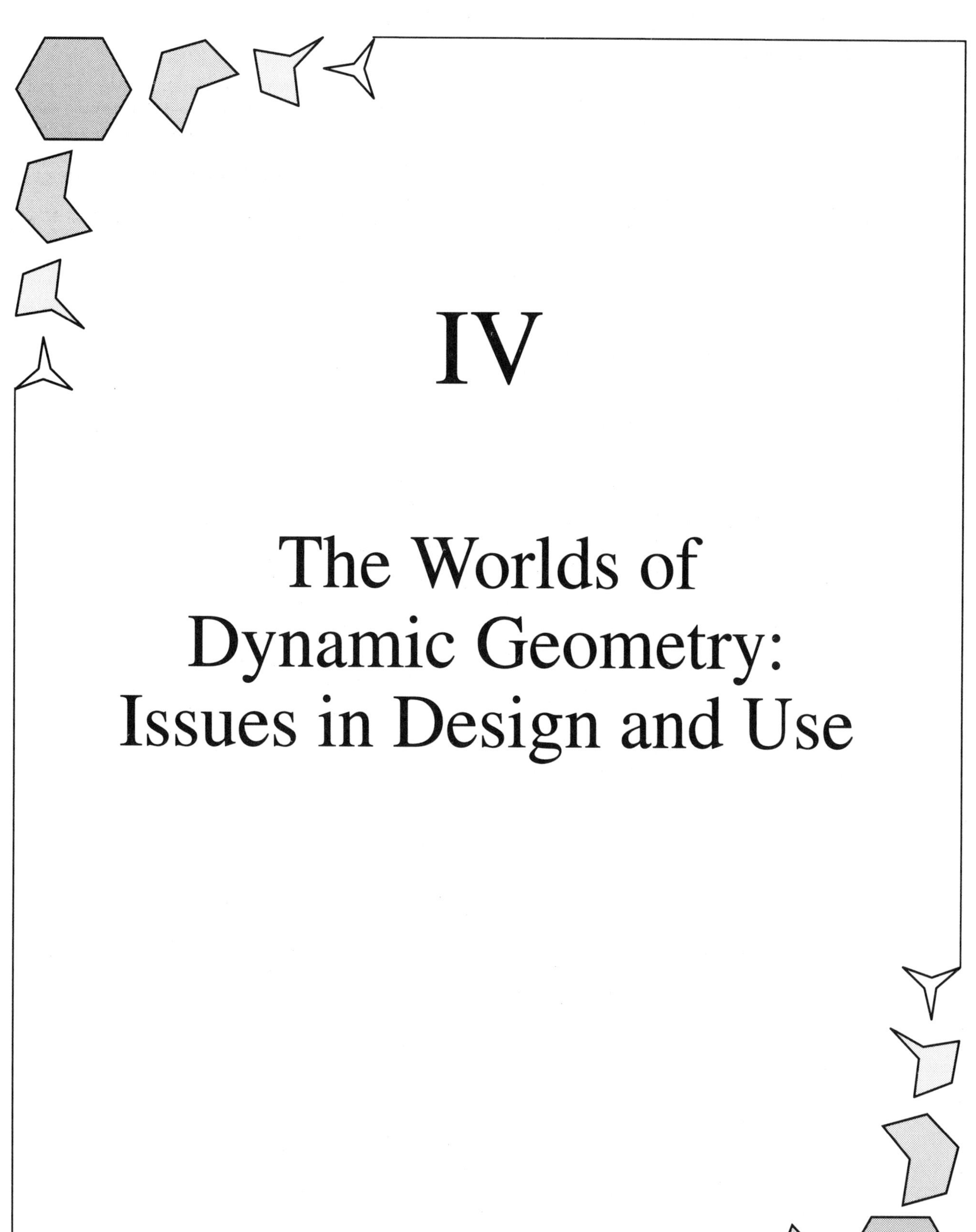

IV

The Worlds of Dynamic Geometry: Issues in Design and Use

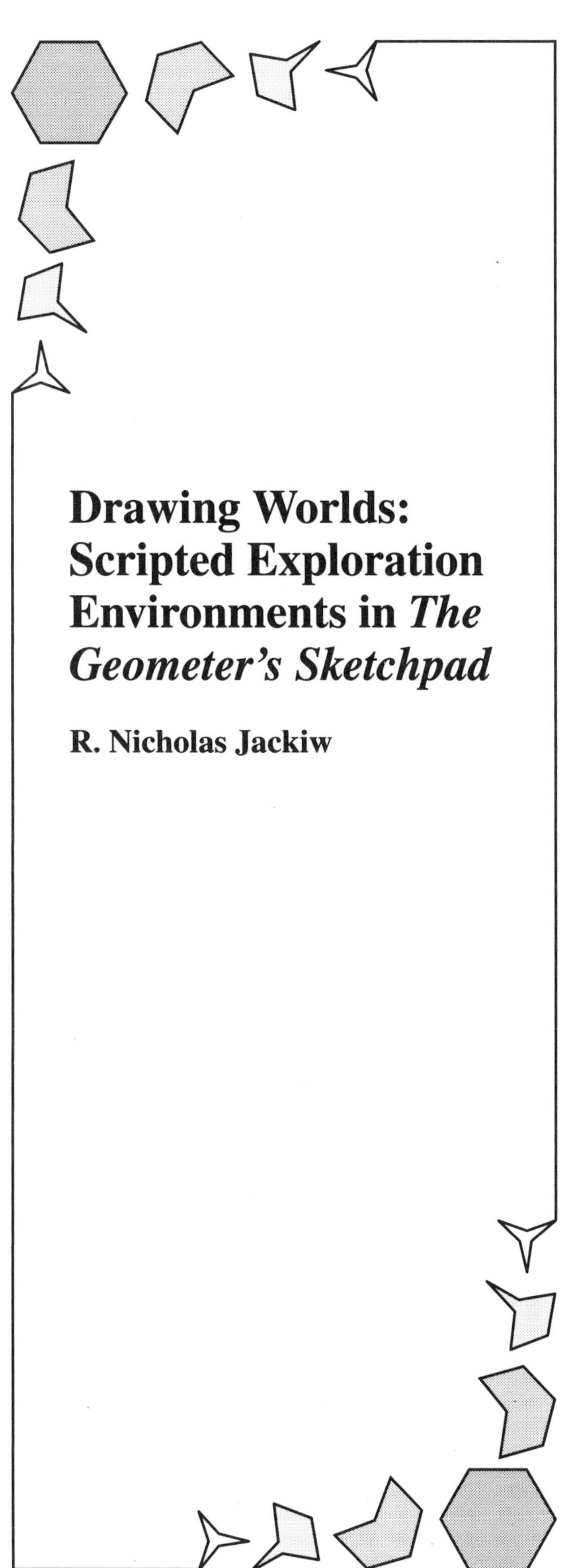

Drawing Worlds: Scripted Exploration Environments in *The Geometer's Sketchpad*

R. Nicholas Jackiw

A sign of maturity in the larger dynamic geometry programs (such as *Cabri II* and *The Geometer's Sketchpad*) is the emergence of meta-geometric authoring tools, which allow sophisticated users—teachers and curriculum developers—to customize the general-purpose environment to a particular student population's specific needs. By restricting problem contexts, customized environments enable students to explore given geometric topics using a variety of relevant tools, while at the same time isolating students from the diverse, partially irrelevant, and potentially intimidating remainder of the software system's feature set. Viewed this way, customization or authoring tools can respond to a variety of pedagogic issues, ranging from those of mathematical inquiry (e.g., what becomes of compass-and-straightedge geometry if we eliminate the compass?) to those of classroom mechanics (e.g., how can my students meaningfully use the software before being introduced to the menu bar?).

This paper describes the idea, development, and deployment of "drawing worlds," an application of the authoring tools found in *Sketchpad* to create special-purpose investigation environments; i.e., micro-microworlds. In drawing worlds, curriculum developers replace *Sketchpad*'s primary compass and straightedge tools with alternate tools of the designers' own invention. From a user's perspective, a drawing world offers an in-depth experience of the geometric context established by the curriculum developer, without requiring any significant amount of intimacy with the general-purpose software environment (or its extended feature set). The drawing world paradigm is relatively neutral to the mathematical nature of the drawing context itself, which lends it broad applicability to the study of alternate (non-Euclidean) geometries, to the investigation of specific application-modeling scenarios, and to the development of geometric intuition and reasoning.

Example Drawing Worlds

This section illustrates two drawing worlds from the perspective of a student. The first depicts a high-school student's investigation of a hyperbolic drawing world published on the World Wide Web.[1] The second documents one of several drawing worlds presented by the author to teachers participating in the 1995 Summer *Sketchpad* Institute (Philadelphia, Pennsylvania).

Poincaré World. In exploring the Poincaré drawing world, the student begins with a *basic model*, a sketch created by the curriculum developer. The basic model serves both as a framework for investigations in this particular world and as

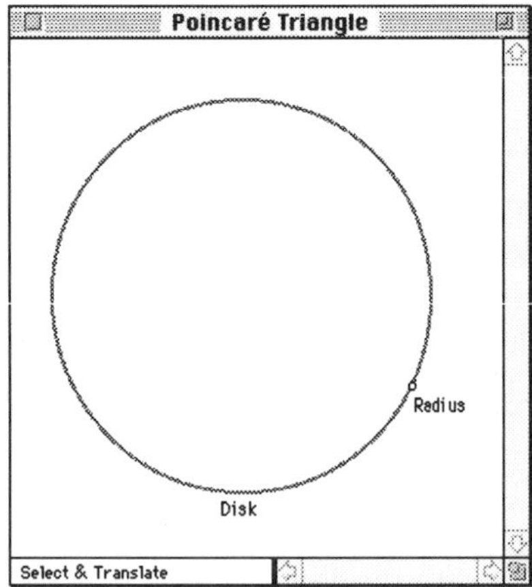

FIGURE 1. Poincaré basic sketch.

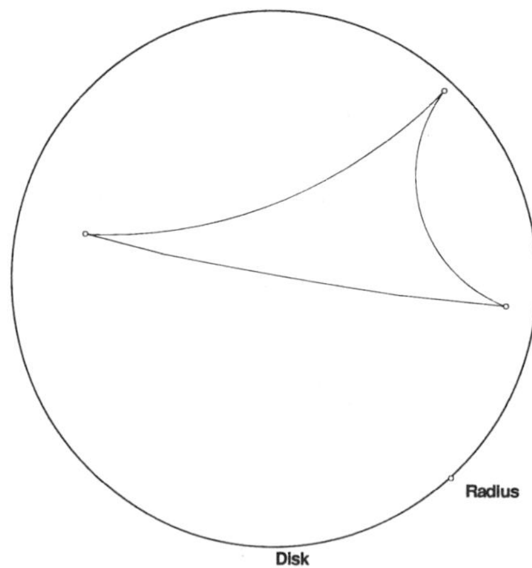

FIGURE 2. A Poincaré Triangle.

a location for the toolset which defines available operations in the world. In this case, the basic model consists simply of a disk (see Figure 1) and the tools: a compass, straight edge, and angle-measuring caliper.[2] A moveable point in the sketch determines the radius of the disk.

In that this world describes an embedding of the Poincaré model of hyperbolic geometry in the Euclidean plane, the student quickly discovers that line segments drawn with the straightedge outside of the central disk are undefined: there, the segment tool simply fails to function. Segments drawn inside the disk, on the other hand, seem subject to a strange curvature, a gravity which bends them toward the disk's center. By dynamically manipulating a triangle constructed with the straightedge, the student realizes that the curvature of the triangle's edges is not constant; instead, it appears to correspond perhaps to the degree to which the edge approaches collinearity with a diameter of the disk (Figure 2). Further investigation reveals that some properties of triangles embedded in the Euclidean plane seem unaltered in this foreign world—certain points of concurrence in the triangle are maintained—but others have been altered. After measuring the triangle's interior angles and dragging its vertices extensively, for example, the student arrives at a conjecture that the sum of the measures of the angles in a Poincaré triangle varies between 0 and π. Returning to the task of descriptively characterizing the degree of curvature of segments, the student varies the radius of the disk itself, and notes a correspondence between a segment's curvature and its length in relation to the radius of the disk. By extending the disk's radius indefinitely outward, the student's curved triangles all "straighten out," confirming the interpretation of the Euclidean plane as a highly local region of the Poincaré disk.

A Mystery World. In this example, teachers working in pairs confront several "mystery worlds." With no prior experience manipulating or exploring drawing worlds, they are given a blank sketch (as a basic model), alleged compass and straightedge tools, and the task of identifying the mechanics which govern drawing in their assigned world. Many begin exploring by drawing triangles. In one particular exploration, the first triangle drawn has an entirely pedestrian appearance (Figure 3, top). However, the second triangle breaks apart as it is drawn and, instead, forms an irregular pentagon (Figure 3, bottom).

Exploring this rupture, the teachers draw several more segments, and realize that each bounces off an invisible barrier running horizontally through the sketch. The compass produces similar results, and reveals the presence of a vertical barrier as well. After verifying these bounds with several other drawings (Figure 4), one of the teachers begins to describe the world as dominated by partial reflection—anything extending past one of two invisible perpendiculars gets mirrored back across it. The other teacher—noting that the mirrors are perpendicular and that drawing can only

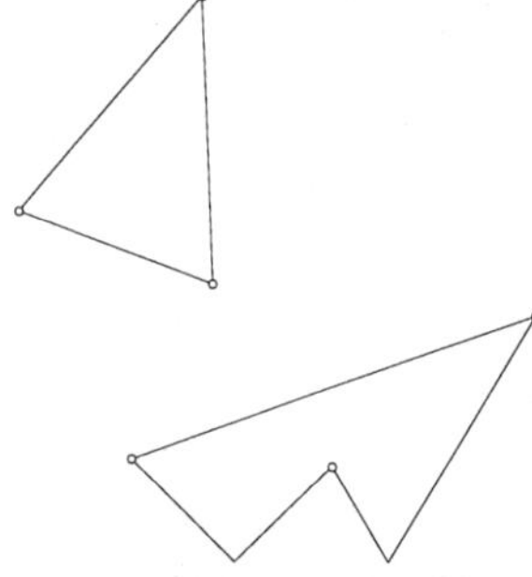

FIGURE 3. Two Triangles in a Mystery World.

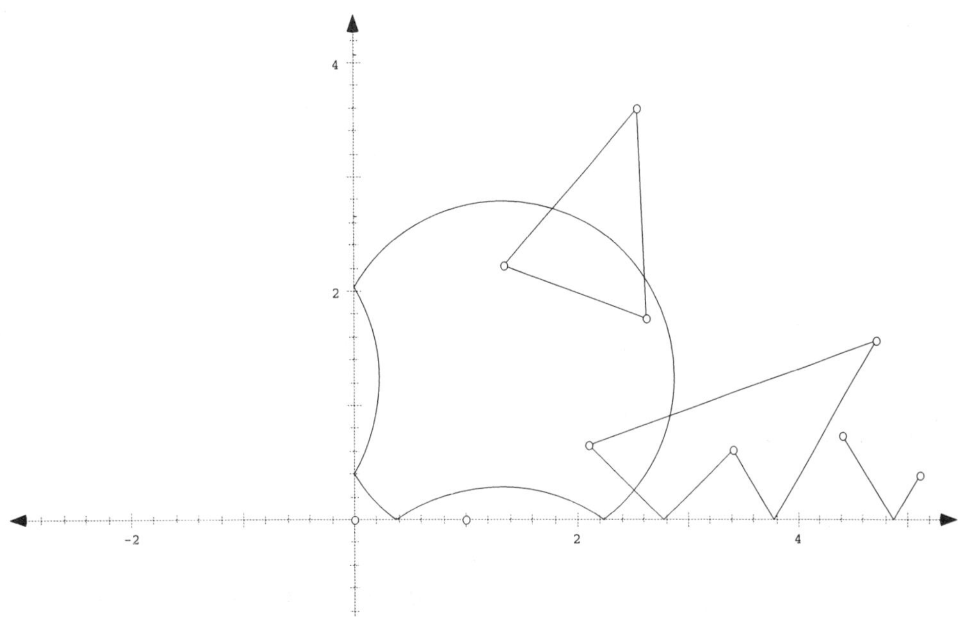

FIGURE 4. Circles, Segments, and the Implicit Coordinate System of The Mystery World.

transpire above the horizontal and to the right of the vertical mirror—identifies the world as a plane mapped through the absolute value function: any "negative extent" of an object gets mapped back to a positive extent in the first quadrant.

Drawing Worlds' Usefulness and Pedagogy

Drawing worlds afford a number of improvements over the standard learning model presented by dynamic geometry environments. While the previous examples do not introduce new mathematical functionality (or "native" understanding) to the *Sketchpad* program, the drawing world paradigm enables both a concentrated focus upon, and an immediacy of interaction with, existing mathematical functionality. The Poincaré disk example, for instance, uses only Euclidean constructions—standard *Sketchpad* mathematics—to generate the arcs which represent segments on the disk. However, these techniques have been consolidated in, and abstracted by, the tools offered in the drawing world. The onus of understanding the Euclidean implementation of drawing lines and circles in the Poincaré disk *a priori* (and in general, of being fluent in construction techniques) is thus isolated from the student's exploration of the model. Similarly, and equally important from a classroom perspective, drawing worlds—by locating the essence of an investigation in the behavior of directly manipulable tools—reduce the amount of preliminary exposure to program menus, dialog boxes, and interface mechanisms that students require before they can interact with the world's mathematics.

Secondly, drawing worlds introduce students to the notion of construction, and to the deterministic behavior that constructed constraints impose on a drawing. A widely-noted characteristic of *Cabri, Sketchpad*, and similar environments, is that students begin using the software by drawing rather than constructing.[3] (For example, a student asked to create a right triangle may draw an arbitrary triangle, and drag it to approximate a right triangle, rather than construct a triangle that has a right angle.) Without external incentive to move beyond the drawing stage to interact with constructions, students may miss a large portion of the potential impact of dynamic geometry. The tools in a drawing world, however, represent constructions themselves, and thereby introduce students to constructed behavior almost immediately. Because drawings produced in the world no longer mirror the gestures students use to draw them, students must develop internal models of the (constructed) behavior of the tool, and then test these models through directed deployment of the drawing tool. In general, drawing tools are the first and most obvious components of geometry software that attract students' attention. By making these tools configurable, drawing worlds allow curriculum developers to shape students' experiences at an early and fundamental level.

Thirdly, drawing worlds provide an excellent mechanism for exploring the concept of a (mathematical) function. In the Mystery World example above, the students explore a mapping between domain (the segments and circles they attempt to draw) and range (the "bent" segments and circles the program creates in response) without encountering either the standard number line or the perpendicular projection of domain onto range associated with more traditional investigations of graphs of functions. On one hand, because neither algebra nor coordinate geometry is required to work in such a drawing world, the approach is well suited to elementary and middle-school classes. On the other, due to the relative novelty of introducing functions through planar mappings as opposed to through $y = f(x)$ graphs, the approach is useful in teacher training contexts as well: it is no coincidence that the "students" in the Mystery World example were teachers themselves!

Anatomy and Origins of Drawing Worlds

While the obvious central aspect of a drawing world is its problem domain (whether explicitly articulated, as in the first example above; or hidden, as in the second), the task of the problem-poser—the teacher/author or curriculum developer—is to create an arena for the problem's exploration using available technology. In *Sketchpad*, one can create a drawing world through a basic model sketch (which may be simply empty) and a set of scripts which express the desired functionality of the world's tools. These scripts are then installed in the toolbar by marking them as "script tools." Once associated together, the scripts and model sketch form the basis of a drawing world fit for dissemination.[4]

From the software developer's perspective, it is interesting to note the causal derivation of the drawing world concept. Scripts in *Sketchpad*—or macros, in *Cabri*—have traditionally been the most versatile feature of the software's vocabulary. By allowing users to consolidate a multi-step construction into a new single-step program operation, scripts theoretically provide the ability to widen the scope of one's explorations indefinitely, while keeping complexity a constant. For instance, once an equilateral triangle has been laboriously constructed using strict compass and straightedge techniques, that triangle can be consolidated into a script, and abstracted into a new primitive. Subsequent needs for equilateral triangles can then invoke the script, rather than require the student to wield once again only a compass and straightedge. Despite this theoretical appeal, however, scripts and macros have remained largely beyond the scope of most students' interaction with the software, perhaps by fault of having no "real world" correspondents easily accessible to the user's imagination. When designing the third version of *Sketchpad*, the idea of installing and deploying scripts as first-class drawing tools came to us largely as an enticement—a possible means to inspire students to begin authoring scripts. While it remains to be seen whether "script tools" lower the experience threshold at which students begin to write their own scripts, once script tools became available, it was ironically apparent that they provided a device for widening *Sketchpad*'s pedagogic utility among student audiences with absolutely no prior familiarity with the software.

Creating Scripts for a Sample World. Techniques for creating a world are largely dependent on the nature of the world's modeled domain. While the drawing world concept and its potential problem space are quite broad, the second example illustrated previously may serve as representative of a typical drawing world. This world visualizes a mapping of the Euclidean plane known to *Sketchpad* through some arbitrary function imposed by the world's author. If a point in the *Sketchpad* world is denoted $P(Sketchpad)$ and $|P(Sketchpad)|$ denotes the absolute value of the coordinates of that point, then

$$P(World) = |P(Sketchpad)|.$$

The choice of available tools in this world is of course arbitrary, but the world's author chose to duplicate *Sketchpad*'s existing tools—the point, compass, and straightedge tools—in order to provide a frame of reference which could stimulate investigation of the properties of the mysterious mapping function. The point tool's script can be conveniently expressed using the software's analytic capabilities:

Given:
1. Point $P(Sketchpad)$

Steps:
1. Let [*m1*] = Coordinates of Point $P(Sketchpad)$.
2. Let Measurement [*m2*] = *abs*[*x*[*m1*]].
3. Let Measurement [*m3*] = *abs*[*y*[*m1*]].
4. Let $P(World)$ = Point defined by (Measurement [*m2*], Measurement [*m3*]) on the rectangular grid.

When used as a drawing tool, this script locates $P(Sketchpad)$ at the current coordinates of the computer mouse, and derives $P(World)$ as its image through the mystery function. One could proceed to author unique scripts for the straightedge and compass tools, but in that the mysterious function describes a transformation of the entire plane, it's tempting to re-use this world's point tool in a generalized locus construction. For example, the compass:

Given:
1. Point A
2. Point B

Steps:

1. Let $[c1]$ = Circle with center at Point A passing through Point B.
2. Let $P(Sketchpad)$ = Random point on Circle $[c1]$.
3. Let $[m1]$ = Coordinates of Point $P(Sketchpad)$.
4. Let Measurement $[m2] = abs[x[m1]]$.
5. Let Measurement $[m3] = abs[y[m1]]$.
6. Let $P(World)$ = Point defined by (Measurement $[m2]$, Measurement $[m3]$) on the rectangular grid.
7. Let Locus $[P(World)]$ = Locus of Point $P(World)$ while Point $P(Sketchpad)$ moves along Circle $[c1]$.

Here, *Sketchpad* matches A and B to the mouse positions at the center and on the circumference of a circle in the world, and derives the world's "circle" as a locus image of a point traveling the Euclidean circle, but transformed by the same construction defined previously in the Point tool script. A similar approach can define tools for straightedges, polygons, or other objects.

Automatic Matching with Script Tools. In some cases, the world's mapping function involves certain constants, the value of which depend upon parameters in the basic model sketch. For instance, in the Poincaré world, the world's tools require knowledge of the circle used as a Euclidean representation of the Poincaré disk. In the process of writing a Poincaré script tool, these parameters are automatically registered as prerequisites (Givens) of the script. However, in this state, the script cannot yet be used as a drawing world tool. Generally, scripts wielded as tools create new objects to match each of their prerequisites. A single invocation of this script would therefore create a new disk, and using the tool naively would result in a basic model rapidly littered with different circles representing different disks, one for each invocation of the tool.

In circumstances like this, one requires the script tool to match certain prerequisites with pre-existing objects in the basic model sketch, rather than to create a new object corresponding to each prerequisite. *Sketchpad* accommodates this requirement, though the script author must declare it at the time of writing the script. In general, if a scripted prerequisite should automatically match an existing object labeled X, then the label of the prerequisite must be stated in the script as *Auto-X*. Thus, in the script corresponding to a Poincaré tool, the first prerequisite will commonly be a given Circle *Auto-Disk*. When the tool is subsequently wielded, *Sketchpad* automatically matches this prerequisite with any circle labeled Disk in the sketch. (If no such circle exists, then the tool forces one to be created when the tool is used, as with conventional script tool prerequisites.)

Automatic matching can be employed in script tools for a variety of novel effects, which move beyond duplicating the functionality of a Euclidean compass or ruler. If all of a script's prerequisites are automatically matched, then the script's construction takes place the moment the tool is invoked. (That is, such scripts require no interactive drawing.) The following script, when associated with a model sketch containing a rectangle $ABCD$, instantly constructs G as a random point inside the rectangle:

Given:

1. Point Auto-A
2. Point Auto-B
3. Point Auto-C

Steps:

1. Let $[j]$ = Segment between Point Auto-B and Point Auto-A.
2. Let $[E]$ = Random point on Segment $[j]$ (hidden).
3. Let $[k]$ = Segment between Point Auto-C and Point Auto-B.
4. Let $[F]$ = Random point on Segment $[k]$ (hidden).
5. Let $[n]$ = Perpendicular to Segment $[j]$ through Point $[E]$ (hidden).
6. Let $[o]$ = Perpendicular to Segment $[k]$ through Point $[F]$ (hidden).
7. Let G = Intersection of Line $[n]$ and Line $[o]$.

Subsequent invocations of the script tool create new random points. In this case, the tool functions less as a drawing device than as a program *command*. Drawing worlds can exploit both sorts of extension to the program's native vocabulary.

Drawing tools can also maintain state—some form of durable memory—across multiple invocations through creative use of automatic matching. In the event that *Sketchpad* attempts to match (automatically) a prerequisite *Auto-P* in a sketch that contains more than one object *P*, *Sketchpad* will always match the most recently created *P*. This most recently created P may in turn have been created by a scripted tool—perhaps even the same tool on a previous invocation. Stated another way, a given invocation of a script can construct some object P, which will automatically match to a prerequisite of the same script on its next invocation—regardless of the presence of other, less recently created Ps. In a basic model containing a Point Turtle, the script

Given:

1. Object *Auto-Turtle*

Steps:

1. Let *Turtle* = Image of Object *Auto-Turtle* translated by 1.00 inches at 90.00 degrees.

will "move" the turtle up an inch on the screen (leaving behind an image of where the turtle began). Subsequent invocations will keep the turtle moving, trailing images of its past locations. In each invocation, the prerequisite turtle is matched to the location of the turtle defined by the script's previous invocation. Several scripts can share a common state, claiming as prerequisites, and creating as constructed objects, the same set of objects. (A suite of scripts automatically matching two points, *TurtlePosition* and *TurtleHeading*, are sufficient to implement a complete turtle geometry drawing world.)

Conclusion

Together, these techniques can combine to describe countless possible drawing worlds. Where the aim of a world is to challenge students to discover some unrevealed central transformation, patterning the world's tools on conventional tools (such as the compass and straightedge) provides a useful point of departure for their exploration. In a world in which the subject domain is explicitly stated, the tools can of course be designed to provide the appropriate operators used in analysis or exploration of that subject: in a taxi-cab geometry environment, these may remain "compass" and "straightedge;" in a perspective rendering world, they may be replaced by sphere and polyhedra tools. Both types of worlds can profit by automatic matching of certain tool parameters to existing quantities in the sketch. For example, a mystery world which maps the polar plane (r, θ) into $(r, k\theta)$ is made both more interesting and more general if k represents some dynamically-adjustable angle in the basic model (made available to tools through automatic matching) than if k is a numeric constant fixed in each of the scripts. Likewise, the perspective-rendering world's tools can usefully infer the horizon line, station point, and projection plane—central characteristics of a perspective model—from global parameters declared in the basic sketch. A collection of worlds demonstrating these techniques is available on the World Wide Web site associated with this volume. Because they eliminate much of the overhead of students' learning to drive a new software package, while at the same time enabling curriculum designers to focus students' activities on well-bounded problem domains in geometry and a variety of related mathematical contexts, drawing worlds can both concentrate and diversify the impact of dynamic geometry on student learning.

Endnotes

1. http://www.forum.swarthmore.edu/dynamic.html. This world, designed by William Finzer, evolved from a collection of *Sketchpad* scripts written in 1992 by Michael Alexander, then at the University of Washington.
2. As of this writing, a limitation in the current *Sketchpad* implementation prevents a drawing world's custom tools from appearing as icons directly in the tool palette (on the left edge of the screen). Instead, each of a world's custom tools is accessible through a pull-down menu associated with the generic "script tool" icon in the main toolbox. Prototype versions of *Sketchpad* allow scripted tools to be installed with icons directly in the main toolbar. Over time, this functionality will likely migrate to the commercial version of the software.
3. For example, see Bennett, Dan and Finzer, William F. "From Drawing to Construction with *The Geometer's Sketchpad*," *The Mathematics Teacher*, NCTM, 1995; Hoyles, Celia, Richard Noss, "Dynamic Geometry Environments: What's the Point?", *The Mathematics Teacher*, NCTM, 1995.
4. The mechanism for creating drawing worlds explained here, as well as the details about customizing worlds which follow, correspond to features available in *Sketchpad* Version 3 (April 1995). Less interactive implementations of the "world idea" can be approximated through conventional deployment of scripts (*Sketchpad* Version 1, and *Geometry Inventor*) and macros (*Cabri*), though such implementations mitigate the benefits of drawing worlds because they require the student to have much greater facility with the environment, and do not offer the immediate feedback of a drawing tool. *Sketchpad* Version 3's ability to create a fully autonomous drawing world is itself limited by the fact that, in the commercial version of the software, one cannot completely eliminate the "primary" toolbox, but rather only extend it with custom drawing tools. Thus one must establish a convention that the primary tools go unused throughout an encounter with a drawing world.

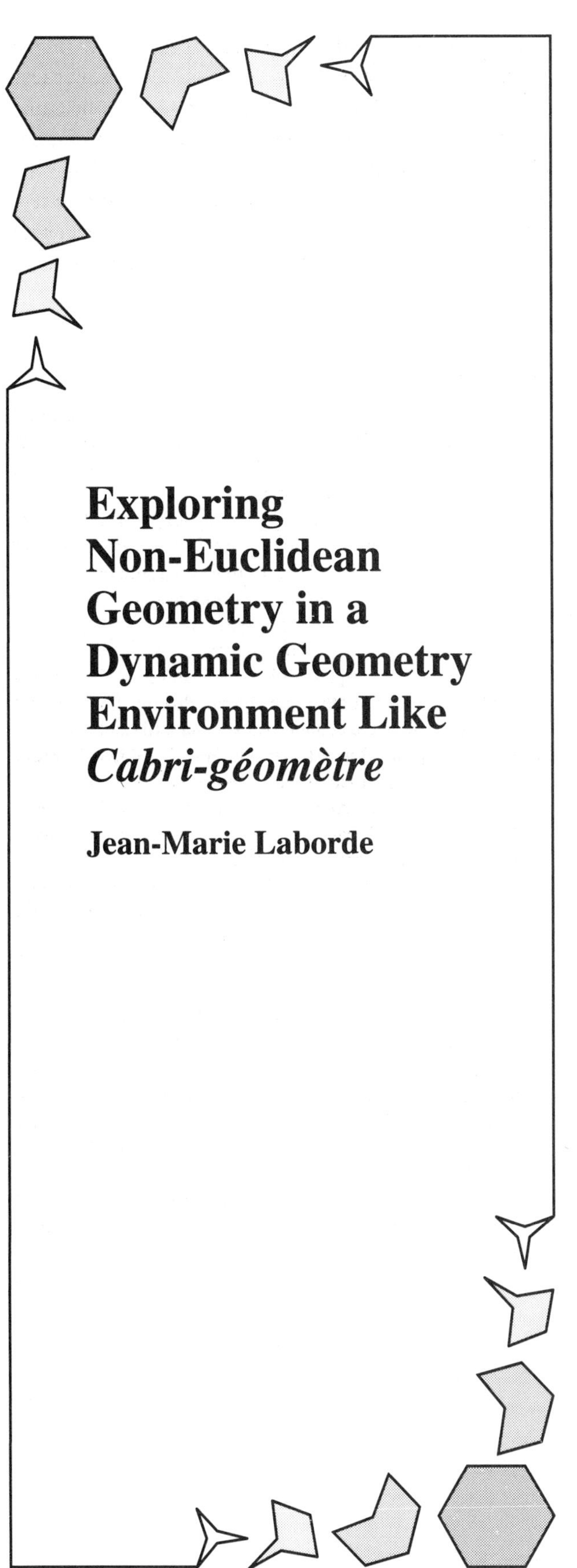

Exploring Non-Euclidean Geometry in a Dynamic Geometry Environment Like *Cabri-géomètre*

Jean-Marie Laborde

It is possible to replace the parallel postulate of Euclid with its negation without introducing more contradictions than are present in standard Euclidean geometry ($\mathbb{R}^2$). In these geometries, through any point not lying on a line d, either there is more than one line parallel to d (hyperbolic type) or there is no parallel at all (elliptic type). We say that a model of one of these geometries is based on standard Euclidean geometry if the model maps the basic elements of the geometry, points and lines, to subsets of $\mathbb{R}^2$ in such a way that the axioms of the geometry are still satisfied inside the set of images of elements.

For hyperbolic geometry, one possible way to do this is given by the *Poincaré model.* Points in the model are all points strictly inside a standard circle h of $\mathbb{R}^2$, called the *horizon*. Lines in the model are arcs of circles (or segments of lines) that cross the horizon orthogonally. The orthogonal circle or line is called the *support* of the hyperbolic line, and the hyperbolic line in the model is the arc (or segment) consisting of the points of the support that lie strictly inside h.

With dynamic geometry at our disposal, by means of by software such as *The Geometer's Sketchpad* [Jackiw 1990–1995] or *Cabri-géomètre* [Baulac, Bellemain, Laborde 1988–1995][1] it is tempting to build such models to show in a more convincing way the situation inside non-Euclidean geometry. Thibault and Labarre [89] proposed the first constructions with *Cabri* in this direction.

I. The *discovery* of a Way to Construct Hyperbolic Lines in the Poincaré Model

Once one is convinced that there exist circles that intersect orthogonally, the question is how to construct them. With a macro that constructs a circle going through three points, it is possible to check that, given any two fixed points A and B, we find one (and probably only one) circle through A and B that intersects a given circle (the horizon h) orthogonally. Here is how that exploration might play out.

A first locus for the center. Every circle that passes through A and B has its center on the perpendicular bisector of the segment AB.

A second locus for the center. We are looking for a circle that satisfies three properties: it passes through A, it passes through B, and it intersects h orthogonally. We have just considered the locus of all circles that pass through two points. We will consider now all circles that pass through one point (say A) and intersect h orthogonally at a point P. In order to do that we consider a (movable) point P on the horizon h. The center C of the desired circle will be at the intersection of the perpendicular bisector of AP and the

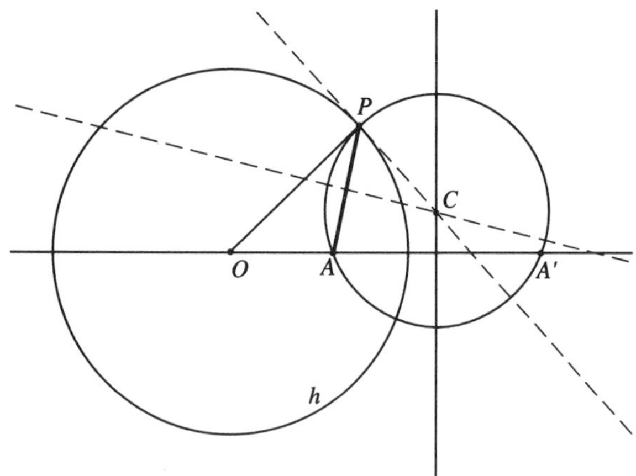

FIGURE 1. A *second* locus for the center of circles orthogonal to h.

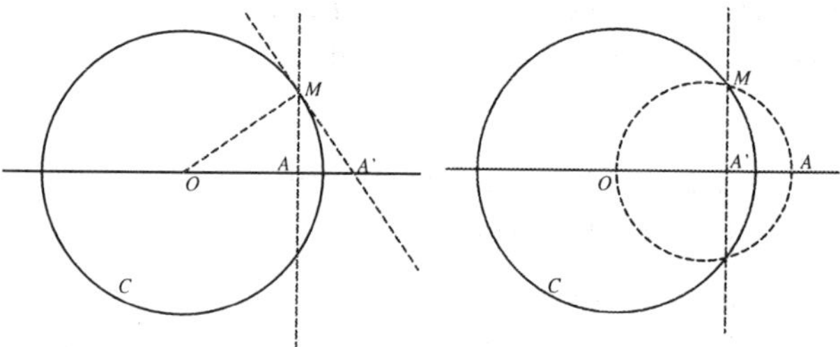

FIGURE 2. Constructions of the inverse A' of A when A lies inside circle c (left) and A lies outside circle c (right).

tangent line to h at P. We now ask what is the locus of C. This locus turns to be a line (Figure 1).

When we play with this situation and move P around, we observe that the line is not only perpendicular to OA but is a diameter of the constructed circle. The intersection A' of OA with the circle varies outside of h if A varies inside and conversely. This suggests that A' is the image of A under inversion in h. This idea can be checked directly in different cases. (With *Cabri II* it is possible to take advantage of the built-in inversion transformation).

For the purposes of this discussion, we will accept this "real world" evidence from geometrical exploration and will not give a standard proof that the second locus is really the perpendicular bisector of A and its inverse A' with respect to circle h. So we have the following easy construction of the support of a hyperbolic line through points A and B inside the horizon h:

The hyperbolic line through A and B is the subset interior to h of the circle through A whose center is the intersection of the perpendicular bisectors of AB and AA', where A' is the image of A with respect to inversion in h.

Construction of Inverted Images. The inversion transformation is defined by a circle c with center O and radius r. By definition, inversion in c maps point $A \neq O$ to point A' such that $\overline{OA} \cdot \overline{OA'} = r^2$. If we are using *Sketchpad* or the first version of *Cabri*, we need a way to construct A'.

The first idea is to make use of the Pythagorean theorem. We construct the line through A perpendicular to OA, and consider the tangent to the circle at one of the intersection points M. A' is the intersection of OA with that tangent (Figure 2, left). For our purposes here, this works well, because most of the time in the sequel we will consider only inversions of points A that lie in the interior of the inversion circle.

But the inversion transformation itself is more general and is defined for *any* point not equal to O. The construction above fails if we drag A outside of circle c. It is possible to reverse the preceding construction, interchanging the role of A and A'—we construct M as the intersection point of c with the circle with diameter OA. Then A' is the foot of the perpendicular line from M to OA (Figure 2, right). This is a valid construction, but it fails if A is not outside c.

When points can move, we prefer a construction that works in all cases, and will call such a construction *generic*. We will not describe here an exploration activity that can lead to the discovery of such a construction, but will just display in Figure 3 three different constructions that work.

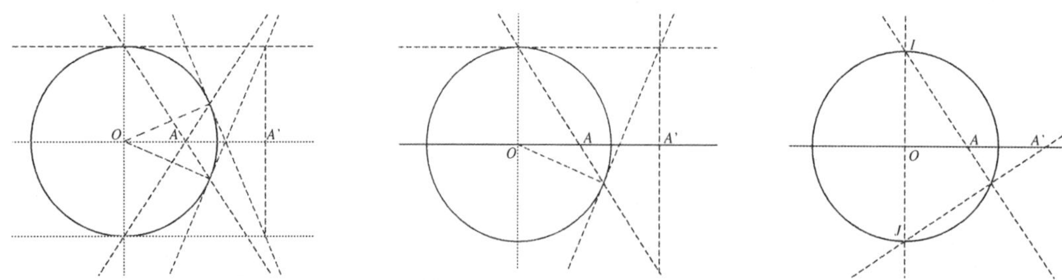

FIGURE 3. Three different generic constructions of the inverse A' of A; these do not depend on the position of A.

One might prefer the last because it uses fewer objects, even though it is probably not as easy to discover.

II. Dynamic Geometry in the Poincaré Model with Elements at Infinity

Orientation. In order to represent a hyperbolic line more accurately, we would like to restrict the drawing of each support circle to its part interior to the horizon, provided the software we are using can draw arcs. If arcs are implemented, we need a way to specify which arc of a support circle we are interested in having drawn.

Cabri II determines an arc using three points, where the first and last points represent the endpoints of the arc, and the third is a point in between. The endpoints U and V of our hyperbolic line are the points of intersection of the support circle with the horizon h (U and V are called limit points). We could try to take the in-between point I as the intersection of the support circle with the perpendicular bisector of the segment UV. This will probably work, but there is a chance that the software will not preserve the correct intersection and may replace it by its opposite for some configurations. This is because there is no canonical way of distinguishing the two intersection points of a line with a circle.

In order to provide a more precise way to construct a point that always lies on the hyperbolic line, we need to introduce an explicit orientation for the support circle. This can be done by means of a ray. The intersection of the ray eminating from the center of the support circle and passing through the midpoint of the segment UV is an acceptable in-between point (Figure 4).

We note that in *Cabri* the "implicit" orientation of lines is as follows:

Line constructed from A to B,	oriented from A to B,
Perpendicular from P to AB,	left turn of $\pi/2$ from the orientation of AB,
Perpendicular bisector of AB	the same.

These orientations are useful in some constructions.

Some More Advanced Situations: The Extended Plane. After constructing the hyperbolic line through A and B, we could try to play with the hyperbolic line and observe it for different positions of A and B. The first interesting case is when A, B, and O lie on a line. Mathematically, in the Poincaré model, the hyperbolic line is a diameter of the horizon. If you try the previous construction of the hyperbolic line using dynamic geometry software, with A and B on a line through O, the construction may fail because it involves the intersection of lines that are now parallel.

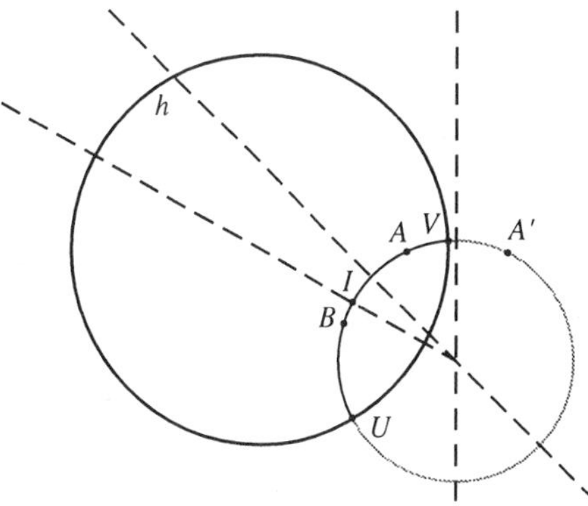

FIGURE 4. The hyperbolic line through A and B is represented by the arc UIV, where I is the intersection of the support circle of the hyperbolic line with the ray from the center of that circle through the midpoint of UV.

With *Cabri II* we began to implement an extension of Euclidean geometry to overcome this type of annoyance. In this extension, when a person makes a construction that involves intermediate objects that do not exist in standard Euclidean geometry, then under certain circumstances the construction will continue to have meaning.

For instance, suppose we construct the circle through three points A, B, C by finding its center as the intersection of the perpendicular bisectors of AB and AC. In the case A, B, and C are collinear, this same construction will construct the line containing A, B, and C. In Figure 5 we show the construction where $C = A'$. The construction works because the software will continue to give meaning to the intersection point, considering it to be at infinity in the direction d of the perpendicular bisectors. Then it constructs a circle through a point in the finite plane which is centered at infinity

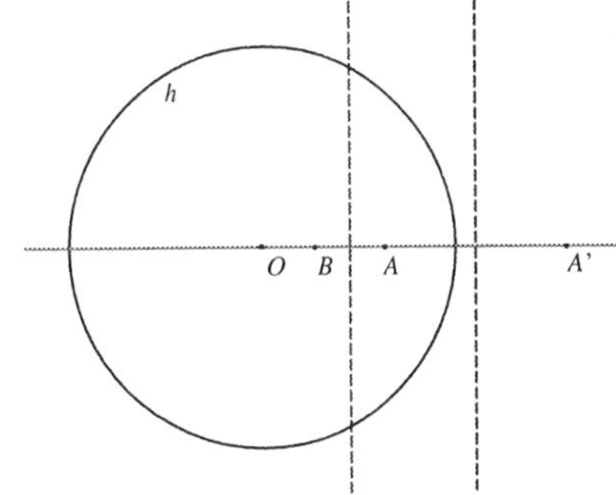

FIGURE 5. The construction of the hyperbolic line through A and B when O, A, and B are collinear. Because *Cabri* constructs the intersection of two lines even if they are parallel, the circle centered at the intersection point of the two parallel perpendicular bisectors of AB and AA' will be shown as the line going through O and A (a diameter of h).

in direction d. The result will appear as the line through the three points. This is acceptable based on the *implicit* continuityof the situation.[2] We are not going to review here all the extensions needed to obtain a coherent system. This is part of the specification of *Cabri II* and will be published elsewhere.

In the extended system of *Cabri II*, the points and lines are the points and lines of the projective plane. These are standard points of the Euclidean plane plus a point at infinity in each vector direction (u, v), and a line at infinity that is the set of points at infinity. In the extended plane, circles are defined by a center O and a peripheral point P (the distance $|OP|$ is the radius of the circle). Their geometric definitions differ according to the finite/infinite character of the points O and P:

Center O	Peripheral point P	Circle
normal	normal	standard Euclidean circle
normal	∞	line at infinity
∞	normal	Euclidean line through P perpendicular to the direction of the center O,
∞	∞	1) O and P have same direction: center O 2) O and P have distinct directions: line at infinity.

Most of the constructions are extended in this model to optimize coherence and satisfy an implicit *rule of continuity*, as illustrated in the following example.

An object will not show on the screen if it is possible to reach it as a degenerate position produced by distinct configurations. For example, consider a line segment AB with A fixed and B going to infinity in direction d (which, without loss of generality, may be assumed to be close to horizontal). For B at infinity, the line segment AB will not be drawn because this situation can be obtained by having B go to infinity to the right of A, producing in the limit a half ray to the right, and the same object can also be produced by having B approach infinity to the left of A.

Although indeterminate objects are not displayed, there is an advantage in keeping some form of existence for objects we consider to be undefined. Suppose we consider the inversion defined by a circle h. In this case, the inversion of its center is not defined in the normal sense. In our extension we define the inverted image of the center as an "undetermined" point at infinity. This concept allows constructions to work smoothly in more cases, as shown the next example.

A problem arises with our earlier construction of the hyperbolic line AB when A (or B) coincides with O. The construction requires the inverse image of O, so it should fail if carried out normally. In fact, *Cabri II* implements a more sophisticated extension than given by projective geometry. *Cabri II* constructs A', the inversion of A, as a point at infinity with undetermined direction. The perpendicular bisector of AA' is then the perpendicular bisector of an ordinary point and A' at infinity, which is the line at infinity. The center of the support circle is the intersection of this line with the perpendicular bisector of AB; this is the point at infinity on the bisector. This constructs the support for the hyperbolic line OB as the Euclidean line OB. Thus the construction (given in Figure 6) still works in this case.

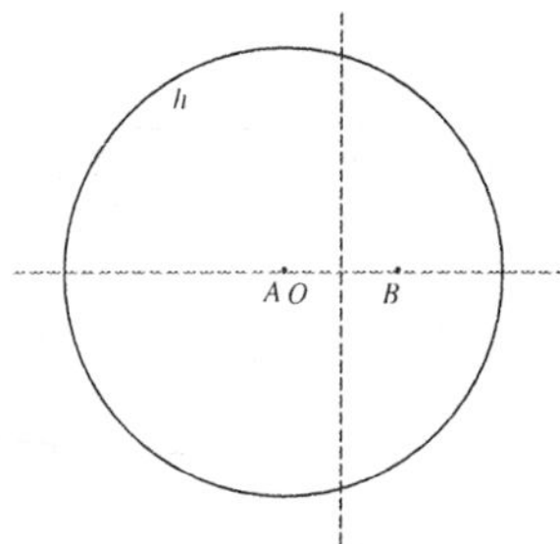

FIGURE 6. The construction of the hyperbolic line through A and B when O, A, and B are collinear and $O = A$. Because *Cabri* constructs A', the inverse of O, as an undetermined point at infinity, the perpendicular bisector of OA' is the infinity line, and its intersection with the perpendicular bisector of OB is the point at infinity in the direction perpendicular to OB. The hyperbolic line through AB appears as a diameter of the horizon h.

III. Stretching the Model

The Poincaré model for hyperbolic geometry relies on a circle that represents the horizon h. If we increase the radius of h, the sides of a given hyperbolic triangle will be arcs with less and less curvature. What happens if we increase the radius to infinity?

Suppose the horizon h is the circle with center O and passes through point H. There are two ways to make OH infinite—send O and/or H to points at infinity. We will first send H to infinity, defining it as the intersection point of two parallel lines. We obtain the picture in Figure 7(a), where the sides of the triangle are now rectilinear, as are the altitudes of the triangle. One can check the rectilinearity by using the "Check property" function available in *Cabri*.

Now let us go back to the situation in which h has a finite radius. We now send the center O to infinity, defining it as the intersection of two parallel lines.[3] In this situation our Poincaré model for hyperbolic geometry turns into the well-known other model, namely the half-plane model. In this

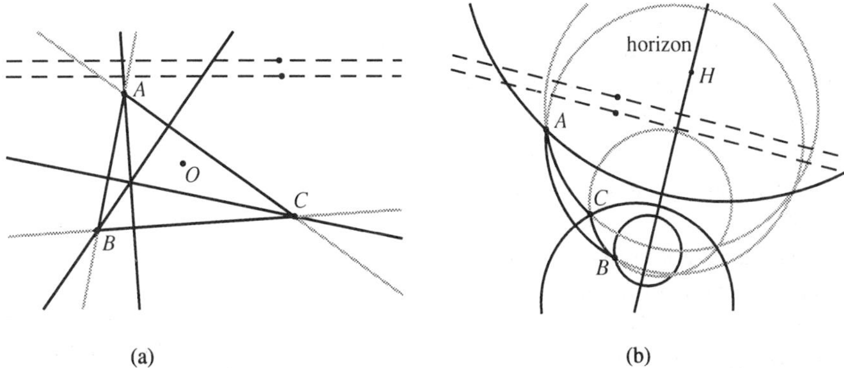

FIGURE 7. (a) This drawing is obtained when one point of the horizon is sent to infinity (as the intersection of the two horizontal parallel dotted lines). (b) The Poincaré model turns into the half-plane model when the center of the horizon is sent to infinity (as the intersection of the two parallel dotted lines).

model, points are the points of an open half-plane (the half-plane without its boundary), and lines are open semicircles centered on the boundary (Figure 7(b)).

Some Questions to be Resolved. In our presentation of constructions of hyperbolic lines, there are three important questions that we have encountered.

1. *Is it true that a generic construction is always possible?*

I would like to illustrate this point with a case studied recently by Pierre Duchet [93]. The following problem of Apollonius is well-known: *construct all the circles (at most eight) tangent to three given circles.* This problem has received a progressive series of answers throughout the history of mathematics. Indeed, by the end of the nineteenth century, mathematicians knew how to construct the solutions in all cases, but different cases (depending on the configuration of the three given circles) were solved using essentially different constructions. Duchet was able to show (with the help of *Cabri*) that all cases fall basically into two different classes: hyperbolic and elliptic. So for this problem, work remains to be done to prove or disprove the following "metaconjecture":

Conjecture. If a Ruler-Compass construction problem has different solutions for different configurations of the givens, a unique solution can be described which covers all cases, producing a so-called generic construction.

2. *How can one take account of implicit orientation of objects to find some reasonable way to identify points of intersection?*

Consider the problem: Given a line m and a segment AB, construct the reflection of AB with respect to m. We are interested in a step-by-step construction, involving only basic primitives of geometry, such as "line segment between two points," "circle with a given center through one point," "intersection of these objects," etc. We will not employ built-in constructions such as point symmetry or line reflection.

A common solution to the problem, especially for students, is to construct the line p from A perpendicular to m, and its intersection I with m. Next, construct the circle c centered at I and going through A. The desired reflected point A' is the intersection point of c and p which is different from A (assuming that A is not on m). See Figure 8.

This construction highlights the need to internally distinguish the two intersections of a circle with a line. Here are two possible ways to carry this out:

(1) Make an internal distinction between the two intersections of a line with a circle by taking the first intersection as the point computed with a $+$ sign in front of the square root in the analytic expression of the coordinates of both intersection points.

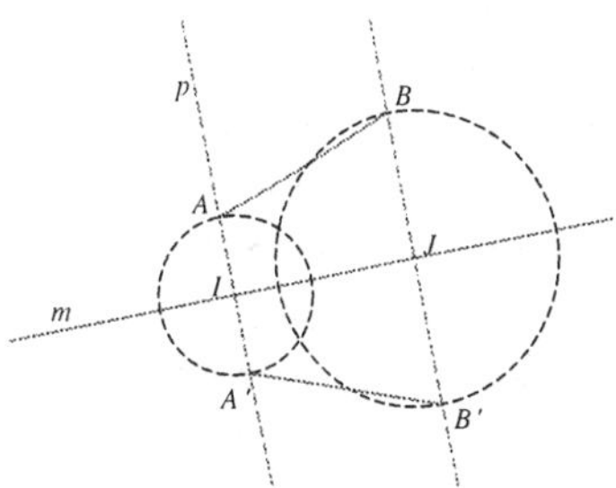

FIGURE 8. Construction of the reflected image of a line segment AB.

(2) Orient implicitly all lines: for instance, orient a "free" line by the direction from the first point used to define it to some other point. Once free lines are oriented, then a line that depends on another line l can be oriented by various means. For example, a line p drawn from a point P perpendicular to a line l could be oriented by (a) rotating the "direction" of l by, say $\pi/2$, or (b) p could be oriented from P to the foot of the perpendicular (where p intersects l), supposed for simplicity not to coincide with P.
In both cases (a) and (b), we define the first intersection I_1 of a circle with p to be the first crossing point when we move on p according to its direction, and I_2 to be the second crossing point.

In our construction in Figure 8 of the reflection of AB in line m, let us assume that m is oriented with direction from right to left. Let us examine how the point A' behaves when we move A around the plane. As long as A does not cross m, everything is OK: $A = I_1$ and $A' = I_2$. When A lies on m, then in case (1) or (2)a we will still have $A' = I_2$ but $A = I_1$ now shares the same location as I_2. When A crosses to the other side of m, $A = I_2$ and $A' = I_1$. In case (2)b, the situation will be better: $A = I_1$ and $A' = I_2$ no matter where A is located.

Nevertheless, most software implementations choose to implement the intersection of a line with a circle according to scheme (1) or (2a), and we can understand why. In the implementation (2b), a first annoyance arises because the orientation of a line p drawn from a point P perpendicular to a line l depends on which side of l the point P lies on. The orientation of the line p if the point P lies on l seems to be possibly not canonical. A second difficulty in the implementation of (2b) comes from some "discontinuities" in the transformation of a figure which may be introduced when we drag elements of the figure around the plane.

It is interesting to test the robustness of dynamic geometry software with respect to the preceding construction. *Cabri* shows acceptable behavior here, because it tests whether a point defining the line also lies on the circle, and in that case the intersection is obtained by constructing only one "new" point, the intersection of p and c different from the point already known.

A similar ambiguous situation occurs when circles intersect. This situation is not handled properly by most dynamic geometry software, as can be checked by the following construction of a rhombus $ABCD$ (see Figure 9). Consider the circle with center A and radius AB. Construct a point on this circle and call it D. The fourth point of the rhombus is then the intersection of the two circles C_1 (center B, radius BA) and C_2 (center D, radius DA). Without care, the rhombus will probably appear for some position of point D relative to B as "two" line segments AB and AD, C coinciding with A.

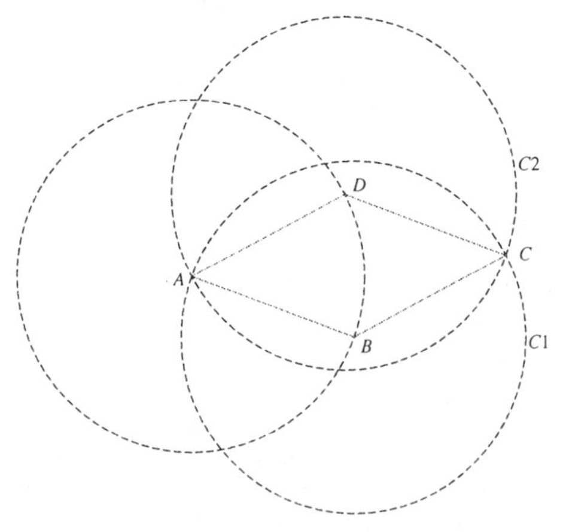

FIGURE 9

3. *Does there exist a "complete" model for dynamic geometry, embracing all degenerate cases in a reasonable way?*

In our second section we discussed some improvements in the behavior of dynamic geometry software brought about by extending the geometry with elements at infinity. The implementation given in *Cabri II* is probably not entirely satisfactory, and it is surely possible to point out cases where its behavior will not fulfill all the expectations of the mathematically literate user. I think we need a real mathematical treatment of all the consequences of stretching geometry in some way to a wider system. This system cannot be simply the projective one if we want to maximize the way the environment takes into account the special characteristics of non-static objects which are at the core of dynamic geometry.

Endnotes

1. *Cabri-géomètre* is sold in the United States as *Cabri Geometry II*. It is available from Texas Instruments.
2. This feature can be switched off for situations where people prefer a "purely Euclidean" behavior from the environment. This might be necessary, especially in a classroom setting, where a teacher designs an activity with special didactic expectations in mind [Laborde 1994].
3. This can be done in a progressive way by constructing two extra lines meeting at O. We first redefine O as this intersection and then manage to force one of the lines to be geometrically

parallel to the other. In *Cabri,* it is also possible to construct a line and one of its parallel lines. In redefining an object, we then choose intersection of two objects and select, successively, the first line and its parallel.

Bibliography

Baulac, Y. (1990). *Un micromonde de géométrie, Cabri-géomètre,* thèse de l'Université Joseph Fourier, Grenoble.

Baulac, Y., Bellemain, F., and Laborde, J. M. (1988). *Cabri-géomètre, un logiciel d'aide à l'enseignement de la géométrie.* Software and User's guide, Cedic-Nathan Paris.

Bellemain, F. (1992). Conception, réalisation et expérimentation d'un logiciel d'aide à l'enseignement de la géométrie: *Cabri-géomètre,* Thèse de l'Université Joseph Fourier, Grenoble.

Bellemain, F., and Laborde, J.-M. (1992). *Specifications of Cabri-geometry II®*, Internal Research Report, LSD2-IMAG Grenoble.

Duchet, P. (1993). *Sur le problème d'Appolonius,* séminaire, private communication (LSD2-IMAG).

Jackiw, N. (1990–95). *The Geometer's Sketchpad,* Key Curriculum Press, Berkeley, CA.

Laborde, J.-M. (1995). "Des connaissances abstraites aux réalités artificielles, le concept de micromonde *Cabri,*" in *Environnements Interactifs d'Appreentissage avec Ordinateurrs,* D. Guin, J.F. Nicaud, D. Py, eds., pp. 29–41 Eyrolles Paris.

Laborde, C. (1994). "Designing Tasks for Learning Geometry in a Computer-based Environment, the Case of Cabri-géomètre," in *Technology in Mathematics Teaching—a Bridge between Teaching and Learning,* L. Burton and B. Jaworski, eds, pp. 35–68, Chartwell-Bratt, Bromley, UK, London 1995.

Laborde, J.-M. and Bellemain, F. (1994). *Cabri-geometry II®*, Texas Instruments, Dallas TX.

Thibaut, M.-F. and Labarre, R. (1989). "Some Hyperbolic Geometry with *Cabri-géomètre,*" in *Intelligent Learning Environments, the Case of Geometry,* J.-M. Laborde, Ed., ASI Series F, vol. 117, Springer Verlag, pp. 218–230.

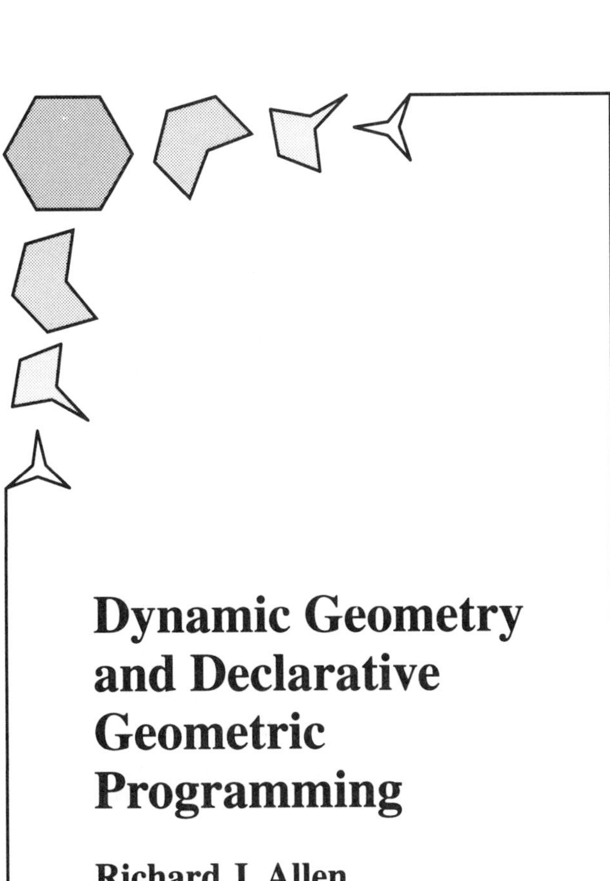

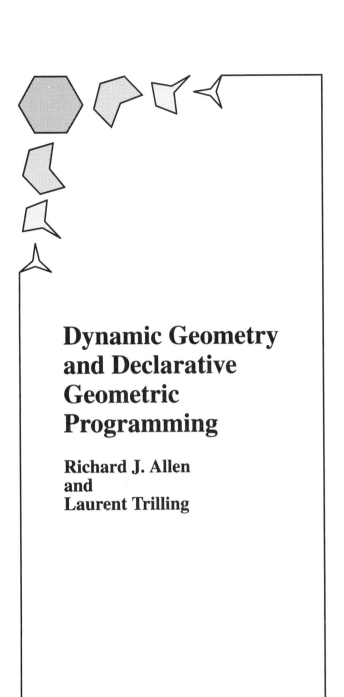

Dynamic Geometry and Declarative Geometric Programming

Richard J. Allen
and
Laurent Trilling

Declarative geometric programming provides a tool with which you can specify a figure by giving the objects and properties of the figure in a logical specification. With such a system you can construct the figure automatically from this specification and a list of base elements. The user can displace the figure dynamically by dragging any one of the base elements. If, on the other hand, you take a procedural approach for constructing the figure, as is the case when you use *Cabri* or *Sketchpad*, there are fewer base elements with which you can displace the figure by dragging; this is because the order of construction of the objects in a figure is important in the procedural case but not in the declarative one.

For example, let ABC be a triangle with H the point of intersection of its altitudes. If you use *Cabri* or *Sketchpad* to construct the triangle and then construct H, the points A, B, C are free to be dragged (and the position of H changes as the shape of ABC changes), but H is not free. In other words, when H is dragged, the triangle ABC moves rigidly with the motion of H, not changing its shape. On the other hand, using a declarative geometric programming system, you can create a specification for the triangle containing the three segment sides, the three altitudes and their point of intersection H, and the three vertices A, B, C. You may then construct any three of the four points, including H, and the system will automatically construct the fourth point as well as the rest of the figure. In addition, and what is important here, you may drag on any one of the four points and the three other points move, changing the shape of triangle ABC in order to preserve all the specified relationships. Thus with such a program, you can often create (through dragging) a wider class of figures having certain properties than with a procedural system such as *Cabri* or *Sketchpad*. The significance of this capability for education is that students can use dragging as a means to discover geometric invariants by dynamically changing a figure into other ones that have the same specifications as the original.

GéoSpécif is an example of a *declarative programming system* and, as such, provides more ways to drag a figure and thus more opportunities to find geometric invariants. Since our system is intended for use in an educational context, we must provide the system and its interface with semantics that are as clear as possible. In the case of *GéoSpécif*, we intend to find all possible constructions of a figure from a given specification that use only the objects found in the specification. The overall organization and syntax of the interface should be as close as possible to the context of geometry pedagogy. We see the use of this system in the classroom as a tool to bridge the gap between the more concrete and dynamic activities of figure construction and property exploration and the use of logic for representing and communicating geometric information.

The latest version of *GéoSpécif* runs on both Macintosh and UNIX platforms and allows direct manipulation of the specification, thus providing the user with a capability for dynamically exploring effects on a figure caused by changes to the specification. The system, which is written in Prolog III and exists in prototype form, has developed as part of continuing research on intelligent tutors for teaching geometry in the Laboratoire de Génie Informatique at IMAG in Grenoble, France. The second author is one of the members of the *Cabri* Project which is also located at IMAG. Some high school students have used the system on an experimental basis.

ITS and Geometry

Intelligent tutors (ITS) for learning secondary school geometry provide tools designed (1) to aid in the construction of geometric figures that conform to a specification provided by a teacher, (2) to stimulate discovery of important geometric properties represented in such figures, and (3) to guide the organization of proofs. In previous work, four main cognitive components of such systems along with related learning activities have been identified (Allen 1990):

- Figure acquisition: The student first constructs a figure which conforms to a specification provided by the teacher.
- Figure appropriation: The student can graphically transform the figure while its logical properties are preserved.
- Property exploration: The student reacts to system-proposed properties by using theorems furnished by the teacher or the tutor.
- Proof organization: The student, utilizing facts discovered during the previous steps and theorems furnished by the teacher or the tutor, constructs a proof which is verified by the system.

All three dynamic geometry systems, *Cabri*, *Sketchpad*, and *GéoSpécif*, focus mainly on providing figure acquisition and figure appropriation capabilities. In addition, *GéoSpécif* provides the user with two principal system parameters, namely, the figure specification and the base element list.

Organization of *GéoSpécif*

The system interface of *GéoSpécif* is a combination of text editor, graphics window, and button panel with two principal menus, one for creating objects and one for specifying properties between objects. The syntax used by the editor to express a logical specification of a figure is close to that found in secondary geometry textbooks. For example, the syntax to specify a triangle, its altitudes, and their intersection could be the following:

point(A), point(B), point(C), point(H),
line($l1$), line($l2$), line($l3$),
segment($s1$, A, B), segment($s2$, A, C),
segment ($s3$, B, C),
$A \in l1$, $B \in l2$, $C \in l3$, $H \in l1$, $H \in l2$, $H \in l3$,
$l1 \perp s3$, $l2 \perp s2$, $l3 \perp s1$.

The other parameter (besides the figure specification) to be communicated to the system is a set of base elements; in this particular example, any three of the four points A, B, C, H will do. Our interface expects the user to place the base points in the graphics window by clicking with the mouse. The teacher might expect the students to determine which points in the specification can be used as base elements; in fact, students may be expected to formulate the entire specification as well. Obviously, there are a number of educational activities that can be focused on the specification and the base element list. After the specification and the base elements have been communicated, the system automatically draws and labels the triangle and its altitudes (Figure 1). Any base element may be used to displace the figure by dragging on it.

For *GéoSpécif*, a geometric figure can be viewed as a collection of pairs of numbers that satisfy a set of constraints expressed as a system of equations and inequalities. This provides a natural pedagogical opportunity to explore the relationship between euclidean geometry and analytic geometry, where points also are represented as pairs of numbers, lines as linear equations, and intersections of lines as solutions of linear equations. Furthermore, use of *GéoSpécif* naturally provides teachers with opportunities for discussion of the relationship between linear algebra and geometry (es-

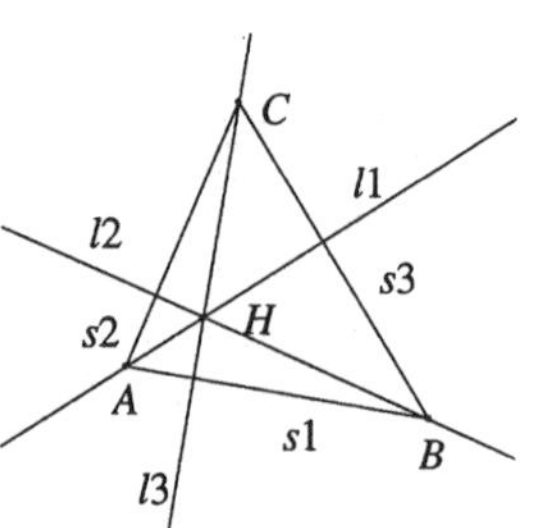

FIGURE 1

pecially solutions of equations), and provides a natural setting for comparing procedural and declarative approaches to problem solving. We propose that declarative geometry systems such as *GéoSpécif* be part of the new dynamic geometry. They add to the dynamic geometry provided by *Cabri* and *Sketchpad* by connecting the visual aspects of figure displacement to formal algebraic methods and systems necessary to support automated and semi-automated construction of figures.

We believe that *GéoSpécif* is an excellent example of systems supporting dynamic geometry since it provides automatic construction of figures from a specification, direct manipulation of the constructed figures as well as direct manipulation of the specification, property exploration and property discovery, and discovery of the largest class of geometrically equivalent figures where all figures in the class have the same object(s) held fixed. Automatic construction has the following potential for users: (1) to obtain geometric figures even if the user does not know how to construct them with straightedge and compass (or their electronic equivalents); (2) to change the specification interactively and observe corresponding visual changes immediately; (3) to construct an entire geometric pattern from only the specification of the pattern and a few elements of the pattern. The user does not need to discover or to know beforehand a step-by-step algorithm to construct a figure before exploring it, but only be able to describe it in a specification.

Dynamic Constructions

GéoSpécif provides a tool for carrying out what we term dynamic constructions. It links automatic figure construction and rapid figure displacement (through dragging) to logical specification. In using *GéoSpécif*, students first obtain a construction and then explore its properties. This can often lead them to discover a ruler and compass construction. Several examples follow that introduce a different kind of (dynamic) geometry we can do in class.

Example 1. Given lines A and B and a point p, construct points a and b on A and B, respectively, such that p is the midpoint of the segment between a and b.

To solve this problem, you can first have *GéoSpécif* construct a point p given A, B, a, and b. Then, freeing a, you can drag b and investigate the properties of the locus of a (the dots in Figure 2).

Example 2. Given three lines, construct a triangle for which the lines are the perpendicular bisectors of its sides.

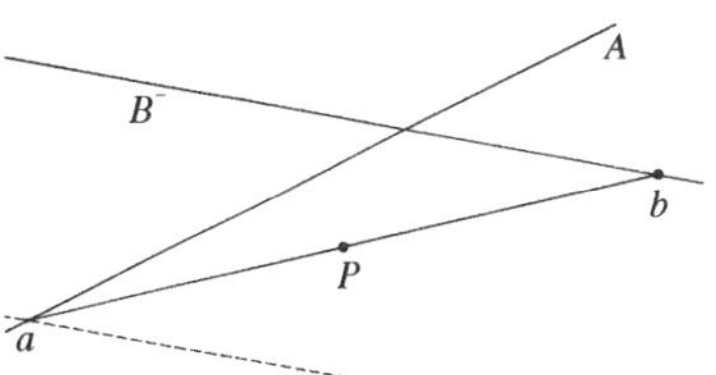

FIGURE 2

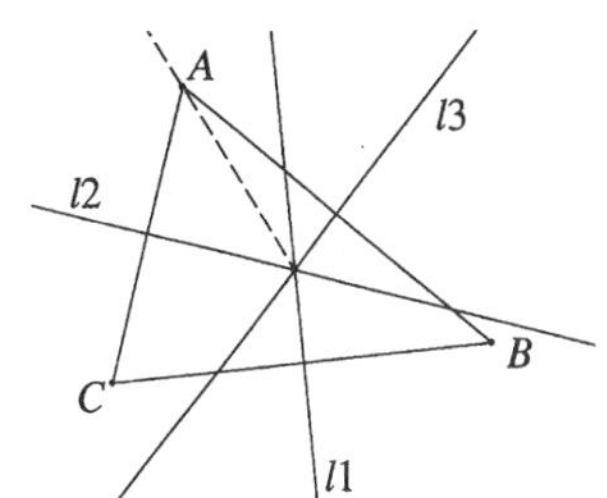

FIGURE 3

The construction of such a triangle is difficult, but exploring with dynamic geometry can help students discover a construction. One approach is to enter into *GéoSpécif* the specification for a triangle with the points A, B, C, as its vertices and with the lines $l1, l2, l3$, as the perpendicular bisectors of its sides. First, take the vertices as base elements and have *GéoSpécif* construct a triangle. Then fix $l1, l2, l3$ (the conditions of the given problem) and drag on A. Watch the locus of A (dashed line segment in Figure 3) and observe that it is a line passing through the intersection of $l1, l2, l3$. Students should then try to discover properties of this line. The problem could also require that A be on a given line D.

Example 3. Given four distinct points A, A', B, B', construct the point P on the line (A', B') such that the line (A, P) is perpendicular to the line (B, P).

First, construct by fixing A', B', A and P. Then free P and fix B. In this case, P cannot be constructed by *GéoSpécif* (note that the circle of diameter AB is not given). But, by weakening the specification so that P does not belong to $A'B'$ and by fixing and dragging line AP, you can discover the locus of P (Figure 4).

Examples 1, 2, and 3 illustrate how *GéoSpécif* can be used to solve construction problems by dragging on different base points. First have the system automatically construct a figure satisfying the specification. Then temporarily remove a constraint in the specification, drag on a judiciously chosen point, and watch the locus of that point, observing relationships helpful for a ruler and compass construction. This process is known as the *method of loci.*

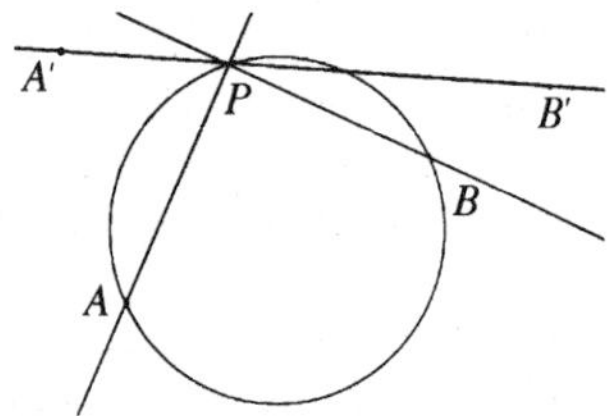

FIGURE 4

The next two examples exhibit the power of *GéoSpécif* to provide, dynamically and automatically, solutions to geometry construction problems that involve distances and directions and are nonlinear (if you think of constraints as equations and inequalities).

Example 4. Given lines A and B, a distance d, and a slope s, construct points a and b on B and A, respectively, such that the line through a and b has slope s and the distance between a and b is d (see Figure 5).

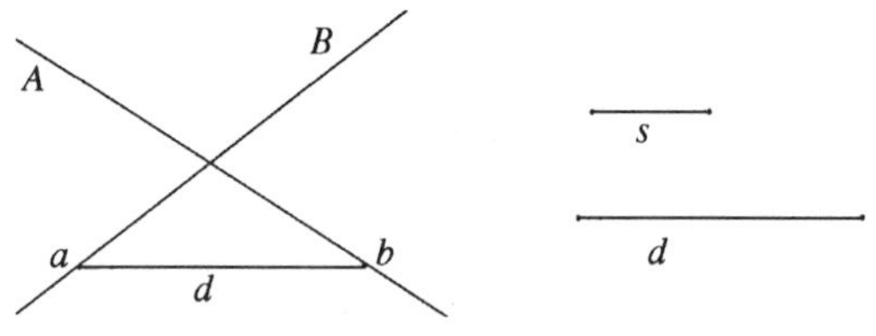

FIGURE 5

Example 5. Given two triangles *ABC* and *DEF*, construct a triangle *GHI* inscribed in triangle *ABC* such that each side of triangle *GHI* is parallel to a side of triangle *DEF* (Figure 6).

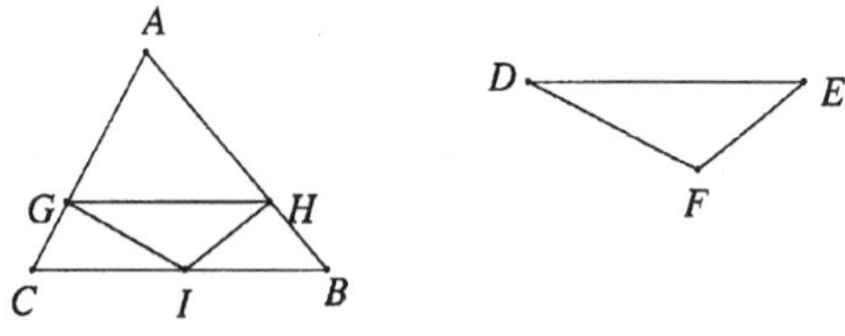

FIGURE 6

The next example shows how *GéoSpécif* can be used for error diagnosis in a construction problem.

Example 6. The scenario goes as follows: Students are to construct point H from a set of base points containing C. But what happened is that they have constructed point H' which is graphically very close to H. Working with the incorrect figure, move C to a position C' which forces the distance of H' from H to be significant, say 1 cm apart. Now change the specification by freeing C and keeping the distance between H' and H equal to 1 cm. Then have the students re-explore the figure. An example where this scenario could easily occur is when H is the orthocenter and H' is the centroid of a triangle that is nearly isosceles.

Conclusion

It is natural to ask questions about the use of *GéoSpécif* in teaching geometry. If the objective is to help a student find a construction, why, therefore, use a system which is able itself to find a construction? The reasons occur on several levels but include the following points.

(1) First and foremost we want students to be able to find ruler-and-compass constructions for the cognitive impact this activity has on them. It is a mechanism for finding, understanding, and remembering important geometric relationships and properties. It is really the same reason why we ask students to be able to do elementary factoring and simplification of algebraic expressions even though we now have automatic symbolic systems that will do it for us.

(2) A system like *GéoSpécif* allows students both to hold parts of the figure fixed and to modify the figure specification quickly and easily, thus producing new, yet related, figures to explore. The more (guided) figure explorations the students have, the greater the potential for discovering those geometric properties and relationships useful in carrying out a ruler-and-compass construction. It really comes down to the suggestive power of visualization, especially when combined with real-time movements. The more possibilities available for effecting such animations, the more likely a student is to perceive something different and important.

(3) The machine-generated construction is often different from the one asked of the student (who usually is working in a cognitive geometry; i.e., within a certain context of definitions, theorems, and exercises which do not include all of geometry) and is useful as a metaphor relating Euclidean geometry to analytic geometry and linear algebra.

Why use a system that has as one of its two principal parameters the logical specification of the figure? Dynamic geometry already uses rapid displacement to empower and extend the understanding of geometric concepts. Such animation is often controlled through dynamically manipulating different objects in the figure. Our approach adds dynamic manipulation of the figure's specification as a means of controlling the geometry of the figure.

References

Allen, R., Idt, J., and Trilling, L. (1993). "Constraint Based Automatic Construction and Manipulation of Geometric Figures," *Proceedings of the 13th International Joint Conference on Artificial Intelligence*, Chambéry, France, vol. 1, pp. 453–458.

Allen, R., Nicolas, P., and Trilling, L. (1990). "Figure Correctness in an Expert System for Teaching Geometry," *Proceedings of the eighth biennial conference of the Canadian society for computational studies of intelligence*, Ottawa, May 22–25, 1990, pp. 154–160.

CABRI Geometry II (1994). Texas Instruments, Dallas, Texas.

Colmerauer, A. (1990). "Prolog III," *Communications of the ACM*, vol. 33, no. 69.

Laborde, C. and Capponi, B. (1994). *Cabri-Classe, apprendre la géométrie avec un logiciel*, Edition Archimède, 11b, Allée Wallon, 95100 Argenteuil, France.

Laborde, J.M., (ed.) (1994). *Intelligent Learning Environments, the Case of Geometry*, NATO ESI Series, Springer Verlag.

About the Authors

Susan Addington is an associate professor of mathematics at California State University, San Bernardino. Her interests include geometry and algebra, the use of computers for mathematical visualization and communication, and math education. Addington is involved in enriching the mathematical education of elementary and middle school students, teachers, and parents through several projects, including The California Math Show, a traveling hands-on exhibit on the idea of symmetry. Her phone is (909) 880 5362, fax (909) 880-7119, email susan@math.csusb.edu, or check her World Wide Web home page http://www.math.csusb.edu/faculty/susan/home.html.

Richard Allen is professor of computer science and mathematics at St. Olaf College in Northfield, MN. One of his major professional interests consists in the development of tutoring systems for learning geometry. At the same time he has worked with teachers and two St. Olaf colleagues during the past ten years in two National Science Foundation funded projects to bring geometry microworlds and geometry tutors into the classroom. A recent publication is "Constraint Based Automatic Construction and Manipulation of Geometric Figures." His phone number is (507) 646 3117 and email address is allen@stolaf.edu.

Benjamin T. Backus is a graduate student in the Vision Science Program at the University of California at Berkeley. He taught mathematics in middle and high schools from 1989 to 1993 in Oakland, California. Together with his advisor, Martin S. Banks, he has published work in binocular vision and the determination of heading from optic flow. He can be contacted at the U.C. School of Optometry, 360 Minor Hall, Berkeley, CA 94720-2020, email: ben@john.berkeley.edu. For those interested in teaching vision through dynamic geometry, he recommends a visit to the web site http://john.berkeley.edu/IndividualPages/MartyPages/OSP.html.

Dan Bennett is a mathematics editor at Key Curriculum Press, publishers of *The Geometer's Sketchpad.* He is author of *Exploring Geometry with The Geometer's Sketchpad* and *Pythagoras Plugged In: Proofs and Problems for The Geometer's Sketchpad.* Before beginning work at Key, he was a high school math teacher in San Francisco. He can be reached at Key Curriculum Press, P.O. Box 2304, Berkeley, CA 94702; e-mail: dbennett@keypress.com.

Kathryn Boehm currently teaches mathematics at Harpeth Hall, a 5–12 independent school for girls in Nashville, TN. She has been teaching geometry for 8 years and has constantly sought ways, such as the use of dynamic geometry software, to make the ancient subject come alive for her students. She is a member of NCTM and resides at 600 Vosswood Dr., Nashville, TN 37205. E-mail: boehmk@ten-nash.ten.k12.tn.us.

Doug Brumbaugh teaches teachers and believes if he is telling others how to do it, he should be out there doing it. This year he will be teaching two geometry classes using *Geometer's Sketchpad* in a local school for at-risk students. Doug recently completed a secondary mathematics methods textbook, *Teaching Secondary Mathematics*, published by Lawrence Erlbaum & Associates. His telephone number is (407) 823 2045; email: brumbad@pegasus.cc.ucf.edu; and web site: http://pegasus.cc.ucf.edu/~mathed.

Jere Confrey is an associate professor of mathematics education at Cornell University. Her research has focused on articulating the potential in mathematical ideas of students. History has been an important tool in establishing these alternative approaches. Her recent work with elementary students has focused on the construct of "splitting," establishing it as an independent but complementary idea to "counting." Currently she is co-authoring a multimedia precalculus course incorporating investigations and computer-based software tools. She is a senior editor for *Computers and Mathematics Learning* in which she co-authored the paper, "A Critique of the Selection of 'Mathematical Objects' as a Central Metaphor for Advanced Mathematical Thinking." Her phone number is (607) 255 1255 and e-mail is jc56@cornell. edu.

Al Cuoco is Senior Scientist and Director of the Mathematics Initiative at Education Development Center, 55 Chapel Street, Newton MA 02158 (alcuoco@edc.org) where he is working on a high school geometry curriculum, Connected Geometry, that makes central use of dynamic visualization. From 1969 until 1993, he taught high school mathematics. His mathematical interests and publications have been in algebraic number theory, but he has slowly learned to appreciate the role of geometry and visualization. His favorite publication in this area is his article in the *American Mathematical Monthly* (98, 4), described by his wife as "an attempt to explain a number system no one understands with a picture no one can see."

Michael D. de Villiers is associate professor of mathematics education at the University of Durban-Westville, South Africa. His major research areas are geometry, the nature and philosophy of mathematics, and applications of school mathematics. Two recent publications are "An alternative introduction to proof in dynamic geometry," *MicroMath*, Spring 1995, 14-19 and *Some adventures in Euclidean geometry* (212 pp.), 1996, University of Durban-Westville. Since 1988 he has been editor of *Pythagoras*, the journal of the Mathematical Association for Mathematics Education of South Africa (AMESA). His fax number is 027-31-8202866 and e-mail is mdevilli@pixie.udw.ac.za.

David Dennis is a professor of mathematics at the University of Texas at El Paso. His research focuses on the history of mathematics and mathematics education. Recently he conducted an extended study of the history of curve-drawing devices and their role in the genesis of calculus. He is currently involved in the Partnership for Excellence in Teacher Education (PETE), an NSF pre-service teacher collaborative to improve the mathematical and scientific preparation of K-12 teachers. Contact: Dept. of Mathematical Sciences, University of Texas at El Paso, El Paso, TX 79968-0514; Tel: (915) 747 6775, Fax: (915) 747 6502, e-mail: dennis@math.utep.edu.

Tim Garry has been teaching secondary mathematics in international schools in Scotland, Switzerland and Norway for the past eight years. In Fall 1996 he began teaching at the University School of Milwaukee, 2100 W. Fairy Chasm Rd., Milwaukee, WI 53217 (fax: (414) 352 8076). His major interests are integrating graphing calculators and geometry software into the math classroom and sharing ideas with teachers at workshops. For the past two years he has been a member of the European Council of International Schools' (ECIS) Mathematics Committee and editor of *The Centroid*, the ECIS mathematics newsletter. His e-mail address is: 100114.2524@compuserve.com.

E. Paul Goldenberg, Senior Scientist at Education Development Center, Inc. (EDC), received his doctorate from the Harvard Graduate School of Education. Author of numerous books and articles on mathematics learning and uses of computers in education, his background includes mathematics coordination for the University of Chicago Laboratory Schools, and extensive teaching in primary, secondary, and graduate school. At EDC, Dr. Goldenberg has been principal investigator of Connected Geometry, a secondary mathematics curriculum project, as well as for a research effort "The Epistemology of Dynamic Geometry," both funded by the NSF. He is also engaged in other research and development efforts in undergraduate mathematics learning. pgoldenberg@edc.org

Catherine A. Gorini teaches mathematics at Maharishi University of Management. Her main interest is in investigating the connections between Maharishi's Vedic Science and mathematics and applying the principles of the development of consciousness to mathematics education. One aspect of her work in this area is in teaching mathematics visually—using art, computer graphics, and models—as a way of appealing to the whole student. Her publications include "An Art Research Project for a Geometry Course" in PRIMUS, December, 1994, and "Symmetry: A Link Between Mathematics and Life in the *Humanistic Mathematics Network Journal*, Number 13, May, 1996. She can be reached by telephone (515) 472 1107, fax (515) 472 1123, or email: cgorini@mum.edu.

David Green is Reader in Mathematical and Statistical Education at Loughborough University, UK, where he has taught for just over 20 years. He has a B.Sc. degree in Mathematics and an M.Sc. degree in Computing both from Manchester University, an MEd degree in Science Education from London University and Ph.D. in Probability Concepts from Loughborough University. He is an editor of the journal *Teaching Statistics* and is the author of many papers and books, mostly on probability concepts and the use of IT in mathematics teaching. His current main interest is the use of dynamic geometry, particularly *Cabri*. His address is: Department of Mathematical Sciences, Loughborough University, Loughborough, Great Britain, LE11 3TU; Tel. 01509-222-864; Fax 01509-223-969; email: D.R.Green@lut.ac.uk.

Tony Hampson is based in the Faculty of International Education at Cheltenham and Gloucester College of Higher Education, Cheltenham, Glos., England (Tel & Fax 01242 532877), where he advises on quality assurance procedures relating to overseas programmes. He teaches mathematics within the undergraduate programme in Cheltenham and also to teachers in Hong Kong as part of the College's overseas In-Service BEd programme. His professional interest lies in undergraduate curriculum development in Mathematics and, in particular, the design and delivery of courses appropriate for intending teachers.

Fadia Harik is a senior scientist at BBN Systems and Technologies. BBN develops simulations and networking environments for education. Her current work and research is in the area of mathematics teaching/learning from a constructivist perspective with particular focus on classroom cultures. She has written curricula for teacher education courses in modern algebra and in dynamic geometry. She is currently directing an NSF teacher enhancement project where a set of video case studies are being developed for use by teachers and teacher educators to unravel the processes of inquiry as well as the obstacles to inquiry in the mathematics classroom at the middle school level. She can be reached by telephone (617) 873 3009, fax (617) 873 2455, and by email: fharik@bbn.com.

Douglas Hofstadter is a professor of cognitive science at Indiana University, with connections to several departments including Computer Science and Psychology. A math major at Stanford, he did his Ph.D. in physics at the University of Oregon. He explored number theory using computers in the early 1960's, and fell in love with "experimental mathematics". He never studied geometry at all, but somehow fell in love with it purely on his own in 1992. His love affair with Euclidean and other geometries (including Euclidual, its dual with respect to the self-dual projective geometry), mediated in large part through *Geometer's Sketchpad*, has been most ardent. He is deeply involved with triangle centers and their complex interrelationships, and has taught several courses on geometry and discovery, called "Circles and Triangles: Diamonds of Geometry" (or "CaT:DoG"). He is working on a book on these topics. Fax: (812) 855 6966; email: dughof@cogsci.indiana.edu or helga@cogsci.indiana.edu (administrative assistant).

R. Nicholas Jackiw is a software designer at Key Curriculum Press, where he directs ongoing development of *The Geometer's Sketchpad* software and print projects, and participates in other software development projects. Jackiw's current research focus is on deploying dynamic geometry principles in the area of student-constructed graphs and related visualizations of quantitative information. He may be reached at njackiw@keypress.com.

Zhonghong Jiang is an assistant professor of mathematics education and computer education at Florida International University. His major research area is the use of technology in mathematics curriculum and instruction. Recent publications include "A Computer Microworld to Introduce Students to Probability" (Jiang & Potter, 1994, in *The Journal of Computers in Mathematics and Science Teaching*, Vol. 13, No. 2) and "A Brief Comparison between the U.S. and Chinese Middle School Mathematics Curricula" (Jiang & Eggleton, 1995, in *School Science and Mathematics*, Vol. 95, No. 4). He can be contacted by phone (305) 348 3790, fax (305) 348 2086, and e-mail: jiangz@solix.fiu.edu.

Michael Keyton has been a teacher of mathematics and music history at St. Mark's School of Texas since 1977. He has contributed to several reference works in music; he was an assistant to Nicolas Slonimsky in the production of the 7th and 8th editions of *Baker's Biographical Dictionary of Musicians*. In addition to his interests in mathematics and music, he was a professional golfer for several years. He studied at Louisiana State University (B.S., 1965; M.S., 1976), Indiana University (M.A., 1967), and the University of North Texas (Ph.D., 1986). He is the author of *92 Geometric Explorations on the TI-92* (Texas Instruments, 1996). His addresses are: 8580 Banff, Dallas, TX 75243; email: mkeyton@tenet.edu.

Jim King has a Ph.D. in mathematics from the University of California at Berkeley and has been teaching at the University of Washington since 1975. Since 1991, he has been an organizer and instructor in the Park City Mathematics Institute. His research areas include geometry of complex algebraic manifolds and other areas of geometry. He began teaching geometry using computers about ten years ago in a course based on problem-solving with *Logo*. Since then he has become involved in the use of dynamic geometry software, writing, teaching and organizing courses, workshops, user groups and sessions for college and high school teachers. He is author of *Geometry Through the Circle with The Geometer's Sketchpad*, Key Curriculum Press, and is currently working on a dynamic geometry project on the World Wide Web. Check out his web page: http://www.math.washington.edu/~king. He can be reached at the University of Washington, Box 354350, Seattle, WA 98195-4350, or by telephone (206) 543 1915, fax (206) 543 0397, or email: king@math.washington.edu.

Jean-Marie Laborde, Research Director at CNRS, founded the Laboratoire de Structures Discrètes et de Didactique (LSD^2) in 1982; this is a research laboratory within IMAG. He graduated in Mathematics at École Normale Supérieure in Paris in 1969. He earned a Ph.D. (Thèse d'Etat) in computer science at the University of Grenoble in 1977. His work on the *Cabri* project began in 1981; it was originally envisioned as an environment for graph theory. He has devoted his research efforts to the use of geometric methods for the study of different classes of graphs, especially hypercubes. He is currently involved with several of his PhD students in new developments of *Cabri-géomètre*. He can be reached at Laboratoire Leibniz—IMAG-Campus, Université Joseph Fourier—CNRS, BP 53 38041 Grenoble cedex 9 FRANCE; by telephone (33) 76 51 46 10 (sec 76 51 46 16), fax 76 51 45 55, or email: jean-marie.laborde@imag.fr.

Stuart Levy serves on the technical staff of the Geometry Center at the University of Minnesota, and among other computer-related activities enjoys programming for interactive computer graphics. He is one of the authors of the Geometry Center's *Geomview* graphical software package. His phone is (612) 624 1867, fax is (612) 625 8083, and prefers e-mail at slevy@geom.umn.edu.

Edwin McClintock, professor of mathematics and computer education at Florida International University in Miami, FL, is writing and testing units of mathematics curriculum for middle school through graduate school utilizing *Sketchpad* in the visualization process. This development relates to a set of grants for which he is principal investigator or co-principal investigator. His phone is (305) 348 2087, fax is (305) 348 3205, and e-mail address is: mcclinto@solix.fiu.edu or EDDMAC@aol.com.

James Morrow, who received his Ph.D. in functional analysis from Florida State University under the direction of W.J. Stiles, directs Mount Holyoke College's SummerMath program, along with his colleague and companion, Charlene Morrow. He is currently working on ways that computer technology, combined with reflection on problem solving, can be used to make sense of mathematics. He enjoys traveling with his dynamic family and watching his daughter, Hannah, learn and grow. He may be reached at SummerMath, Mount Holyoke College, 50 College Street, South Hadley, MA 01075-1441, or by phone (413) 538-2069, or email: jmorrow@mhc.mtholyoke.edu.

John Olive, an associate professor of Mathematics Education at the University of Georgia, Athens, has been working with students and teachers from kindergarten through college for the past 28 years. He received his Ph.D. from Emory University in 1985. He currently is co-director of two NSF-funded projects: Project LITMUS, a five-year Teacher Enhancement Project, and a research project investigating children's construction of fractions in the context of computer microworlds. He is author of a chapter on teaching and learning with *The Geometer's*

Sketchpad in the book *New Directions in Teaching and Learning Geometry*, Lawrence Erlbaum Associates, and also has several journal articles on technology and school mathematics. Email jolive@coe.uga.edu or find his web page at http://jwilson.coe.uga.edu/olive/welcome.html.

James M. Parks is a professor of mathematics at SUNY-Potsdam, and is currently a visiting lecturer for the MAA. Recent publications include *Generic Topology*, McGraw-Hill, rev. 1994, and "On Proof and Dynamic Software," *Math. in College J.*, CUNY, 1995–96. His current interests are in geometry and applications of dynamic geometry software. His email address is parksjm@potsdam.edu.

Arnold Perham is a teacher of mathematics and computer science at St. Viator High School, 1213 E. Oakton St., Arlington Hts., IL, 60004. He is particularly interested in using geometry construction software in an experimental setting. He coauthored the text *Topics in Discrete Mathematics, Computer Supported Problem Solving*, Addison-Wesley, 1993. He also coauthored "Discrete Mathematics and Historical Analysis: A Study of Magellan," *Mathematics Teacher* 88 (February 1995), pp. 106–112. He can be reached by telephone (847) 392 4050, ext. 226, fax (847) 392 4101, and email: aep@svhs.viator.k12.us.

Bernadette H. Perham, until her death in April, 1996, was a professor in the Department of Mathematical Sciences, Ball State University, Muncie, Indiana. Her main interests included curriculum development and assessment. She was president-elect of the Muncie Branch of the American Association of University Women. In 1991, she received the Ball State University Outstanding Faculty award. She coauthored the text *Topics in Discrete Mathematics, Computer Supported Problem Solving*, Addison-Wesley, 1993 and also coauthored "Discrete Mathematics and Historical Analysis: A Study of Magellan," *Mathematics Teacher* 88 (February 1995).

Doris Schattschneider has a Ph.D. in mathematics from Yale University and has taught at Moravian College for over 25 years. Her main research interests are in discrete geometry, especially tiling, and in the visualization of mathematical ideas. She was Senior associate on the Visual Geometry Project, which included the development of the software *The Geometer's Sketchpad*. She has led workshops for teachers at all levels on the use of dynamic geometry software, and been active as co-organizer and presenter in sessions on dynamic geometry. She is author of more than 40 articles and several books, of which her favorite is *Visions of Symmetry: Notebooks, Periodic Drawings, and Related Work of M.C. Escher*, W.H. Freeman. She is the recipient of the MAA award for distinguished College or University Teaching of Mathematics. Contact her at Moravian College, 1200 Main St., Bethlehem, PA 18018-6650, or by telephone (610) 861 1373, fax (610) 861 1462, or email: schattdo@moravian.edu.

Heinz Schumann is senior lecturer in the Department of Mathematics and Informatics at the Pedagogical University (PH) Weingarten, D-88250, Germany. He is guest lecturer at the Technical University of Karlsruhe and was a visiting lecturer at the University of Josef Fourier, Grenoble (France). He has written several articles and books on geometry teaching and learning (with or without computers), and has produced educational software. He introduced *Cabri-géomètre* to German schools, teacher training institutions and universities and with David Green (Loughborough University/England) coauthored the book *Discovering Geometry—using Cabri-géomètre*. He is guest editor of the *International Reviews on Mathematical Education* (ZDM). His recent didactical research concerns computer algebra and the use of computers in spatial geometry at secondary school level. Fax: (49)75150200; e-mail: schumann@ph-weingarten.de.

Laurent Trilling is professor of computer science at the Université Joseph Fourier and head of the PLIAGE group (Programmation Logique Intelligence Artificielle Génie Informatique) in the LSR laboratory (Logiciel Système Réseau) of IMAG, both institutions in Grenoble. His professional research interests include artificial intelligence, constraint logic programming, and intelligent tutoring systems. A recent publication is "Figure Correctness in an Expert System for Teaching Geometry" and "Programmation géométrique, impérative et logique" in *Journées francophones des langages applicatifs, collection didactique*, INRIA (1996). His phone number is 33-76-82-72-13 and email address is trilling@imag.fr.

Information on Software and Interactive Sketches

Information Sources on the World-Wide Web

Geometry computer files for many of the papers in this volume can be found on the World-Wide Web at:

http://forum.swarthmore.edu/dynamic/geometry_turned_on

A general source for materials and discussion of dynamic software for geometry is the Corner for Interactive Geometry Software on the Math Forum:

http://forum.swarthmore.edu/dynamic/

Software Contact Information

We have listed below contact information for software mentioned in articles in this volume *Geometry Turned On*. We have tried to include telephone telephone numbers, addresses on the World-Wide Web, and some information about the computer platforms on which the software runs. Pricing information is not included.

We hope that this information will prove useful. Because of the rapidly-changing world of computer software, some of this information may be out of date by the time you read this book, so be sure to check for current information.

Cabri Geometry II. This software was developed at the University of Grenoble by Yves Baulac, Franck Bellemain, and Jean-Marie Laborde.

For information about *Cabri Geometry II* in the United States, contact Texas Instruments, 1-800-TI-CARES. Information on Cabri can also be found on the Texas Instruments World-Wide Web site at

http://www.ti.com/calc/docs/cabri.htm.

Cabri Geometry runs on Macintosh OS and MS-DOS. A version of *Cabri Geometry* can also be found on the TI-92 calculator.

An earlier version of *Cabri* is published as *CABRI: The Interactive Geometry Notebook for Macintosh OS* (ISBN: 0-534-17059-5) or *MS-DOS 3.1 or later* (ISBN: 0-534-17586-4) by Brooks/Cole Publishing Company, 511 Forest Lodge Road, Pacific Grove, CA 93950-5098; telephone (408) 373-0728; e-mail: info@brookscole.com, and World-Wide Web:

http://www.thomson.com/brookscole/default.html

Cabri is available in a number of languages and versions. Information on *Cabri* outside the United States can be found on a World-Wide Web site at the University of Grenoble:

http://www-cabri.imag.fr.

Geometry Inventor. *Geometry Inventor* is available from Logal Software, Inc., 125 Cambridge Park Dr., Cambridge, MA 02140, Telephone (800)-564-2587, 617-491-4440, Fax 617-491-5855.

Geometry Inventor runs on Macintosh OS and Windows 3.0 or higher. Systems and other information is available on the World-Wide Web at

http://www.logal.com/.

The Geometer's Sketchpad. *The Geometer's Sketchpad* was developed by the NSF-funded Visual Geometry Project, directed by Eugene Klotz at Swarthmore College. The program's designer and programmer is Nicholas Jackiw.

Contact Key Curriculum Press, 1-800-995-MATH (outside the U.S. call 510-548-2304) for information about licenses and pricing. This information can also be found on the Key Curriculum Press World-Wide Web site at

http://www.keypress.com.

The Geometer's Sketchpad runs on Macintosh OS and Microsoft Windows 3.1 or later. Sketchpad has been translated into several languages. For information about international distributors and non-English versions, see the Key Curriculum Press World Wide Web site or send email to sales@keypress.com.

The Geometric Supposers and SuperSupposer. Several versions of the *Geometric Supposer* and also the *Super-Supposer* are available from Sunburst Communications, 101 Castleton Street, Pleasantville, NY 10570, U.S.A. Telephone (800) 321-7511; International: (914)747-3310; Fax: (914) 747-4109. The World-Wide Web site is

http://www.nysunburst.com/

Geomview. Geomview is a 3D object viewer. The software is free from the Geometry Center at the University of Minnesota. Several versions of *Geomview* can be downloaded from the web site

http://www.geom.umn.edu/software/download/

The following versions are available: SGI, Sun, NeXT, HP, Linux, IBM RS/6000, Dec Alpha, source.

Mathematical Bouncing Ball Lab. This software is presently available from its author, Fadia Harik, at BBN Systems and Technologies, 70 Fawcett St., Cambridge, MA 02138; Telephone (617)-873-3009.

Physics Explorer-One Body. Contact Logal Software Inc., 125 Cambridge Park Dr., Cambridge, MA 02140. Telephone (800)-564-2587, 617-491-4440; Fax 617-491-5855.

Physics Explorer runs on Macintosh OS and Windows 3.0 or higher. Systems and other information is available on the World-Wide Web at

http://www.logal.com/.

Trademarks

Cabri Geometry II and Cabri-géomètre are trademarks of Université Joseph Fourier.

Geometry Inventor and *Physics Explorer* and *Logal* are trademarks of Logal Software, Inc.

The Geometer's Sketchpad and *Dynamic Geometry* are registered trademarks of Key Curriculum Press, Inc. *Sketchpad* is a trademark of Key Curriculum Press.

Apple and *Macintosh* are registered trademarks of Apple Computer, Inc.

IBM is a registered trademark of International Business Machines Corporation.

Microsoft and *MS-DOS* are registered trademarks, and *Windows* is a trademark of Microsoft Corporation.